Earthworm Ecology
FROM DARWIN TO VERMICULTURE

MAN·IS·BVT·A·WORM·

Frontispiece Cartoon from *Punch*, December 6th, 1881

Earthworm Ecology

FROM DARWIN TO VERMICULTURE

Edited by

J. E. Satchell

Institute of Terrestrial Ecology
Merlewood Research Station
Grange-over-Sands
Cumbria, UK

1983

LONDON   NEW YORK

CHAPMAN AND HALL

First published 1983 by
Chapman and Hall Ltd
11 New Fetter Lane, London EC4P 4EE
Published in the USA by
Chapman and Hall
733 Third Avenue, New York NY10017
© 1983 Chapman and Hall Ltd
Printed in Great Britain at the
University Press, Cambridge

ISBN 0 412 24310 5

British Library Cataloguing in Publication Data

Earthworm ecology.
  1. Opisthopora
  I. Satchell, J. E.
  595.1'46   QL391.04

  ISBN 0-412-24310-5

Library of Congress Cataloging in Publication Data

Main entry under title:

Earthworm ecology.

  Bibliography: p.
  Includes index.
  1. Opisthopora—Ecology.   2. Earthworm culture.
I. Satchell, John E.
QL391.A6E27   1983   595.1'46045   83-1928
ISBN 0-412-24310-5

Contents

Preface   xi

Contributors   xiii

DARWIN'S CONTRIBUTION TO EARTHWORM ECOLOGY

1 Darwin's *Formation of Vegetable Mould* – its philosophical basis
   M. S. Ghilarov   I

2 Darwin on earthworms – the contemporary background and what the critics thought
   O. Graff   5

EARTHWORMS AND ORGANIC MATTER

3 Darwin's 'vegetable mould' and some modern concepts of humus structure and soil aggregation
   M. H. B. Hayes   19

4 Organic matter turnover by earthworms
   J. D. Stout   35
   (with a section by B. J. O'Brien)

5 Effect of earthworms on the disappearance rate of
   cattle droppings
   P. Holter   49

6 Soil transport as a homeostatic mechanism for stabilizing the earthworm environment
   A. Kretzschmar   59

vi   Contents

EARTHWORM ECOLOGY IN GRASSLAND SOILS

7   Earthworm ecology in grassland soils
     J. K. Syers and J. A. Springett   67

8   Effect of earthworms on grassland on recently re-
     claimed polder soils in the Netherlands
     M. Hoogerkamp, H. Rogaar and H. J. P. Eijsackers   85
     (with Appendix by J. H. Oude Voshaar and H. J. P.
     Eijsackers)

9   The activities of earthworms and the fates of seeds
     J. D. Grant   107

EARTHWORM ECOLOGY IN CULTIVATED SOILS

10   Earthworm ecology in cultivated soils
     C. A. Edwards   123

11   Nitrogen turnover by earthworms in arable plots
     treated with farmyard manure and slurry
     N. C. Andersen   139

12   Earthworm population dynamics in different agricul-
     tural rotations
     A. Lofs-Holmin   151

EARTHWORM ECOLOGY IN FOREST SOILS

13   Earthworm ecology in forest soil
     J. E. Satchell   161

14   Earthworm ecology in deciduous forests in central
     and southeast Europe
     A. Zicsi   171

EARTHWORM ECOLOGY IN TROPICAL AND ARID SOILS

15   Earthworms of tropical regions – some aspects of
     their ecology and relationships with soils
     K. E. Lee   179

16   The ecology of earthworms in southern Africa
     A. J. Reinecke   195

17   Effects of fire on the nutrient content and microflora
      of casts of *Pheretima alexandri*
      M. V. Reddy   209

EARTHWORMS AND LAND RECLAMATION
18   Earthworms and land reclamation
      J. P. Curry and D. C. F. Cotton   215

19   Earthworm ecology in reclaimed opencast coal
      mining sites in Ohio
      J. P. Vimmerstedt   229

20   Development of earthworm populations in aban-
      doned arable fields under grazing management
      H. J. P. Eijsackers   241

EARTHWORMS AND SOIL POLLUTION
21   Heavy metal uptake and tissue distribution in
      earthworms
      M. P. Ireland   247

22   Heavy metals in earthworms in non-contaminated
      and contaminated agricultural soil from near
      Vancouver, Canada
      A. Carter, E. A. Kenney, T. F. Guthrie and
      H. Timmenga   267

23   Earthworms and TCDD (2,3,7,8-tetrachloro-
      dibenzo-*p*-dioxin) in Seveso
      G. B. Martinucci, P. Crespi, P. Omodeo, G. Osella
      and G. Traldi   275

UTILIZATION OF WASTES BY EARTHWORM CULTURE
24   Earthworms as a source of food and drugs
      J. R. Sabine   285

25   Assimilation by the earthworm *Eisenia fetida*
      R. Hartenstein   297

26   The culture and use of *Perionyx excavatus* as a protein resource in the Philippines
      R. D. Guerrero   309

27   Utilization of *Eudrilus eugeniae* for disposal of cassava peel
      C. C. Mba   315

28   Cultivation of *Eisenia fetida* using dairy waste sludge cake
      K. Hatanaka, Y. Ishioka and E. Furuichi   323

29   The earthworm bait market in North America
      A. D. Tomlin   331

30   A simulation model of earthworm growth and population dynamics: application to organic waste conversion
      M. J. Mitchell   339

EARTHWORMS AND MICROFLORA
31   Earthworm microbiology
      J. E. Satchell   351

32   The effects of fungi on food selection by *Lumbricus terrestris* (L.)
      A. Cooke   365

33   Introduction of amoebae and *Rhizobium japonicum* into the gut of *Eisenia fetida* (Sav.) and *Lumbricus terrestris* L.
      J. Rouelle   375

34   *Enterobacter aerogenes* infection of *Hoplochaetella suctoria*
      B. R. Rao, I. Karuna Sagar and J. V. Bhat   383

EARTHWORMS IN FOOD CHAINS
35   Predation on earthworms by terrestrial vertebrates
      D. W. Macdonald   393

36   Predation on earthworms by the Black-headed gull
     (*Larus ridibundus* L.)
     G. Cuendet   415

37   *Agastrodrilus* Omodeo and Vaillaud, a genus of car-
     nivorous earthworms from the Ivory Coast
     P. Lavelle   425

EARTHWORM EVOLUTION AND DISTRIBUTION
PATTERNS
38   The establishment of earthworm communities
     M. B. Bouché   431

39   The structure of earthworm communities
     P. Lavelle   449

TAXONOMY AND NOMENCLATURE
40   The scientific names of earthworms
     R. W. Sims   467

41   A guide to the valid names of Lumbricidae
     (Oligochaeta)
     E. G. Easton   475

     Systematic index   489

     General index   492

Preface

'**Darwin cleared: official**' This 1982 *Times* (7 January) headline of a first leader, reporting the astonishing case brought in Arkansas against compulsory teaching of a biblical account of creation, hopefully set at rest doubts about Darwin in the minds of a public confused by media presentations of such unfamiliar concepts as punctuated equilibria, cladism and phenetics. Mud sticks, but Darwin's perturbed ghost may have found some consolation in the concurrent celebrations at Grange-over-Sands, a modest township in Cumbria, UK, of the centenary of the publication of his less controversial book *The Formation of Vegetable Mould through the Action of Worms*. In the form of a symposium on earthworm ecology, this attracted some 150 participants, predominantly adrenalin-charged research workers in the full heat of peer-group interaction. This book comprises a selection of the more ecologically oriented papers contributed to the symposium, brutally edited in the interests of brevity and thematic continuity.

The book opens with an appraisal of Darwin's earthworm work in its historical and philosophical context and relates his views on 'vegetable mould' to current concepts of humus formation. Thereafter, quotations from Darwin made out of piety have been rigorously excluded. Subsequent sections each comprise a review chapter and two or three 'case studies' presenting new data on a related topic. It has always been difficult to formulate a balanced judgement on how far earthworm activity creates fertile soils and how far such soils create a favourable environment for earthworm activity. The early chapters of this book now facilitate a clear discrimination between the chicken and the egg and fully vindicate

Darwin's views on earthworm pedogenesis. Subsequent sections cover earthworm biology in natural and man-made ecosystems including polluted land and commercial vermiculture. No attempt has been made to impose uniformity in species nomenclature but the final chapters analyse this problem and propose a list of valid names as a basis for further attempts to resolve the current nomenclatural chaos.

Each of the twelve review chapters cites some 20–100 published papers and together provide a rather comprehensive coverage of the earthworm ecology literature. They do not include references on earthworm physiology, enzyme chemistry, subjects like wound healing where the earthworm is of interest only as experimental material, or papers primarily of local distributional interest. A more comprehensive bibliography of the earthworm literature was prepared for distribution at the 1981 symposium and an updated supplement to this is now available. Stephenson's monograph 'The Oligochaeta' reviewed the earthworm literature up to 1930 and around 2500 papers have been published in scientific journals since. This does not include popular articles or the profuse manifestations of excessive drilophilia generated by commercial interests or organic faddists.

Despite his well-known tag '*Natura non facit saltus*', Darwin was quite clear that evolutionary rates are non-constant. Darwin's contemporaries thought of earthworms as pests to be eradicated. The publication of his book produced a great *saltus* in popular opinion, approaching modern concepts of the role of earthworms in soil formation, and with the subsequent doubling of the number of earthworm research papers every decade his ideas have now been fleshed out.

Currently, the role of earthworms in many ecosystems is appreciated as being of much wider significance and of potential in such practical applications as pollution control and protein production. It is hoped that the publication of this book will provide a befitting tribute to a great pioneer of soil science and a base from which earthworm ecology can make a further leap forward.

J. E. Satchell

Contributors

N. C. Andersen — Zoology Department, Royal Veterinary and Agricultural University, Bulowsvej 13, DK-1870 Copenhagen V, Denmark.

J. V. Bhat — Laboratory of the Director of Research, Kasturba Medical College, Manipal, Pin-576119, India.

M. B. Bouché — Centre d'Etudes Phytosociologiques et Ecologiques, Laboratoire de Zooecologie du Sol, Route de Mende, BP 5051, F-34033 Montpellier Cedex, France.

A. Carter — University of British Columbia, Department of Soil Science, Suite 248, 2357 Main Hall, Vancouver, BC, Canada V6T 2A2.

A. Cooke — Liverpool Polytechnic, Department of Biology, Byrom Street, Liverpool L3 3AF, UK.

D. C. F. Cotton — School of Science, Regional Technical College, Sligo, Ireland.

P. Crespi — Universita degli Studi di Padova, Istituto di Biologia Animale, Via Loredan 19, 35100 Padova, Italy.

G. Cuendet — Institut de Génie de l'Environnement, EPFL-Ecublens, 1015 Lausanne, Switzerland.

J. P. Curry — Department of Agricultural Zoology, University College, Belfield, Dublin 4, Ireland.

E. G. Easton — Department of Zoology, British Museum (Natural History), Cromwell Road, London SW7 5BD, UK.

C. A. Edwards	Entomology Department, Rothamsted Experimental Station, Harpenden, Herts AL5 2JQ, UK.
H. J. P. Eijsackers	Research Institute for Nature Management (RIN), Kemperbergerweg 67, 6816 RM Arnhem, The Netherlands.
E. Furuichi	Technical Research Institute, Snow Brand Milk Products Co. Ltd, Kawagoe, Saitama, 350 Japan.
M. S. Ghilarov	Academy of Sciences of the USSR, Moscow V-71, 33 Lenin Ave., USSR.
O. Graff	Karl-Sprengel-Strasse 10, 3300 Braunschweig-Völkenrode, West Germany.
J. D. Grant	School of Plant Biology, University College of North Wales, Memorial Buildings, Bangor LL57 2UW, Gwynedd, Wales, UK.
R. D. Guerrero	Blis II Task Force, Technology Resource Center, Buendia Avenue, Makati, Metro Manila, Philippines.
T. F. Guthrie	University of British Columbia, Department of Soil Science, Vancouver, BC, Canada V6T 2A2.
R. Hartenstein	State University of New York, College of Environmental Science and Forestry, Syracuse Campus, Syracuse, New York 13210, USA.
K. Hatanaka	Technical Research Institute, Snow Brand Milk Products Co. Ltd, Kawagoe, Saitama, 350 Japan.
M. H. B. Hayes	Department of Chemistry, University of Birmingham, PO Box 363, Edgbaston, Birmingham B15 2TT, UK.
P. Holter	Institute of General Zoology, University of Copenhagen, 15 Universitetsparken, DK-2100, Copenhagen Ø, Denmark.
M. Hoogerkamp	Centre for Agrobiological Research (CABO), P.O. Box 14, 6700 AA Wageningen, The Netherlands.
M. P. Ireland	University College of Wales, Department of Zoology, Penglais, Aberystwyth, Dyfed SY23 3DA, Wales, UK.

Y. Ishioka — Technical Research Institute, Snow Brand Milk Products Co. Ltd, Kawagoe, Saitama, 350 Japan.

E. A. Kenney — University of British Columbia, Department of Soil Science, Vancouver, BC, Canada V6T 2A2.

A. Kretzschmar — Institut National de la Recherche Agronomique, Station de Faune du Sol, 7 rue Sully, 21034 Dijon Cedex, France.

P. Lavelle — Laboratoire de Zoologie, École Normale Supérieure, 46, Rue d'Ulm, 75230 Paris Cedex 05, France.

K. E. Lee — Commonwealth Scientific and Industrial Research Organization, Division of Soils, Private Bag No. 2, Glen Osmond, South Australia 5064, Australia.

A. Lofs-Holmin — Swedish University of Agricultural Sciences, Department of Ecology and Environmental Research, S-750 07 Uppsala, Sweden.

D. W. Macdonald — Oxford University, Department of Zoology, Animal Behaviour Research Group, South Parks Road, Oxford, OX1 3PS, UK.

G. B. Martinucci — Universita degli Studi di Padova, Istituto di Biologia Animale, Via Loredan 19, 35100 Padova, Italy.

C. C. Mba — Department of Soil Science, University of Nigeria, Nsukka, Nigeria.

M. J. Mitchell — State University of New York, College of Environmental Science and Forestry, Department of Environmental and Forest Biology, Syracuse, New York 13210, USA.

B. J. O'Brien — Institute of Nuclear Sciences, Department of Scientific and Industrial Research, Lower Hutt, New Zealand.

P. Omodeo — Universita degli Studi di Padova, Istituto di Biologia Animale, Via Loredan 19, 35100 Padova, Italy.

G. Osella — Museo Civico di Storia Naturale, Lungadige Porta Vittoria 9, 37100 Verona, Italy.

Contributors

B. R. Rao — Department of Zoology, Mahaveera College, Moodabidri-574 227, Karnataka State, India.

M. V. Reddy — North Eastern Hill University, College of Agriculture, Medziphema – 797 106, Nagaland, India.

A. J. Reinecke — University of Potchefstroom, Institute for Zoological Research, Department of Zoology, Potchefstroom 2520, South Africa.

H. Rogaar — Agricultural University, Department of Soil Science and Geology, P.O. Box 37, 6700 AA Wageningen, The Netherlands.

J. Rouelle — Institut National de la Recherche Agronomique, Station de Faune du Sol, 7 rue Sully, 21034 Dijon Cedex, France.

J. R. Sabine — University of Adelaide, Waite Agricultural Research Institute, Department of Animal Physiology, Glen Osmond, South Australia 5064.

I. Karuna Sagar — Department of Microbiology, College of Fisheries, Mangalore, Pin-575002, India.

J. E. Satchell — Institute of Terrestrial Ecology, Merlewood Research Station, Grange-over-Sands, Cumbria LA11 6JU, UK.

R. W. Sims — Department of Zoology, British Museum (Natural History), Cromwell Road, London SW7 5DB, UK.

J. A. Springett — Ministry of Agriculture and Fisheries, Private Bag, Palmerston North, New Zealand.

J. D. Stout — Dr Stout's chapter was published *post mortem*. All correspondence to: G. Yeates, Department of Scientific and Industrial Research, Soil Bureau, Private Bag, Lower Hutt, New Zealand.

J. K. Syers — Massey University, Department of Soil Science, Palmerston North, New Zealand.

H. Timmenga — University of British Columbia, Department of Soil Science, Vancouver, BC, Canada V6T 2A2.

A. D. Tomlin

Research Centre, Agriculture Canada, University Sub PO, London, Ontario, Canada N6A 5B7.

G. Traldi

Istituto di Patologia Generate Generale Veterinaria, Via Celoria 10, 20133 Milano, Italy.

J. P. Vimmerstedt

Ohio Agricultural Research and Development Center, Department of Forestry, Wooster, Ohio 44691, USA.

J. H. Oude Voshaar

Institute Toegepast Natuurwetenschappelijk Onderzoek for Mathematics, Information Processing and Statistics, P.O. Box 100, 6700 AC Wageningen, The Netherlands.

A. Zicsi

Department of Systematic Zoology and Ecology of the Eötvös Loránd University, H-1088 Budapest, Puskin ucta 3, Hungary.

Chapter 1

Darwin's *Formation of Vegetable Mould* – its philosophical basis

M. S. GHILAROV

The name of the great naturalist Charles Darwin is connected in the mind of humanity primarily with the theory of natural selection and with his strictly materialistic theory of organic evolution. Using consistently materialistic premises, Darwin has shown the possibility of the origin of the multitude of plants and animals and of the rise of new species as a result of the accumulation of small hereditary changes of ancestral forms. Darwin's methodology was a spontaneously dialectic one; his theoretical constructions and conclusions may serve as brilliant illustrations of the philosophical law of transition by quantitative changes into the new quality.

Correct methodological premises, ability to observe subtle distinctions and the astonishing diligence of the great scholar enabled him to make a permanent impact not only on biology but also on other branches of natural history. In geology, especially famous are his views on reef formation (1847, 1874) by the slow growth and skeleton deposition of coral polyps (and other reef-forming organisms). The concept is based on the same methodological premise, i.e. on the recognition that small 'insignificant' quantitative changes, on accumulation, provoke significant qualitative change, a qualitative jump: certainly a philosophical jump for Darwin who used to stress that *natura non facit saltus*. This methodological approach is clearly evident in Darwin's last work on 'The formation of vegetable mould through the action of worms' (1881). It appeared in two Russian editions, translated simultaneously by M. Lindeman and M. A. Menzbier in 1882, just after Darwin's death.

Darwin's last book was of paramount significance for the development of soil science. Not in vain Sir E. J. Russell (1927) evaluated it as 'the most interesting book ever written on the soil'. A. Yarilov, a famous Russian historian of soil science, the organizer and for many years editor of

Pochvovedenie (Pedology), the first magazine dedicated to this science, founded in 1899, wrote in 1936 that the author of *The Formation of Vegetable Mould* is to be recognized as one of the founders of modern soil science.

Earthworms attracted the attention of the great English naturalist throughout his life. His first communication on their activity was made in 1837 to the Geological Society of London; he showed that pieces of chalk and other materials laid on a meadow surface became covered with earthworm castings and after some years were found under the turf. He had already concluded that all the vegetable mould, as he called the humified soil layer, had passed through earthworms' intestines. Being seriously ill after his 'Beagle' voyage and staying almost all the time in his home at Downe, Darwin continued observations and gathered literature upon earthworms, counting their castings on pathways and lawns. I was happy to be invited in 1958 to Downe by Darwin's family and to look at the places of his regular systematic observations. Darwin also reared earthworms in pots when studying their behaviour and activity in more detail. *The Formation of Vegetable Mould*, the result of a lifetime's work, was the great naturalist's last accomplished work – his 'swan song'.

Shortly after its publication the famous Russian naturalist Vasiliy Dokuchaev issued his classical book *Russian Chernozem* (1883) which is regarded as the very foundation of the modern concept of the soil as a 'fourth natural body' in addition to the three discriminated by Linnaeus – mineral, plant and animal.

Darwin was the first to characterize and evaluate quantitatively the role of biological agents in soil formation; this enables us to regard the great naturalist as the founder of biological direction in soil science. In the preface to his work, Darwin remembered his first communication on the subject made to the Geological Society and mentioned that in the 'Gardener's Chronicle' (1869) Mr. Fish rejected his conclusions on the participation of earthworms in vegetable mould formation 'merely on account of their assumed incapacity to do so much work'. Mr. Fish remarked that 'considering their weakness and their size the work they are represented to have accomplished is stupendous'. Darwin adds with bitterness: 'Here we have an instance of that inability to sum up the effects of a continually recurrent cause, which has often retarded the progress of science, as formerly in the case of geology, and more recently in that of the principle of evolution'.* This note reflects Darwin's profound certitude of the correctness of his methodological approach, which he applied when examining reef formation and when explaining evolutionary changes under the action of natural selection resulting in the formation of new

---

* pp. 5–6 in the preface to the 1904 edition.

species. The methodology of scientific investigation, originating from Lyell's actualistic concepts in geology, reached its sharpest point in the *Origin of Species*. When finishing his book on earthworms, Darwin wrote: 'It may be doubted whether there are many other animals which have played so important a part in the history of the world, as have these lowly organized creatures. Some other animals, however, still more lowly organized, namely corals, have done far more conspicuous work in having constructed innumerable reefs and islands in the great oceans; but these are almost confined to tropical zones' (p. 288).

It is a matter of real sadness that the founder of genetical soil science, Dokuchaev (1883), who based his soil classification on the ways and modes of soil formation, underestimated the role of earthworms, though he had thoroughly studied and cited Darwin's book on the subject. Dokuchaev considered counts made by Darwin as either exaggerated or having only local significance. Dokuchaev's position is explained by the fact that when studying chernozem (black soil) profiles he worked in steppe regions where the activity of earthworms is not so evident, especially in summer.

Later, the immediate disciple of Dokuchaev – G. Vysotskii (1930), who worked in the steppes in all seasons for many years, explained on the basis of his observations and experiments the high stability of the granular structural units of chernozems, characterizing earthworm activity as one of the main factors of this water-stable granular structure. At the centenary of the first Darwin communication on earthworms, Yarilov (Jarilow) called him not only one of the founders of modern pedology, but the father of biological soil science and of archaeological soil science.

In Darwin's approach to the characteristics of the soil-forming activity of earthworms, especially essential is the fact that, when studying soil, he does not draw attention to its static state, nor does he describe its horizons etc., but he investigates 'vegetable mould' from the viewpoint of its dynamics, the process and the development of the soil layer.

At the time of Darwin's *Origin of Species*, ecology as a branch of biology was not yet formed. The term 'ecology' and its definition were introduced later by the adept and passionate fighter for Darwinism – the famous German zoologist Ernst Haeckel (1870). The concept of the biocenose was formulated still later by Möbius (1877). But Darwin himself, in the third chapter of the *Origin of Species*, when illustrating the struggle for existence, used changes in biocenoses as examples, whereas in his *Earthworms* his attention was drawn to that aspect of ecology which at that time and still many years later was neglected by ecologists.

Ecology, according to its very definition, studies interaction and interrelationships of organisms and their environments. Up to a short time ago, ecologists only studied dependence of organisms on their environment. Darwin in his *Earthworms* has shown brilliantly the other side of the medal – the influence of organisms on their environment, i.e. the dependence of the milieu, of the environment, on their activity.

He also recognized the quantitative approach to the evaluation of the effect of the activity of organisms in changing their environment; in particular he estimated the velocity of changes in the earth's surface under the influence of earthworms. The broad generalization of various observations on earthworms led him to conclusions on their geological significance, on their participation in denudation processes and in transformation of hilly relief into a peneplain.

The last book by Charles Darwin may also be regarded as one of the foundations of invertebrate ethology. It is interesting that Darwin stressed the non-stereotypic, adaptive character of earthworm activity, e.g. when closing their holes with fallen leaves and other objects. Though some of his descriptions of earthworm behaviour are rather anthropomorphic, they are, as are all Darwin's descriptions, very clear and detailed so that it is possible to discriminate the strictly objective facts he observed from their interpretation.

Comparison of the methodological premises of such different works by Darwin as *Origin of Species*, *Coral Reefs* and *Earthworms* reveal the philosophical concepts of the great scholar. Being an adversary of catastrophism, supporting Aristotle's position that *natura non facit saltus*, Darwin developed in his works the dialectic philosophical concept of qualitative changes resulting in the accumulation of small quantitative changes, and showed it brilliantly in the theory of evolution, in geology and in soil science.

REFERENCES

Dokuchaev, V. V. (1883) *Russian Chernozem*. St. Petersburg. 376 pp. [In Russian].

Haeckel, E. (1870) Über Entwicklungsgang und Aufgabe der Zoologie. *Jen. Z. Naturw.*, **5**, 353–370.

Möbius, K. (1877) *Die Auster und die Austerwirtschaft*. P. Parey, Berlin. 126 pp.

Russell, E. J. (1927) *Soil Conditions and Plant Growth*. 5th edn. Longmans, London.

Vysotskii, G. N. (1930) Agents of soil structure. *Trudy Naukovo-Doslidn. stancii Gruntoznavstva* (Proc. Exp. Stn. Pedol., Kharkov, Ukraine). **1**, 1–37.

Yarilov, A. A. (1936) Charles Darwin – the founder of soil science. *Pochvovedenie* **4**, 17–23. [In Russian].

Chapter 2

Darwin on earthworms – the contemporary background and what the critics thought

O. GRAFF

2.1 INTRODUCTION

From December 1831 until October 1836 Charles Darwin participated in the famous journey round the world on the British sailing ship 'Beagle'. During 1837 he worked intensively on the material he had collected in the last five years and on the manuscripts and notes of his studies of the coasts and countries he had visited. By now his health was somewhat affected and, as he wrote to a friend, the doctors urged him strongly in 1837 to 'knock off all work and go and live in the country for a few weeks'. These weeks he spent at the country seat of his uncle Josiah Wedgwood in Maer, Staffordshire. Here Wedgwood told him about his own observations on earthworms, which he had made previously on his pastures.

Darwin's son Francis (1887) writes of this: 'He was indebted to his uncle Josiah Wedgwood, who suggested that worms, by bringing earth to the surface in their castings, must undermine any objects on the surface.' Wedgwood had observed that burnt marl, cinders and pieces of brick, which were spread on the pastures as fertilizer some years before had disappeared from the surface and could be found some inches beneath.

2.2 DARWIN'S PAPER ON THE FORMATION OF MOULD

On November 1st 1837, Darwin read his paper before the Geological Society of London. He acknowledges the role of Mr Wedgwood's observations as the basis of this communication. The paper was published twice: in the *Proceedings of the Geological Society*, **2** and in the *Transactions* *of the Geological Society*, **5**. The two drafts are somewhat different. The first is a report of Darwin's lecture presented by someone else and already printed in 1838. The second was evidently written by Darwin himself and was printed in 1840 (Barrett, 1977).

The writer of the first draft cites a remark of Darwin which does not appear in the second: 'Darwin has found in Chile, 1300 feet above sea level, *Acephala* shells lying on the soil surface as well as included in vegetable earth.' Perhaps Darwin at first drew a parallel with the observations on Mr Wedgwood's land but he dropped this idea later. He appears to have made no observations on earthworms during the Beagle journey.

In the first decades of the last century the famous work of Gilbert White *The Natural History of Selborne* was well known to educated people in England. His famous letter **XXXV**, reporting his observations on earthworms and his ideas about their importance as 'the intestines of the earth' is still famous today, but was it known to Darwin?

The Natural History of Selborne was first printed in 1789. Two further editions of 1825 (two volumes) and 1843 (two volumes in one) are in Darwin's private library, today kept in the Cambridge University Library. In the edition of 1825 we find Darwin's pencil-notes on several pages, but not on the earthworm letter. The edition of 1843 bears a dedication 'Charles Darwin Esq. from the editor with many regards' and below on pages 281 and 282 a reference is printed to Darwin's lecture of 1837. However, we have to assume that in 1837 Darwin did not know White's observations on earthworms, otherwise he would have cited them.

In J. C. Loudon's *Encyclopaedia of Gardening* first published in 1820 (German translation, Loudon 1826) earthworms are listed under the heading of noxious animals, with snails, caterpillars and other insects. Poisons for eradicating them are quoted. In the English edition of 1835 it states on the contrary that 'Earthworms, unless existing in great numbers, cannot be ranked among injurious animals . . . They perforate the earth in every part and accelerate the process of vegetation'. Although this change may have resulted from White's letter, which had become better known in the meantime, it seems that Darwin had no knowledge of Loudon. We must therefore conclude that Darwin's lecture of 1837 is based – simply and solely – on the suggestion of his uncle (and later father-in-law) Josiah Wedgwood.

In the second version of his paper (1840), Darwin cites some new information about earthworm activity which he had received since 1837. In this some mistakes were printed. Darwin therefore wrote another short note 'On the origin of mould', which came out in No. 14 of the *Gardeners Chronicle and Agricultural Gazette* (April 6th 1844).

The Darwin papers in the Cambridge University Library include a handwritten note by Darwin dated May 26th 1842, which sets right a mistake in his essay of 1838. He had described a field, of which he had interpreted the effects of earthworm activity, as having been marled 80 years ago. The correct reading should have been '28 years ago'.

How was Darwin's paper reviewed? Evidently it was acknowledged by the Geological Society, otherwise they would not have published it twice. The agricultural experts also found their own observations corroborated but we have very little knowledge of what Darwin's scientific contemporaries said.

A French geologist, E. de St. Simon, Vicomte d'Archiac (1847), criticized Darwin's statements as too far-ranging. At most earthworm activity could have importance in low-lying and humid pastures.

In 1849, Morren, the author of a famous dissertation in 1829 on the anatomy of *Lumbricus*, published a later but favourable review of Darwin's early paper and subjoins some observations of his own made in the meantime in Belgium. An equally good opinion was expressed by Samuelson (1860) who quoted complete sentences of the paper. However, as noted by Ghilarov (1974), not all the reviews were favourable, Mr. Fish in the Gardeners Chronicle of 1869 rejecting his views on the grounds that earthworms were too small and weak to move the quantities of soil Darwin had described.

In the following decades Darwin focused his attention on his other publications which established his fame. Earthworms continued to be widely regarded as injurious animals, the agricultural and horticultural literature – newspapers and textbooks alike – recommending methods of exterminating them (Graff, 1979).

2.3 EARTHWORM PUBLICATIONS BETWEEN WHITE (1789) AND DARWIN (1837)

What was written about the real value of earthworm activity after White and before Darwin (1837)? Krünitz (1812), in a short notice under the key-word 'Regenwurm', wrote only a single sentence: 'They are useful for the soil by perforating it, so that the humidity can penetrate easier'. De Montègre (1815) published some observations on the mating behaviour and feeding of earthworms and stated that they don't eat earth alone. He opened the intestines of some worms and found various organic residues. Cuvier (1826) presented a paper written by the young French zoologist, J. C. Savigny, to the Physical Department of the Royal Academy of Science in Paris. Savigny was the first to state that there are many different earthworm species, not only the one which Linnaeus (1758) in the famous *Editio Decima* of his *Systema Naturae* had listed. Dufour (1829) wrote about 'The cocoons or eggs of *Lumbricus terrestris*' but there was total obscurity about cocoon-formation.

Ihrem Werk über den Regenwurm sehe ich mit grosser Spannung entgegen, dies ist es auch, was mich zu der unmittelbaren Beantwortung Ihres freundlichen Briefes veranlasst.

Der dänische Zoologe P. E. Müller hat in Tidsskrift for Skovbrug Bd III I u. II Hft. Kjobenhavn 1878 in den „Studier over Skovjord“ (Walderde) die Thätigkeit des Regenwurms weiter untersucht, ich selbst habe mir diese Arbeit <u>noch nicht</u> verschaffen können, da Sie aber grade über den Gegenstand schreiben und gewohnt sind in ausgedehntem Maasse die Literatur zu benutzen und zugeben, wollte ich Sie für den Fall, dass Sie die Arbeit noch nicht kennen sollten, auf ihr Vorhandensein aufmerksam machen.

Mit dem Ausdruck ausgezeichnetster

Hochachtung

Ihr ganz ergebener

V. Hensen.

I am eagerly looking forward to your publication on the earthworm and this is also the reason for my prompt reply to your letter.

The Danish zoologist, P. E. Müller, has a paper in *Tidsskrift for Skovbrug*, Vol III, I and II, Copenhagen 1878 in the 'Studies over Skookord' (forest humus) where he examined the activities of the earthworm; I myself have not been able to obtain this article, but as you are just writing about the subject and will utilize and cite the literature to a very wide extent, I wanted to bring this to your attention in case it had escaped your notice.

Yours very sincerely,
V. Hensen

Fig. 2.1 Letter (opposite), with English translation (above), from Victor Hensen to Charles Darwin dated 25th May, 1881. (By permission of the Syndics of Cambridge University Library.)

2.4 EARTHWORM PUBLICATIONS BETWEEN DARWIN (1837) AND DARWIN (1881) WITH REFERENCES TO ECOLOGY

Bridgman (1849), in a letter to the editor of *Zoologist* about his own observations wrote that 'earthworms don't feed on living parts of plants but only on such ones in which decay had already commenced'. This was corroborated by Edward Newman, the commentator of *Zoologist*. Bridgman's conclusion was so surprising in the opinion of the common reader that his letter received great attention. It was printed the following year (Bridgman, 1850) in German translation.

Pontallié (1853) published his own observations on earthworms which cleared up some earlier errors about their habits but in various German publications of the fifties and sixties I found mainly recipes for the expulsion or eradication of earthworms and nothing about their utility (e.g. Lenz (1852), Schnee (1862), Glaser (1867)). The famous work of Brehm (1869) *Illustrated Animal Life (Illustriertes Tierleben)*, which devotes three pages to earthworms, contains nothing about their importance to the soil.

The investigations of Hensen, Professor of Physiology at the Faculty of Medicine in Kiel, brought greater progress in estimating the activity of earthworms. In 1871, he read a paper on 'The Relations of the Earthworm to the Cultivation of the Soil' before the annual meeting of the German Naturalists and Physicians. A short report of this was printed in many newspapers.

Six years later, Hensen (1877) published his observations in a scientific journal. In the meantime more correct evaluations of earthworm activity were printed in German journals (Fückel (1871), von Lengerke (1872), Haberlandt (1875)) and the collection by earthworms of small stones on garden paths was described by Robert (1873). Hensen's 1877 paper was highly approved of by the Russian zoologist Grimm (1877). In England the paper was reviewed by Wilson (1877) in *Nature* and Cooper Key (1877) and Philipps (1877) wrote letters on the subject to the editor. Hensen also received non-acceptance and mockery, especially from von Homeyer (1879), although in the same year, Hüttig (1879), probably without any knowledge of Hensen's work, came to conclusions similar to Hensen's.

Naturally Hensen's paper was read by Darwin. On May 25th 1881, Hensen thanked him for a letter in which Darwin had announced that he would bring out his earthworm book in the near future. In his answer Hensen pointed out the work of the Danish forestry scientist, P. E. Müller (1878) which is cited in a later edition of *The Formation of Vegetable Mould . . .* (Fig. 2.1).

2.5 DARWIN'S PREPARATIONS FOR *THE FORMATION OF VEGETABLE MOULD . . .*

Darwin's preoccupation with the famous works which earned him the title of the 'Copernicus of the Organic World' did not prevent him from observing the life and habits of creatures regarded as the humblest. At the beginning and at the end of his scientific life we find him occupied with the earthworm theme. He came to it as a geologist and saw the problems always from a geologist's point of view.

I saw only part of the complete collection but from the years 1842 to 1870 I found no reports by Darwin concerning earthworms. There are three letters from his niece Lucy Wedgwood showing that on the recommendation of her uncle, she made some observations on earthworms, e.g. she writes on May 5th 1870 ' . . . My worms have not turned up any earth since I enclosed them . . . '.

Beginning from the winter 1870/71 there is an accumulation of handwritten notes about worms. Some of them are quick scribbles, many others look like final copies. More details are noted in 1872. From a letter of Archibald Gerkie, University of Edinburgh, dated October 10th 1881, we learn that ten years previously – that is in 1871 – he had corresponded with Darwin about 'the action of the earthworm'. Many other people helped Darwin by collecting observations on earthworms and three of his sons assisted him with his experiments.

All the material on 'Earthworms for the Formation of Vegetable Mould . . .' still existing has been collected and is well stored by the Manuscript Department of the Cambridge University Library

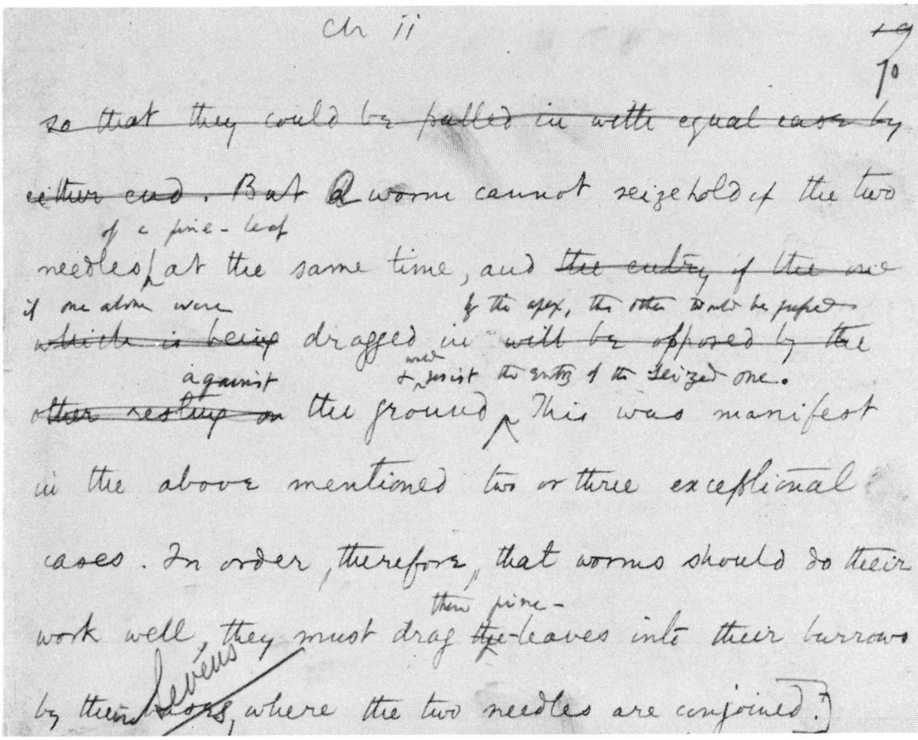

Fig. 2.2 Part of the manuscript of 'The Formation . . .' Chapter II, p. 68. (Corrections in Darwin's hand). (By permission of the Syndics of Cambridge University Library.)

(Classmark: 'Darwin papers' vol. 63–65). The manuscript of the worm book is almost complete and forms volumes 24 and 25 of the Darwin papers (Fig. 2.2).

Darwin was very cautious in the formulation of his manuscript. He was not sure whether it was permissible to call 'intelligent' the manner in which earthworms seize leaves and draw them into their burrows. He therefore sent the relevant pages of his manuscript to George John Romanes, FRS, for comment. Encouraged by Darwin early in his scientific career and subsequently a life-long friend, Romanes had delivered a lecture on 'Animal Intelligence' to the British Association in 1878 and published an extensive collection of data under the same title in 1881. On March 7th 1881 he replied to Darwin ' . . . there may be intelligence without self-consciousness . . . '. Romanes concludes the letter: 'Again thanking you very much for letting me see the MS and also for the notes of references'. Before completing his manuscript, Darwin had studied the non-ecological papers on earthworms published after his first paper of 1837, e.g. Hoffmeister (1845), Claparède (1869), Eisen (1871), Perrier (1874), Krukenberg (1877) and Frédéricq (1878). All these authors are quoted frequently.

2.6 PUBLICATION

The manuscript was finished at the end of March 1881 and correspondence began with Robert Cooke, clerk of the publisher John Murray and with Messrs Clowes, the printers. On April 10th, Darwin wrote to Cooke 'I do not care whether it is published by you on commission or on our former terms of two thirds profit . . . But unless Mr Murray can make up his mind in a day, I wish to publish on commission so that no time may be lost before the MS is in Messrs Clowes' hands'. Murray preferred to wait until October because he had agreed with the American publisher, D. Appleton of New York, that the American and English editions should come out on the same day. Appletons could not get the illustration blocks until the printing in England was finished. On July 29th, Darwin wrote again to Cooke ' . . . I thought you understood that I wished to publish at once . . . , it destroys all my satisfaction in the book as I hate to have the publication hanging over me, for until published, I cannot banish the subject from my mind'. Cooke immediately wrote a calming reply and announced on September 10th 'I send you by post an early copy of worms for your approval. Keep it quiet. You need not return it. The price we propose is 9/-.'

On October 3rd Cooke writes that he will send the author's copies in the next few days and adds: 'I hear of a mysterious copy of the work having been seen in the hands of a reviewer and I tremble in case a premature notice may appear before our copies go out to the Press generally.' On October 7th Cooke wrote that they had decided to print 500 more copies, and on the next day a post-card followed: 'We have made up our number to 1300 and shall perhaps have to print off 1000 extra. What about the German edition?' On October 25th. the second thousand was sold out and the third in preparation. The last sheet of this interesting correspondence dated November 5th brings a sensation: 'We have now sold 3500 worms! ! !' Probably a bookfair was held in the meantime, earlier works of Darwin having been sold in this way.

2.7 THE SUCCESS OF THE WORM BOOK

At about the same time as the English and American editions were prepared, German, French, Italian and Russian editions were being translated. The German Publishing house Schweizerbart in Stuttgart had engaged Victor Carus, Professor of Physiology in Leipzig, as translator of

Darwin's work 'because this gentleman has spent some years at the University of Oxford and knows the English language perfectly.' Two of Carus' letters are in the Cambridge collection: one (24th June 1881) dealing with the timetable and conditions of translation; the other (6th December 1881), after the book had appeared, reporting many misprints which were corrected in later editions. The German edition came out early in 1882, possibly while Darwin was still alive.

A French translation by M. Levêque was published by C. Reinwald, Paris in 1882, a letter among the Darwin papers of 24 November 1881 announcing that translation was in progress. In the same year translations were published in Italian by M. Lessona and two in Russian, one by M. A. Menzbir printed in Moscow (Freeman, 1977) and one by M. Lindeman printed in St Petersburg.

Darwin received 100 free copies of his book before it was on sale and sent them to friends and colleagues immediately. I found 24 letters of thanks in Cambridge, eleven of them dated during October, five in November, three in December, the others later.

Many writers related also their own experiences with earthworms, some of them somewhat strange. Douglas J. Wintle of Newnham, Gloucestershire, writes in December 1881: 'I have been a volunteer private for 10 years, and after a few minutes volley or file fixing with blank cartridges by a company or battalion, I have seen many large earthworms crawling about on the surface with great rapidity as if alarmed – and quit of their burrows. I have seen this on many occasions . . . '.

Mary A. Peek of Tilney, perhaps a relative of Darwin, wrote to him on March 6th 1882 that in the year 1862, her father's farmland was damaged by an inundation: 'My father claimed £1100 but did not make any special claim for the loss of the worms though things were mentioned . . . '.

2.8 THE CRITICS

The letters written by people who had read the book were mostly enthusiastic or at least in agreement. The *Country Gentleman* of Albany, New York, in its issue of February 2nd 1881 rejected Darwin's book totally on the grounds that earthworms damage plants in flowerpots. In Belgium, van Hulle (1882) also wrote a cutting rejection of Darwin's new book, similarly based on criticism of earthworms in flower pots. Some pages further on in the same journal, de Duren (1882) judged a little more objectively on the 'wormbook', but misunderstood Darwin's remarks about denudation and considered that if earthworms bring fine soil particles to the surface they may cause further erosion. Another scientist from Belgium, Errera (1882), closes his review – in opposition to his countrymen – with the following words: 'By these fine observations and arguments of Darwin the earthworm is raised from his humble position and we learned to regard it as a non-despicable factor of geology and agriculture.'

Extensive and positive reviews appeared in many agricultural journals in Germany and Austria, e.g. that of Medicus (1882) who took the opportunity also to cite Hensen (1877). Hensen himself published a long paper in 1882 about 'The fertility of soil as dependent on the actions of worms living in the earth-crust'. He considers Darwin's statements critically, partly approving and partly trying to disprove them. The two authors had different working methods, Darwin mostly examining farmland and pastures, and Hensen, gardens. The areas they studied have climatic differences which are important for the life and development of earthworms. For this reason, Hensen sometimes reached conclusions different from Darwin's. Darwin having quoted Hensen (1877) at nine different points in his book, Hensen (1882) now expressed his gratitude by an extensive appreciation.

At this time, Professor Ewald Wollny in Munich was the most known and best expert in soil science in Germany. He was editor of his own journal *Forschungen auf dem Gebiete der Agriculturphysik*, in which scientific papers and numerous reviews were published. In 1882 he reviewed the newly edited German translation in six pages, ending with the conclusion: 'Summarized, what has been said above, shows that the author has by far overestimated the role which worms have played and are still playing in the formation of vegetable mould. He relies far too much on the prejudices of gardeners and farmers.'

Some pages later in the same volume he disposes of Hensen's newly published paper, using much the same arguments as against Darwin. He ends his review: 'I want to advise every farmer – before engaging earthworms in the amendment of his land as the author proposes – to convince himself about the harmlessness of these animals'.

Wollny, head of a scientific institute in Munich, wanted however to do more than write a bad critique; he proposed to refute Darwin and Hensen by experiments. He cultivated various plants, e.g. rye, oats, peas, beans, sweet peas, linseed and potatos, in pots. In all cases the yield of the pots was higher when earthworms were present in the soil. Wollny published these results in 1890. He remarks, restricting the validity of his experiments, that his findings may not be transmitted to field conditions, 'but one fact may be elucidated from the results reported, namely that the position which the earthworms take in the soil with regard to plant growth is a useful and noteworthy one'.

Three years before Wollny, the Swiss Conrad Keller (1887) had published *Travel Pictures from East Africa and Madagascar (Reisebilder aus Ostafrika und Madagaskar)*. He confirmed that Darwin's observations were – on the whole – valid also in these tropical countries and named a most active earthworm species in honour of Darwin, *Geophagus darwini*. Many papers have been written in the last 100 years about the agricultural, horticultural and silvicultural importance of earthworms but the controversy was settled in favour of Darwin by the publications of Keller and Wollny.

In 1880, the famous French microbiologist Pasteur published a theory about the etiology of anthrax to the effect that, if animals which had died of anthrax were buried in pastures, earthworms might bring the spores to the soil surface with their excrements, facilitating infection of grazing animals. Darwin didn't mention this theory of Pasteur, but some of his critics believed that his observations on earthworm casting provided more evidence for their noxiousness than for their utility (Wanderer, 1882).

2.9 DARWIN ON HIS BOOK

In his autobiography, published by his son Francis in 1887, Darwin expresses all his feelings after finishing and delivering the manuscript but before its success was assured. 'I have now (May 1 1881) sent to the printers the MS. of a little book on 'The Formation of Vegetable Mould, through the Action of Worms.' This is a subject of but small importance; and I know not whether it will interest any readers, but it has interested me. It is the completion of a short paper read before the Geological Society more than forty years ago, and has revived old geological thoughts.'

To the German translator, V. Carus, Darwin wrote: 'The subject (the observation of earthworms) has been to me a hobby-horse, and I have perhaps treated it in foolish detail.'

To S. I. Hooker: 'I am glad that you approve of the "worms". When in old days I was to tell you whatever I was doing, if you were at all interested, I always felt as most men do when their work is finally published.'

To Mr Dyer: 'My book has been received with almost laughable enthusiasm, and 3500 copies have been sold!'

On February 4th 1882, two and a half months before his death he grumbles in a letter to Mr A. Rich: 'I have been plagued with an endless stream of letters on the subject; most of them very foolish and enthusiastic; but some containing good facts which I have used in correcting yesterday the sixth thousand.'

2.10 AN OBITUARY

A very large number of obituaries were written for Darwin. That of de Candolle (1882), famous Swiss botanist who had corresponded with

Darwin since 1840, concluded 'I had no misgivings in walking on the habitations of these lowly creatures, called earthworms, subject of a recent work in which Darwin has shown once more that small causes produce great effects. It occupied him for thirty years, but I ignored it'.

2.11 THANKS

The author wishes to express his thanks to all people who helped him to study and to collect 'Darwiniana'. In

Berlin:	Dr Gerhard Drude, Bibliothek der Technischen Universität, Abt. Gartenbau.
Braunschweig:	Dr Heinz Borkott and the staff of the Institut für Bodenbiologie and the staff of the Zentralbücherei der Bundesforschungsanstalt für Landwirtschaft.
Cambridge:	Mr P. J. Gautrey, Manuscript Department, Cambridge University Library and the staff of the 'Manuscripts Room'.
Dijon:	Mr J. Rouelle, Station des Recherches sur la Faune de Sol.
Downe, Kent:	Mr P. Titheradge, Darwin Memorial, Down House.
Grange-over-Sands, Cumbria:	Dr J. E. Satchell, Merlewood Research Station.
Montpellier:	Dr M. B. Bouché, Laboratoire de Zooécologie du Sol.

2.12 REFERENCES

Barrett, P. H. (1977) *The Collected Papers of Charles Darwin*. University of Chicago Press, Chicago and London. Vol. 1. 277 pp., Vol. 2. 326 pp.

Brehm, A. E. (1869) *Illustriertes Tierleben*. (Hildburghausen). 6, 694–696.

Bridgman, W. K. (1849) On leaves adhering to the casts of worms. *Zoologist*, 2576–2577.

Bridgman, W. K. (1850) Über die Blätter, welche in den Regenwurmlöchern stecken. *Frorieps Tagsber.*, 9(1), 20–21.

Cambridge University Library (1960) *Handlist of Darwin Papers* at the University Library, Cambridge. 72 pp.

Claparède, E. (1869) Histologische Untersuchungen über die Regenwürmer. *Z. Wiss. Zool.*, 19, 603–606.

Cooper Key, H. (1877) The earthworm in relation to the fertility of the soil. *Nature (London)*, 17, 28.

Cuvier, G. (1826) Analyse des travaux de l'Académie Royale des Sciences. Partie physique. *Mémoires de l'Académie Royale des Sciences de l'Institut de France*, 5, 176–184.

Darwin, C. (1837) On the formation of mould. *Proc. Geol. Soc.*, 2, 574–576.

Darwin, C. (1840) On the formation of mould. *Trans. Geol. Soc. London*, II Ser., 5(III), 505–509.

Darwin, C. (1844) On the origin of mould. *Gardeners Chronicle and Agricultural Gazette*, **14**, 218.

Darwin, C. (1881) *The Formation of Vegetable Mould through the Action of Worms with Observations on their Habits*. Murray, London. 298 pp.

Darwin, F. (1887) *The Life and Letters of Charles Darwin*. Murray, London.

d'Archiac, E. J. A. D. St. Simon (1847) *Hist. Prog. Geol.*, **I**, 223–224.

de Candolle, A. (1882) Darwin. *Arch. de Séances de la Bibliothèque Universelle*, 481–495.

de Duren, E. (1882) Formation de la terre végétale par les vers. *Rev. Hort. Belge Étrang.*, **8**, 101–102.

de Montègre, A. F. J. (1815) Observations sur les lombrics ou vers de terre. *Mém. Mus. Hist. Nat. Paris*, **1**, 242–252.

Dufour, L. (1829) Notice sur les cocons ou les oeufs du *Lumbricus terrestris*. *Ann. Sci. Nat.*, **1**, Sér., **5**, 17–21.

Eisen, G. (1871) Bidrag till Skandinaviens Oligochaetfauna. Terricolae. *Öfvers. K. Vetensk. Akad. Förh.*, **27**, 951–971.

Errera, L. (1882) Charles Darwin, The formation of vegetable mould. (A review). *Biol. Zentralblatt.* **2**, 33–37.

Fish, D. T. (1869) A chapter on worms. *Gardeners Chronicle and Agricultural Gazette*, (Apr. 17), 417–418.

Frédéricq, L. (1878) La digestion des matières albuminoides chez quelques invertébrés. *Arch. Zool. Exp. Gén.*, **7**, 391–400.

Freeman, R. B. (1977) *The Works of Charles Darwin*. Dawson (Wm.) & Sons Ltd., Folkestone. 235 pp.

Fückel, L. (1871) Review. *Fühlings Landw. Z.* **20**, 157.

Ghilarov, M. S. (1974) Entstehung und Entwicklung der Bodenzoologie in der UdSSR. *Pedobiologia*, **14**, 61–75.

Glaser, L. (1867) *Landwirthschaftliches Ungeziefer, dessen Feinde und Vertilgungsmittel* Schneider, Mannheim. 11–13.

Graff, O. (1979) Die Regenwurmfrage im 18. und 19. Jahrhundert und die Bedeutung Victor Hensens. *Z. Agrargesch. Agrarsoziol.*, **27**, 232–243.

Grimm, O. A. (1877) Snačenie dla plodorodia pocvy dozdevago cervjaka. *Trudy̆ imperatorskago vol'nago ekonomiceskago obscestva*, 179–195.

Haberlandt, F. (1875) Die Struktur der Ackerkrume. *Mitt. Landw. Lab. K. K. Hochschule Bodenkultur. Wien.*, **1**, 1–9.

Hensen, V. (1877) Die Thätigkeit des Regenwurms (*Lumbricus terrestris* L.) für die Fruchtbarkeit des Erdbodens. *Z. Wiss. Zool.*, **28**, 354–364.

Hensen, V. (1882) Über die Fruchtbarkeit des Erdbodens in ihrer Abhangigkeit von den Leistungen der in der Erdrinde lebenden Würmer. *Landw. Jrb.*, **11**, 661–698.

Hoffmeister, W. (1845) *Die bis jetzt bekannten Arten aus der Familie der Regenwürmer*. Vieweg, Braunschweig, 45 pp.

Hüttig, O. (1879) Der Regenwurm, ein Verbesserer der Ackererde. *Dt. Landw. Presse*, **6**, 515.

Keller, C. (1887) *Reisebilder aus Ostafrika und Madagaskar*. Winter, Leipzig, 341 pp.

Krünitz, J. G. (1812) *Ökonomisch-technologische Encyclopädie*, Berlin, **121**, (see 'Regenwurm').

Krukenberg, C. J. W. (1877) Studien über die Verdaungsvorgänge bei Wirbellosen. *Unters. Physiol. Inst. Univ. Heidelberg*, **2**, 37.

Lenz, H. O. (1852) *Gemeinnützige Naturgeschichte. Gotha.*, **3**.

Linnaeus, C. (1758) *Systema Naturae.* Editio Decima. (Holmiae), 647–648.

Loudon, J. C. (1820) *Encyclopaedia of Gardening*. London, 1st edn. (2nd edn. 1824, 3rd edn. 1825, further editions 1835 and 1850). (see 'Earthworms').

Loudon, J. C. (1826) *Eine Encyclopädie des Gartenwesens enthaltend die Theorie und Praxis etc.*, Aus dem Englischen. Weimar. (VIII. Abtheil. Ungeziefer, Insecten etc. p. 596).

Medicus, W. (1882) Über Darwins Werk:Die Bildung der Ackererde durch die Thätigkeit der Würmer. *Z. Landw. Ver. Bayern*, **72**, 446–454.

Morren, C. F. (1829) *De Lumbrici terrestris Historia Naturalis necnon Anatomia Tractatus.* (Brussels). 280 pp.

Morren, C. F. (1849), De l'utilité des vers de terre pour l'agriculteur, l'horticulteur et le géologue. *Ann. Soc. R. Agric. Bot., (Gand)*, **5**, 273–279.

Müller, P. E. (1878) Studier over Skovjord. *Tidsskr. Skovbrug (Copenhagen)*, 3.

Pasteur, L. (1880), Sur l'étiologie du charbon. *C. R. Hébd. Séances Acad. Sci. Ser. D*, **91**, 86–94.

Perrier, E. (1874) Etudes sur l'organisation des lombriciens. *Arch. Zool. Exp. Gén.*, **3**, 331–350.

Philipps, G. H. (1877) The earthworm in relation to the fertility of the soil. *Nature (London)*, **17**, 62.

Pontallié, (Ns). (1853) Observations sur le lombric terrestre. *Ann. Sci. Nat.*, **1** Ser., **19**, 18–24.

Robert, E. (1873) Sur les moyens employés par les lombrics pour défendre l'entrée de leur galeries souterraines. *C. R. Hebd. Séances Acad. Sci., Ser. D*, **76**, 785.

Samuelson, J. (1860) *The Earthworm and the Common Housefly*. van Voorst, London, 2nd edn., 79 pp.

Schnee, G. H. (1862) *Handbuch der Landwirthschaft in alphabetischer Ordnung.* 2nd edn. Braunschweig (see 'Regenwurm').

van Hulle, H. J. (1882) Les Lombrics (vers de terre). *Rev. Hort. Belge Etrang.*, **8**, 66–68.

von Homeyer, E. F. (1879) Der Regenwurm. *Dtsch. Magazin Gart. Blumenkde (Stuttgart)*, **32**, 34–37.

von Lengerke, (Ns) (1872) Der Regenwurm, ein landwirthschaftliches Nutzthier. *Hannov. land-u. Forstw. Vereinsblatt*, **11**, 148–149 und 161–163.

Wanderer, (Ns) (1882) Les vers de terre. Importance du rôle qu'ils jouent dans la nature. *Rev. Hort.*, **54**, 307–308.

White, G. (1789) *The Natural History and Antiquities of Selborne*. Benjamin White, London. 1825 edition: Rivington, London. 1843 edn.: van Voorst, London.

Wilson, S. (1877) The earthworm in relation to the fertility of the ground. *Nature (London)*, **17**, 18–19.

Wollny, E. (1882a) (Besprechung von Ch. Darwin "Die Bildung der Ackererde . . . etc." deutsch von V. Carus, Stuttgart 1882). *Forsch. Geb. Agrikulturphysik*, **5**, 50–55.

Wollny, E. (1882b) (Besprechung von V. Hensen "Über die Fruchtbarkeit . . . etc." *Landw. Jrb.*, **11**, 1882). *Forsch. Geb. Agrikulturphysik*, **5**, 423–425.

Wollny, E. (1890) Untersuchungen über die Beeinflussung der Ackerkrume durch die Thätigkeit der Regenwürmer. *Forsch. Geb. Agrikulturphysik*, **13**, 381–395.

Chapter 3

Darwin's 'vegetable mould' and some modern concepts of humus structure and soil aggregation

M. H. B. HAYES

3.1 INTRODUCTION

Darwin's observations on earthworms can be regarded as a milestone in our understanding of soil biology and an enormous contribution to some aspects of the genesis of humus and of its role in soils. This chapter outlines Darwin's conclusions on the role of earthworms in the formation of vegetable mould and his interpretations of its nature and properties, especially those of its organic constituents. It discusses modern concepts of the composition, structure and properties of humus substances and the role they play in soil, especially in stabilizing aggregates and retaining water and plant nutrients.

3.2 DARWIN'S 'VEGETABLE MOULD'

Darwin (1881) defined as vegetable mould the uniformly fine soil particles of dark colour 'which covers the whole surface of the land in every moderately humid country'. This mould, which could extend to a depth of 40 cm or more in the soil profile, would have passed many times through the 'intestinal canals' of worms and would, he stated, be more appropriately termed 'animal mould'. He considered the chief work of earthworms to be 'to sift the finer from the coarser (soil) particles, to mingle the whole with vegetable debris, and to saturate it with their intestinal secretions'.

Darwin cited results from a number of experiments which indicated that the soil brought to the surface annually in wormcasts was equivalent to about 0.5 cm of soil profile. Since most earthworms in soil live at depths of less than 25–30 cm he concluded that this surface soil is worked over most intensively. However, some soil is brought as casts from considerable depths, especially when worms burrow deeper during dry or cold periods. Thus, he stated that 'the superficial layer of mould would ultimately attain, though at a slower and slower rate, a thickness equal to the depth to which worms ever burrow, were there not opposing agencies at work which carry away to a lower level some of the finest earth which is continually being brought to the surface by worms'. In one of his measurements of the dry weight of the casts brought to the surface annually, Darwin quoted a value of 18.12 tons of earth per acre. On the basis of the generally accepted value of 2×10^6 kg of soil ha^{-1} in the soil plough layer (15 cm) this corresponds to a depth of about 0.3 cm.

Darwin distinguished between vegetable mould and 'ordinary mould', considering that the colour in the latter might be attributable to small amounts of decaying organic matter. He drew attention to the role of earthworms in consuming, moistening, tearing to small shreds, partially digesting, and intimately mixing the vegetable matter with earth and considered that these processes would give vegetable mould its uniform dark tint. There is no doubt about the importance of the shredding, partial digestion, and mixing processes, and these take place on a relatively short time scale. It is likely, however, that the darkening is slower and is part of the humification process which involves chemical reactions and microbial activity.

Darwin's observations on the effects of 'humus acids' on oxides is very pertinent to modern concepts of the role of humic substances in the chelation of metals and their transportation in the soil profile, as well as to their role in weathering rocks and minerals. He concluded that the acids formed during digestion in the earthworm gut are similar to those in ordinary 'mould or humus' and indicated that such acid materials can dissolve iron oxides far more readily than acetic acid and do so more efficiently than 'even hydrochloric, nitric and sulphuric acids, diluted as in the Pharmacopoeia'. He referred to azohumic acids, described by Thénard (in Julien, 1879), capable of dissolving colloidal silica in proportion to the nitrogen contents of the acids. In the view of some chemists of the time, humus contains 'more than a dozen different kinds' of acids which could 'act energetically on carbonate of lime and on oxides of iron'. Modern evidence indicates that acidic humic materials are polyelectrolytes and it is thus unlikely that the acid substances secreted in the earthworm gut have similar molecular compositions.

Darwin referred to the role of earthworms in mechanically grinding in their gizzards mineral (sand) particles capable of passing through the alimentary tract. This can be regarded as a contribution to soil genesis. He also took account of the way in which earthworm casts can lead to soil erosion, pointing out that castings brought to the surface during rain or shortly before heavy rain are moved short distances down inclined surfaces. He stated that 'much of the finest levigated earth is washed completely away from the castings. During dry weather the castings often disintegrate into small rounded pellets, and these from their weight often roll down any slope. This is more especially apt to occur when they are started by the wind, . . .'. His observation that 'castings when first ejected are viscid and soft; during rain, at which time worms apparently prefer to eject them, they are still softer; so that I have sometimes thought that worms must swallow much water at such times' focuses attention on the need to bring soil particles into close proximity (e.g. by drying) in order to allow aggregate-stabilizing compounds, such as natural and synthetic polymers, to form bridges between adjacent soil particles. A complex of soil plus aggregate-stabilizing or 'cementing' agent formed during passage through the earthworm gut would need to be at least partially dried before it could acquire a resistance to mechanical disruption from raindrops and leaching away of finer (silt and clay-sized) particles.

Darwin made it abundantly clear that earthworms, burrowing animals and insects are essential components of soils because of the role they play in macerating vegetable matter, mixing it with the soil, and sifting finer from coarser particles. Such macro-organisms help prepare media which promote microbial activity, part of which is responsible for the synthesis of the range of polymers found in humus substances and essential to the fertility of most soils.

3.3 MODERN CONCEPTS OF THE COMPOSITION AND STRUCTURE OF HUMUS SUBSTANCES

Humus substances range from relatively simple low-molecular-weight materials such as oligosaccharides to highly complex polydisperse humic polymers. They are composed predominantly of humic acids, fulvic acids and humins. Humic acids are precipitated by acidification to pH 1 of alkali-soluble humic extracts, fulvic acids remain in solution when the alkaline extracts are acidified, and the humins are insoluble in dilute acids and bases though they may be dissolved in concentrated sulphuric acid. Compounds in humus such as polysaccharides, polypeptides and altered lignins which are synthesized by micro-organisms proliferating on humifying substrates, or formed by modifications to parent polymers, are regarded as non-humic substances.

3.3.1 Determination of the structures of humus substances

In order to assign fully a structure to any organic polymer it is necessary to identify the repeating units (*primary structures*), to know the order in which these units are linked together (*secondary structure*) and to determine its shape and size and the arrangement in space of the components (*tertiary structure*). Homogeneous biological polymers, whose synthesis is genetically controlled, have invariant primary structures, linked in an invariant sequence. This almost certainly applies for homogeneous soil polysaccharides, but it is doubtful that it does for humic materials.

The classical procedures for determinations of primary structures in polymers have degraded the macromolecules to identifiable components which could be assigned to structures within the polymer. Such procedures work well where labile bonds, such as the peptide bonds of proteins and the glycosidic linkages of polysaccharides, hold the component molecules together. Because of the high energy input required to cleave some of the linkages in humic substances, the products identified in degradation reaction digests are, in many instances, merely derivatives of the components released from the polymer structures. Progress is being made in non-degradative spectroscopy of humic polymers.

Hayes and Swift (1978) have outlined the principles and procedures involved in the degradation of humic substances by several reagents under a variety of conditions. In their view, most information has been obtained from degradations with permanganate, alkaline cupric oxide, sodium amalgam, sodium sulphide, and phenol plus p-toluene sulphonic acid. The more widespread availability of gas liquid chromatography–mass spectrometry instrumentation has greatly accelerated identification of digest degradation products.

The availability of these non-degradative procedures is having an impressive impact on the advancement of knowledge of humic structures. The most useful information has been obtained using ^{13}C-nuclear magnetic resonance (n.m.r.) instruments incorporating cross polarization and magic angle spinning (CPMAS) facilities (Barron *et al.*, 1980; Worobey and Webster, 1981; Wilson *et al.*, 1981; Schnitzer, 1982).

3.3.2 Conclusions from structural studies of humus substances

There is strong evidence from the structures identified in digests of humic substances to indicate that single ring aromatic compounds are components of humic structures. Additional evidence of aromaticity is being provided in n.m.r. spectra of solid humic materials, the available data suggesting that many of the aromatic structures are di- or poly-substituted.

The fused aromatic structures identified in zinc dust distillation digests of humic substances may have been artefacts of the highly energetic reactions although benzenecarboxylic acids identified in the digests of permanganate-oxidized humic materials could have had their origins in fused aromatic components in the polymer structures. It is more likely, however, that the aromatic acid structures in the digests were derived from the oxidation of aliphatic and/or carbonyl substituents on benzene-type structures.

Many of the benzene structures also contain hydroxy and methoxy substituents, and the positioning of these substituents in the rings suggests possible lignin-type precursors for some of the humic structures. Others of the phenolic structures could arise also from products of microbial metabolism.

Sugars and amino acids released during the hydrolysis of humic materials could arise from polysaccharides and peptides physically sorbed on these materials. Some of these sugars and amino acids might be integral parts of the humic structures because they are released in hydrolysates even after humic substances are carefully fractionated by procedures which would remove physisorbed components. Phenolic glycoside structures could link sugars, oligosaccharides, or even polysaccharides in the humic polymer structures. Peptide bonds could attach amino acids and peptides in the polymer and, additionally, free amino groups could form covalent links with carbons *ortho* to the carbonyl group of quinone structures. It is generally agreed that humic substances contain some heterocyclic nitrogen but little is known about these types of structures in the polymer.

The types of linkages which bind together the primary structures in humic polymers are poorly understood. Data from sodium amalgam cleavage reactions suggest that phenolic ether links could be important. For the most part though, it seems that linkages in the humic 'backbone' or core are through carbon to carbon bonds. However, the activation energies for the cleavage of such bonds is lowered where polar substituents are present on the carbon atoms. Degradation and spectroscopic studies suggest therefore that humic polymers contain saccharide and amino acid or peptide components which may be peripheral to the major structural units. These major structural units could involve single-ring aromatic structures which in many instances are substituted with carboxyl, phenolic, hydroxyl and phenolic ether groups and with hydrocarbon substituents. Quinone structures inevitably arise from the oxidation of phenols. Some of the linking hydrocarbons could be expected to have substantial unsaturation and to be substituted with acidic and with polar functional groups. The saturated hydrocarbons isolated could also have come from cleavage of interaromatic links, or they might merely be products of microbial metabolism which were sorbed by humic substances.

Extrapolation from studies of solution conformations of humic polymers allows predictions to be made about the nature of these polymers in the solid and gel phases which are the states most relevant to field conditions. The evidence for random coil conformations put forward by Cameron *et al.* (1972) was based on studies of the polymers in solution. Their data are excellent but they had to base their interpretations on the properties of the polymers which dissolved in Tris buffer at a pH value of about 9. They were able to predict that polymer branching increased as molecular weight increased. However, substantial amounts of the humic acids were insoluble in the buffer, and it may be assumed that such materials were of higher molecular weight, possibly more heavily branched and/or cross-linked, and/or had lower charge densities than the soluble components.

Humic and fulvic acids exchanged with monovalent cations such as NH_4^+, K^+ and Na^+ would readily be lost in solution in drainage waters. H^+-exchanged humic acids would be retained because the acids are largely undissociated and hence would lack the impetus to dissolve which is provided by charged groups. Equally important is the fact that inter- and intramolecular hydrogen bonding condenses the structures and drives out water from the polymer matrix. In a similar way divalent and polyvalent cations, which bridge two or more charged groups within or between the polymer strands, render humic and even some of the fulvic acids insoluble by shrinking the structures and driving water out. Further drying, as occurs under field conditions, would lead to additional condensation of the structures and eventually to the formation of strong intramolecular bonds, especially between hydrophobic portions of the molecules. This would explain why it is especially difficult to rewet soil organic matter and humic substances that have been dried under laboratory conditions or under drought conditions in the field.

3.4 THE ROLE OF HUMUS SUBSTANCES IN STABILIZING SOIL AGGREGATES

Clays, and oxides and hydroxides, especially those of iron and aluminium, have extensive surfaces and are the most reactive of the inorganic components in soil aggregates. Clays are considered to associate in parallel alignment to form quasi-crystals and domains (Quirk, 1978) of up to 5 μm in lateral extent. These can associate with silt and fine sand to form micro-aggregates in the size range 5–500 μm; micro-aggregates in turn can associate together and with sand particles to form aggregates in the size range 0.5–5 mm. It is generally accepted that humus colloids associate with the soil inorganic constituents to help stabilize aggregates and it is recognized that in certain circumstances oxides and hydroxides also act as aggregate stabilizers.

3.4.1 Binding of humus substances to soil inorganic colloids

All the colloidal humus substances in mineral soils are not necessarily associated with mineral components and there is evidence that substantial portions of the soil inorganic colloids remain free. Considerable attention has been focused on interactions between humic materials and various polysaccharides with clays from commercial deposits. Such model studies are not always relevant to the reactions which take place in the field where the clays are often not well defined, are frequently contaminated with metal oxides and hydroxides, and with indigenous soil organic materials. Theng (1979), Burchill *et al.* (1981) and Hayes and Swift (1981), among others, have reviewed aspects of interactions between humus substances and soil inorganic colloids.

(a) Interactions of humic substances with clay minerals and soil clays

The permanent negative charge on clay minerals is of the same sign as that of dissociated humic molecules. Hence such polymers are not adsorbed from solution by monovalent cation-exchanged clays. H^+-exchanged, water-soluble fulvic acids can be adsorbed even between the layers of expanding clays because such polymers are only weakly dissociated at low pH values and can adsorb by mechanisms similar to those for neutral molecules. Similar studies cannot be undertaken for H^+-humic acids because these polymers are insoluble in water. Nevertheless, humic acids do associate with clays, and one plausible association mechanism would involve the bridging through divalent cations between the polymer and the clay (Theng, 1979; Burchill *et al.*, 1981).

Schwertmann and Taylor (1981) have referred to interactions between oxides and various solid soil constituents and have stressed that there is little known about the affinities of the oxides for other minerals, especially for clay silicates. Aluminium hydroxides can be positively charged in the pH range which predominates in many soils and thus can neutralize the negative charges on the clays and be held by them. In the soil environment, humic colloids may be held by such hydroxides which in turn may or may not be bound to clays. Other possibilities for inorganic colloid-humic associations might involve precipitation in the presence of metal ions of humic materials on the inorganic surfaces.

Turchenek and Oakes (1979) have provided data on the extent to which organic matter and elements such as Al, Ca, Fe, Mg, P, Si and Ti are present in different density fractions of soils. They have shown differences in the properties of the organic materials associated with the different clay-size fractions; for instance, their data suggest that the heavier fine clays contain the more aliphatic humic and fulvic materials, and that these might be adsorbed through physical adsorption forces, whereas the more highly humified and aromatic humic components appear to associate preferentially with the lighter, coarser soil clays. CPMAS n.m.r. spectroscopy can be expected to play an important role in resolving differences between the types of organic materials associated with different inorganic colloids.

(b) Interactions of polysaccharides with clay minerals and soil clays

Adsorption of polysaccharides by clays and soils has been discussed in recent reviews by Hayes and Swift (1978), Cheshire (1979), Theng (1979) and Burchill et al. (1981). These refer to work by Clapp and his associates (e.g. Olness and Clapp, 1975) and by others which show that some but not all dextrans are adsorbed by various clay mineral preparations. Similar variability in adsorptivity appears to apply to the various polysaccharides found in soil. Olness and Clapp's work suggests that β-glycosidic linkages in dextrans are more conducive than α-linkages to binding by clays. This may be explained by differences in the solution conformations of the polymers and by the fact that the β-linkages would allow more points of contact than the α-linkages between the sugars and the clays.

In theory, the higher-molecular-weight polymers should be the better stabilizers of aggregates. To be an effective stabilizer, the polymer should be able to bridge the distances between domains and even between micro-aggregates. This is especially true where newly synthesized polymers would diffuse into preformed aggregates and bridge-adjacent components in the structure. Linear polymers would bridge over larger gaps than random coil structures. Polymer size may not be quite so important where the organic and inorganic polymers are intimately mixed by the soil fauna during the formation of the aggregates.

The evidence for the involvement of polysaccharides in the stabilization of aggregates is strong, although the mechanisms by which the binding occurs are unclear (Cheshire, 1979; Hayes, 1980). Polysaccharides of plant and of microbial origins are generally considered to be involved and mucopolysaccharides from the earthworm gut could also be important. Polymers with substantial amounts of xylose and arabinose are likely to have plant origins while those containing galactose, mannose, rhamnose, fucose, galactosamine and glucosamine are likely to have been synthesized by micro-organisms. Adsorbed polysaccharides within aggregates may be sterically protected from decomposition by micro-organisms. On the other hand, the polysaccharides synthesized by micro-organisms feeding on organic matter within aggregates could help stabilize structures weakened by the removal of sorbed organic polymers.

(c) Aggregate stabilization of humus substances in the presence of water

Water greatly influences the shape and the dimensions of humus polymers and as these shrink during drying, the hydrophilic groups orientate towards the interior exposing the more hydrophobic components to the outside.

Any consideration of bridging of inorganic components by humus substances as a mechanism for the stabilization of aggregates should take into account the vast dimensions (in molecular terms) of the component inorganic structures and of the distances between the particles, compared with those of the individual polymer molecules. Thus molecules with linear or helical solution conformations would bridge longer distances than the same molecules in random coil or in globular conformations. Hayes (1980) has suggested that a molecule of the β-(1 →6)-linked dextran (Polytran) used by Olness and Clapp (1975), with a molecular weight of the order of 2×10^6, could have a linear dimension of up to $7\,\mu m$. Molecules of this type could effectively bridge domains into micro-aggregate structures. The fact that such polymers adsorb from solution on to mineral surfaces indicates that they can compete with water for the adsorbent surfaces. However, components of highly hydrophilic polymer molecules extending away from adsorbent surfaces would have little affinity for each other in a medium containing excess water. Affinity would increase as water is removed during drying. To be effective for stabilizing aggregates, polymers anchored on one clay particle or on one domain of clay particles would have to associate strongly with the unadsorbed molecules in the soil solution or with other molecules in the medium which could be involved in forming a bridge. This gives rise to the concept of a chain of polymer molecules involved in stabilizing soil structures.

Humic molecules in the solid/gel phase in soil have relatively highly condensed structures compared with those in solution. Thus inorganic soil components would have to be very close to be bridged by individual humic molecules. Swelling of the particles could be expected to weaken or break up such aggregates. Where aggregates are stable under conditions of wetting and drying, and where they swell and shrink in the process, it is reasonable to assume that the humic substances bound to one inorganic component are linked through a chain of unadsorbed molecules to those bound to the next, and so on. Under such circumstances the humic molecules would swell and shrink but not dissolve, and so the chain would be preserved in the wetting and drying sequences in the soil. The organic polymers would also act as a reservoir of water in the soil.

These concepts of aggregate stabilization have yet to be proved by experiment. Laboratory studies readily establish the ease and extent of adsorption of polymers by soil components but it is very difficult to confirm what is taking place within the aggregate and micro-aggregate structures important to soil fertility.

Organic soils which lack inorganic colloids can be highly fertile where the organic matter is highly humified and the soils are well drained and fertilized. Such soils can have cation-exchange capacity values of the order of 3000 to 4000 μeq g^{-1}, and thus have a high reservoir potential for plant nutrients. The concepts of aggregate stabilization described for mineral soils do not apply in cases of highly organic soils. However, bridging of polymer chains by divalent and polyvalent cations confers structure to organic materials. These concepts do not apply either to highly sandy soils but these are generally free draining because of their coarse particle size. They are generally of low fertility except where substantial amounts of organic matters are present. Again, the humic substances in the organic matter provide the plant nutrients 'sink'.

3.5 THE ROLE OF EARTHWORMS AND OF SOIL BIOTA IN FORMING SOIL AGGREGATES

The role of biological oxidation in depleting the organic matter content of arable soils is emphasized in most soil texts and the loss of organic matter is well correlated with the degradation of soil structure. Well-aerated grassland soils generally have good structure and such soils are invariably well structured where there is an ample earthworm population. Grassland soils provide an abundant source of organic residues to feed earthworms which mix the soil particles with organic debris and create air and water channels in the profile. In addition, the abundant grass roots push soil particles closer together as they permeate the root zone and in this way promote the formation of aggregates. Living roots can secrete mucilagenous polymers which help bind the soil aggregates and when the roots die they serve as a substrate for micro-organisms; polysaccharides and mucopolysaccharides and humic substances from the microbial metabolism are thought to be active in binding soil components together and in stabilizing aggregates.

There is not sufficient awareness of the importance of earthworms in promoting the formation of soil aggregates, although reviews by Satchell (1967) especially, and by Edwards and Lofty (1972) have clearly summarized evidence which indicates that soil structure is invariably good when there is an adequate earthworm population present.

There is abundant evidence (Edwards, Chapter 10) that change from pasture to arable cultivation causes a significant decrease in the earthworm population, largely by the depletion of plant residues. Satchell (1967), for example, has provided data which show that at Rothamsted, the earthworm population in Park Grass, a permanent mowing meadow, was significantly greater than that of Broadbalk under continuous wheat.

Work by Ehlers (1975) and Dexter (1978) focused attention on the reclamation of compacted soils by earthworms. Ehlers stressed the importance of worms in forming channels to the soil surface in a grey–brown podzol with a high silt content and which crusted after the soil aggregates were broken down by rainfall. He showed that a change from cultivation to no-cultivation, where seeds were direct drilled and the plant residues were not removed, allowed earthworm channels to develop and irrigation water to be transferred to a depth of 180 cm. Dexter showed that *Allolobophora caliginosa* tunnelled through compacted soil by ingesting soil particles.

Satchell's (1967) review refers to the abundant evidence showing that worm casts contain more water-stable aggregates than non-cast soils and that worm-worked soils are more water stable than unworked soils. He summarized the main explanations given up to his time of writing for the enhanced stability of aggregates in worm-worked soils. Some of these explanations considered that the stability arose from mechanical binding by vascular bundles from ingested plant remains or from the growth of fungal hyphae after the casts had been excreted. The majority opinion considered that the stabilizing components originate from the micro-organisms which proliferate in the ingested materials in the gut or develop after the mixture has been excreted, and one theory suggested that soil particles in the worm's intestine are cemented by calcium humate 'formed from ingested decomposing organic matter and calcite excreted by the calciferous glands'.

These various views on the role of earthworms in the formation of stable aggregates highlight the shrewdness of Darwin's observations on the important contribution earthworms make when mingling soil particles with vegetable debris and in saturating the mixture with their 'intestinal secretions'. Darwin's views on the importance of 'intestinal secretions' have not been substantially researched but considerable interest has been focused on the numbers and types of micro-organisms contained in casts and in the soil mixture in the earthworm alimentary canal. The micro-organisms contained in these media are broadly similar to those in soil (Satchell, 1967) but little definitive information is available about the products which they produce which influence aggregate stability. The same applies for the components of the 'intestinal secretions'.

Some interesting preliminary studies have not been followed up. For instance Bhandari et al. (1967) showed that the worm casts they studied had considerably more organic carbon and nitrogen than the parent soils and that the casts were significantly richer in polysaccharides than the soil. However, the work did not investigate the origins or composition of the polysaccharides. A logical approach to a study of the role of secretions by earthworms for the stabilization of soil aggregates might be to obtain the mixtures from intestinal and body wall mucus secretions and to measure the extent to which they interact with soil components to produce stable aggregates.

Gum-like, nitrogenous, mucus materials, probably mucoproteins left behind when earthworms pass over the soil surface, readily adhere to the surfaces with which they come into contact. Such body wall secretions stabilize the walls of the burrows (Edwards and Lofty, 1972) and soil particles would be readily formed into aggregate structures when mixed with them.

Hayes et al. (1981) have carried out preliminary investigations on the influence of earthworm body wall secretions on the stabilization of soil aggregates. These secretions had high nitrogen contents, were probably mucopolysaccharides, and were very sparingly soluble in water. However they could be dissolved by applying gentle heat to the mildly acidified aqueous mixture. Solutions were applied to soil aggregates (< 2 mm) which had been equilibrated at a moisture tension of 6 cm. After drying, disintegration of the aggregates was measured by measuring the extents to which clay-size particles were released from the aggregates. The results clearly indicated that the earthworm secretions had stabilizing effects on the aggregates. Crude fractionations of the polymers gave components which were more active than others for the stabilization process.

In cases where aggregate-forming secretions are resistant to microbial attack, the micro-aggregates and aggregates formed could be expected to be stable. However, it is very unlikely that earthworm secretions would be immune to attack by micro-organisms unless they were sterically shielded in the aggregate structures.

An explanation can now be attempted of Darwin's observation that castings ejected during rain have little stability and that those ejected during dry weather often disintegrate into small rounded pellets. As the castings pass through the intestinal tract the inorganic soil components are intimately mixed with the fragmented plant residues, with secretions from the epithelial lining of the intestine, and with the earthworm faeces. Ammonia, urea and possibly uric acid and allantoin are excreted to the exterior through the nephridiopores, or, in some species e.g.

Allolobophora antipae, into the gut (Edwards and Lofty, 1972) where the nitrogenous compounds would be mixed with the cast materials.

Soil particles are macerated in passing through the earthworm gizzard; aggregates would not survive this process, although domains could, and some very small micro-aggregates might. Clearly a reassociation of aggregate-forming components must take place as the materials pass through the earthworm intestine. The calciferous glands would provide appropriate amounts of calcium to allow divalent cation bridging to take place between the humic substances and the negatively charged inorganic colloid species. Bearing in mind the 'bridging' concepts previously outlined, aggregate formation is unlikely to take place in the presence of excess water in the intestine. Drying of the casts after ejection would bring the particles into close proximity and would allow polymers secreted in the intestine as well as the indigenous soil humic substances and polysaccharides to form the necessary bridges.

Fungal hyphae may or may not contribute to aggregate stability but it is less likely that they are involved in stabilizing micro-aggregates. The earthworm secretions which stabilize aggregates would probably not resist microbial decomposition and degradation of the larger aggregate-size particles may be attributable to biodegradation of the stabilizing polymers which would have little steric protection because of the large pore sizes in aggregates. Such polymers would have greater protection in the more stable micro-aggregates.

The macerated litter materials which are included in the aggregates from earthworm casts would provide a substrate for the proliferation of micro-organisms. With adequate moisture and nitrogen present, this litter would provide a substrate for microbes. Any mechanical binding strength which these residues provide would be lost as humification, resulting from microbial activity, takes place. However, the humus substances would provide appropriate polymers for binding soil particles and stabilizing aggregates. These polymers would have to be of the appropriate size and abundance to bridge the distances between the components of the aggregates. The larger the aggregates, the longer and the more numerous these bridges must be to confer stability. Large aggregates will also incorporate relatively large soil components having small surface areas with little affinities for the organic colloids. These aggregates will have much less resistance than the smaller units to mechanical impacts and will be more readily broken down, as observed by Darwin.

Casts from grassland soils may be more stable than those from arable soils simply because the former contain more humic substances initially. An obvious approach to this question would be to compare the stabilities of aggregates from casts from a single earthworm species, from soils with high contents of humus substances where the earthworms are fed with different substrates, ranging from fresh young vegetation to straw. The results might be compared with those for the same species fed in the same way but in soils with very low contents of humus materials. There is a need in such experiments to carefully identify the amounts and types of inorganic colloids present because their varying affinities for the different organic polymers present might explain variabilities in the stabilization of the wormcast aggregates.

3.6 REFERENCES

Barron, P. F., Wilson, M. A., Stephens, J. F., Cornell, B. A. and Tate, K. R. (1980) Cross-polarization ^{13}C NMR spectroscopy of whole soils, *Nature (London)*, **286**, 585–587.

Bhandari, G. S., Randhawa, N. S. and Maskina, M. S. (1967) On the polysaccharide content of earthworm casts. *Curr. Sci.*, **36**, 519–520.

Burchill, S., Hayes, M. H. B. and Greenland, D. J. (1981) Adsorption. In *The Chemistry of Soil Processes* (eds. D. J. Greenland and M. H. B. Hayes), Wiley, Chichester and New York. 221–400.

Cameron, R. S., Thornton, B. K., Swift, R. S. and Posner, A. M. (1972) Molecular weight and shape of humic acid from sedimentation and diffusion measurements on fractionated extracts. *J. Soil Sci.*, **23**, 394–408.

Cheshire, M. V. (1979) *Nature and Origin of Carbohydrates in Soils*, Academic Press, London and New York.

Darwin, C. (1881, 1927) *The Formation of Vegetable Mould through the Action of Worms with Observations on their Habits*, Faber and Faber, London.

Dexter, A. R. (1978) Tunnelling in soil by earthworms. *Soil Biol. Biochem.*, **10**, 447–449.

Edwards, C. A. and Lofty, J. R. (1972) *Biology of Earthworms*, Chapman and Hall Ltd., London. 283 pp.

Ehlers, W. (1975) Observations on earthworm channels and infiltration on tilled and untilled loess soil. *Soil Sci.*, **119**, 292–299.

Hayes, M. H. B. (1980) The role of natural and synthetic polymers in stabilizing soil aggregates. In *Microbial Adhesion to Surfaces* (eds. R. J. Berkeley *et al.*), Ellis Horwood, Chichester. 263–296.

Hayes, M. H. B. and Swift, R. S. (1978) The chemistry of soil organic colloids. In *The Chemistry of Soil Constituents* (eds. D. J. Greenland and M. H. B. Hayes), Wiley, Chichester and New York. 179–320.

Hayes, M. H. B. and Swift, R. S. (1981) Organic colloids and organo-mineral associations. *Bull. Int. Soc. Soil. Sci.*, **60**, 67–74.

Hayes, M. H. B., Isaacson, P. and Shimane, S. (1981) Stabilization of aggregates with soluble solutions of earthworm body wall secretions. Chemistry Department, University of Birmingham.

Julien, A. A. (1879) On the geological action of the humus-acids. *Proc. Am. Assoc. Sci.*, **28**, 311.

Olness, A. and Clapp, C. E. (1975) Influence of polysaccharide structure on dextran adsorption by montmorillonite. *Soil Biol. Biochem.*, **7**, 113–118.

Quirk, J. P. (1978) Some physico-chemical aspects of soil structural stability – a review. In *Modification of Soil Structure* (eds. W. W. Emerson, R. D. Bond and A. R. Dexter), Wiley, Chichester and New York. 3–16.

Satchell, J. E. (1967) Lumbricidae. In *Soil Biology* (eds. A. Burges and F. Raw), Academic Press, London and New York. 259–322.

Schnitzer, M. (1982) Quo vadis soil organic matter research, *Trans. 12th Int. Congr. Soil Sci.*, **5**, 67–78.

Schwertmann, U. and Taylor, R. M. (1981) The significance of oxides for the surface properties of soils and the usefulness of synthetic oxides as models for their study. *Bull. Int. Soc. Soil Sci.*, **60**, 62–66.

Theng, B. K. G. (1979) *Formation and Properties of Clay–Polymer Complexes*, Elsevier, Amsterdam and New York.

Turchenek, L. W. and Oakes, J. M. (1979) Fractionation of organo-mineral complexes by sedimentation and density techniques. *Geoderma*, **21**, 311–343.

Wilson, M. A., Barron, P. F. and Goh, K. M. (1981) Differences in structure of organic matter in two soils as demonstrated by ^{13}C cross polarisation nuclear magnetic resonance spectroscopy with magic angle spinning. *Geoderma*, **36**, 323–327.

Worobey, B. L. and Webster, G. R. B. (1981) Indigenous ^{13}C-NMR structural features of soil humic substances. *Nature (London)*, **292**, 526–529.

Chapter 4

Organic matter turnover by earthworms

J. D. STOUT

4.1 INTRODUCTION

Many factors affect the turnover of soil organic matter: the kind of organic matter, its origin, its initial distribution within the profile, the temperature and moisture regimes of the soil, its physical and chemical properties, and not least the populations upon whose activity the comminution, mixing and decomposition of soil organic matter eventually depend. In any particular soil these factors are inter-related and two difficulties are paramount in evaluating their relative roles and importance. First, the turnover of soil organic matter may take place over a very long period of time – centuries or even millenia – and second, the quantities of soil organic matter are vast and consist of a great diversity of chemical compounds. Thus any investigation is faced with the problem of diversity of age, kind and distribution of organic matter within the soil profile.

It is not practical to characterize the specific organic components of the soil, and their concentration is commonly expressed in terms of % carbon in the whole soil on a weight basis, which, given the volume–weight of a soil, can be transformed into a unit area and horizon distribution. But the carbon species in soil comprise three isotopes: the two stable isotopes, ^{12}C and ^{13}C, and the long-lived radiocarbon, ^{14}C, with a half-life of 5730 years. Knowledge of the concentration of these isotopes provides a clue to the origin and turnover rate of the soil organic matter (O'Brien and Stout, 1978). The ratio of the two stable isotopes is determined at photosynthesis and a broad distinction obtains between plants using the Calvin photosynthetic pathway and those using the Hatch–Slack pathway. Further discrimination occurs in the synthesis of the different cell components, the proportion of the heavier isotope (^{13}C) being greater in carbohydrates and organic acids than in fats and waxes. During decomposition, the

35

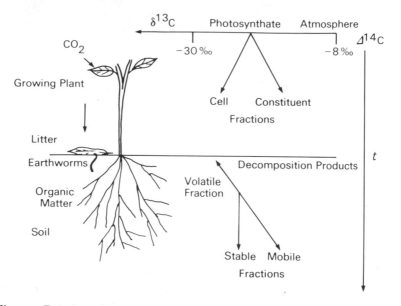

Fig. 4.1 Relation of the stable carbon isotope ratio ($\delta^{13}C$) and the radiocarbon age ($\Delta^{14}C$) of organic matter to the soil organic cycle.

volatile, mobile and stable fractions of the soil organic matter show fractionation of the stable isotopes (O'Brien *et al.*, 1981). Thus the ratio of the two stable isotopes provides a clue to the origin of the organic matter in soil and its state of decomposition (Fig. 4.1). Radiocarbon provides a chronology for, although atmospheric concentration of radiocarbon has varied with time and more recently been sharply elevated as a consequence of thermonuclear explosions, it still provides a guide to the relative age of different organic components in soil and to rates of turnover (Stout *et al.*, 1981). These data have been used to develop a model of organic matter turnover in a New Zealand soil (Fig. 4.2) in which the following factors were distinguished:

1. Annual input of carbon
2. Rate of decomposition
3. Diffusivity of carbon through the soil
4. Turnover time.

This model, however, did not distinguish the relative roles of the agents of turnover. Even within a small area of this soil, e.g. a 70 cm square, differences in the distribution and mixing of organic matter within the profile are enormous, with topsoil depth varying from 12 to 22 cm. These may reflect differences in root distribution or the burrowing of

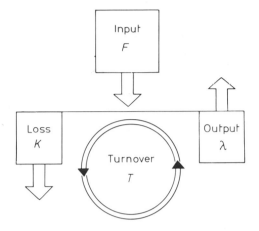

Fig. 4.2 The steady-state model of O'Brien and Stout (1978).

earthworms even when the soil parent material is relatively uniform. This illustrates the difficulties inherent in evaluating the role of different agencies in effecting the translocation, mixing and turnover of soil organic matter.

In this chapter, carbon isotope data from a range of soil sites of known history are discussed as an indication of the importance of earthworms in the turnover of soil organic matter.

4.2 SITES SAMPLED

These included sites with known earthworm populations and sites without earthworms. Because of the slow turnover of soil organic matter, the present vegetation is not necessarily an indication of the origin of plant debris incorporated into the soil and, in interpreting the data, some knowledge of the history of the site is as essential as a knowledge of its present climate, vegetation and soil properties.

4.2.1 Hubbard Brook, New Hampshire, USA

This site has been described in detail by Likens *et al.* (1977) and Bormann and Likens (1979). Their samples were taken from the exposed face of a trench cut across some metres of the catchment. The vegetative cover was natural hardwood forest dominated by beech (*Fagus grandifolia*), birch (*Betula alleghaniensis*) and maple (*Acer saccharum*). The area is not known to have earthworms present and the soil, overlying the granite parent rock, has a podzolic horizon (spodosol) of sharply contrasting organic and mineral horizons.

4.2.2 Brogden's Belt, Wytham Woods, Oxford, England

This is an elevated site with a rendzina-type soil on coral limestone overlain with glacial drift, with beech (*Fagus sylvatica*) the principal woody species and with little under-storey (hornbeam, *Carpinus betulus*) (Phillipson *et al.*, 1975, 1976, 1978). Moss and *Mercurialis perennis* cover part of the ground. The area studied is on sloping ground, so that most of the leaf fall tends to accumulate on the lower part of the slope, while the upper exposed part remains bare. Earthworm activity is confined to surface dwelling species on the exposed upper area and burrowing species occur in the lower moister area where the litter accumulates.

4.2.3 Tihoi soil, North Island, New Zealand

This soil, a loamy sand, has developed at a high altitude (*ca.* 600 m) on Taupo pumice, erupted 1800 years ago, under luxurious rain forest and a high rainfall (1500 mm y^{-1}) over rolling to hilly terrain (Herbert, 1980). It is a coarse textured, strongly leached podzolic soil (spodosol). Native earthworms were associated with the original native forest but where this has been replaced during the last century with exotic permanent pasture, the native acanthodrilid earthworms have disappeared and the pasture has tended to develop a turf mat. European lumbricid earthworms were introduced into the pasture area and these have slowly colonized the soil. The coarse texture of the soil, its acid pH and, initially, the low productivity of the pasture has restricted the multiplication of the populations, but they have steadily increased in number and slowly spread from their points of introduction. Initially, most of the soil organic matter in the pasture soil was the relic of the former forest vegetation and there is a contrast between this old organic matter and that originating from the pasture.

4.2.4 Wehenga soil, South Island, New Zealand

The Wehenga soil, a silt loam, developed on weathered schist under a tussock grassland vegetation at an elevation of 430 m and a mean annual rainfall of 700 mm (Ragg and Miller, 1978). This has been replaced by permanent exotic pastures of ryegrass (*Lolium perenne*) and clover (*Trifolium repens*). With the removal of the tussock grasses, the endemic megascolecid earthworm fauna disappeared but lumbricid earthworms were introduced into the pasture and grew well, multiplying and spreading from the site of colonization. The turf mat which forms in the absence of earthworms rapidly disappears as they colonize the pasture and the topsoil structure alters as the earthworms permeate the soil and mix the organic and mineral components (Stockdill, 1982). The soil which tends to retain moisture has proved to be a favourable site for earthworm activity.

4.3 MATERIAL AND METHODS

$\Delta^{14}C$ and $\delta^{13}C$ values were determined by the Radiocarbon Laboratory of the Institute of Nuclear Sciences, New Zealand Department of Scientific and Industrial Research, Lower Hutt, on whole soil samples collected from the soil horizons but freed of roots. Samples were pretreated with HCl to remove any carbonate present so that the isotope values determined relate simply to the organic soil carbon.

4.4 RESULTS AND DISCUSSION

4.4.1 Hubbard Brook, New Hampshire (Table 4.1)

At this site earthworms are either very rare or absent, and the soil profile has developed relatively undisturbed over a long period of time. A substantial part of the old plant litter remains on the surface of the mineral soil, forming a layer about 20 cm deep with no apparent 'bomb' radiocarbon enrichment. Physical comminution, the activity of arthropods and other fauna, transports part of this litter into and through the upper horizons and eventually into the drainage system of the catchment. There is an accumulation of organic matter in the B_h horizon and this material is very similar in carbon isotope values to the overlying

Table 4.1 Distribution of soil organic carbon below hardwood forest at Hubbard Brook, New Hampshire, USA with no earthworm activity, and at Brogden's Belt, Oxford, England, with and without topsoil worms.

Soil depth (cm)	C (%)	$\delta^{13}C$	$\Delta^{14}C$	Radiocarbon age (y BP)	Modern C (%)
Hubbard Brook					
		Site without worms			
11–10	42	−26.1	− 49 ± 4	390 ± 30	95 ± 0.4
7–10	2.6	−25.9	− 42 ± 7	330 ± 60	96 ± 0.7
24–40	0.16	−24.7	−459 ± 6	5070 ± 100	54 ± 0.6
Brogden's Belt					
		Site with no topsoil worms			
5–0	14	−27.0	+305 ± 11		130 ± 1
0–10	6.5	−25.8	− 9 ± 6		99 ± 1
		Site with topsoil worms			
0–10	9	−26.6	+ 36 ± 7		104 ± 1

surface litter. There is little fractionation of the stable isotopes, indicating very limited decomposition, and the radiocarbon concentration is similar to that of the litter layer, indicating relatively free movement of the more finely comminuted material through the upper podzolic horizon, where virtually no organic matter is retained. In the deep soil horizons (24–40 cm), there is a much lower concentration of total carbon (0.16 %), its mean radiocarbon age is about 5000 years, and the stable isotope ratio ($-24.7\%_o$) shows that the material has undergone significant chemical decomposition.

This site demonstrates how organic matter may move through a soil profile without the intervention of earthworms, but that it will move only through the upper, less compact horizons and not into the deeper layers of the profile, and that it moves and decomposes very slowly.

4.4.2 Brogden's Belt, Wytham Woods, Oxford (Table 4.1)

On the exposed site where only surface earthworms are active, plant litter accumulates on the surface to a thickness of about 5 cm. This accumulated plant material is enriched with 'bomb' radiocarbon ($+305\%_o$) and the stable isotope ratio is that of the living leaf ($-27\%_o$). The underlying mineral soil (0–10 cm) has a much smaller concentration of total organic carbon (6.5 vs 14 %) and shows no significant incorporation of the 'bomb' radiocarbon which had enriched the atmosphere and the growing plant tissues for almost 20 years. But the stable isotope ratio shows a significant fractionation ($-25.8\%_o$), indicating the more strongly decomposed state of the organic matter.

By contrast, the adjacent site on the leeward slope on to which much of the leaf litter tends to be blown has earthworms active in the mineral topsoil and no surface litter accumulation. The fresh plant debris is rapidly incorporated into the topsoil (0–10 cm), which consequently is richer in total carbon (9 %) and in 'bomb' radiocarbon ($+36\%_o$) than the other site but shows less fractionation of the stable isotopes, i.e. decomposition of the soil organic matter. It is clear, in this case, that the presence of burrowing topsoil earthworms has led to a significant level of incorporation of fresh organic matter into the topsoil.

Mineralogical analysis of samples of the topsoils showed that the upper site topsoil has less carbonate (10 %) than the lower site topsoil (40 %), and that the silt fraction of the upper site showed only traces of carbonate ($\sim 1\%$), although it was common in the silt fraction of the lower site (10–30 %). The quartz-rich fraction is derived from quartz-rich metamorphic or sedimentary rocks – presumably contained in the glacial drift – while the carbonate fraction comes from the recrystallized calcitic limestone, which is very nearly pure carbonate. These mineralogical results were confirmed by the carbonate carbon figures and, taken in conjunction with the total organic carbon figures, also indicate very much less mixing of the soil materials at the upper site than the lower.

4.4.3 Tihoi soil, North Island

An area of rolling hilly country at an altitude of 500–600 m was sampled under secondary-growth rain forest and under recently introduced pasture (Table 4.2). The forest site was on a steep slope and had a high concentration of total carbon throughout the soil profile, with slightly less in the eluviated horizon than in the B_{sh}. This reflects continual litter fall of the forest and the deep rooting of the forest trees. At all depths the carbon was enriched with 'bomb' radiocarbon, indicating its recent origin and the free movement of the finely comminuted litter material and extensive distribution of the roots.

Table 4.2 Distribution of soil organic carbon in Tihoi soil, North Island, New Zealand, below native forest and in a recent exotic pasture without earthworms.

	Horizon	Depth (cm)	C (%)	$\delta^{13}C$	$\Delta^{14}C$
Spodosol under native forest	A_1	6–15	7	−28.8	+ 73
	E	15–24	2.7	−27.0	+ 56
	B_{sh}	24–37	3.5	−26.7	+ 49
Spodosol under exotic pasture (without earthworms)	A_1	0–8	10	−26.6	+144
	B_{sh}	8–27	2.4	−25.9	+ 59
	B_s	27–44	0.8	−25.9	+ 1

In the recently established exotic pasture on a gentler slope, the accumulation of organic matter was concentrated in the topsoil because of the shallower rooting of the pasture. The high level of 'bomb' radiocarbon indicates its recent origin. An occasional earthworm was found at the site but a large population had not yet become established. However, there was no surface mat of old plant litter as occurs in the complete absence of earthworms.

Samples were also collected from another area of the Tihoi soil, at an altitude of 450 m, where the pastures had been established for much longer and there was no residual forest (Table 4.3). Lumbricid earthworms had been deliberately introduced about 6 years previously but had spread and multiplied very slowly (Table 4.4), *Allolobophora* caliginosa and *Lumbricus rubellus* being the most successful colonists. *L. rubellus* is a surface-feeding worm surviving the summer drought in egg capsules, while *A. caliginosa*, a topsoil-dwelling species, aestivates coiled up in the soil (Martin, 1978). The depth of horizons in the soil profile differed from the hillier area, there being a much deeper topsoil. At the original site of earthworm introduction, the carbon of the A_1 horizon showed greater enrichment with 'bomb' radiocarbon than an adjacent site more recently invaded by the earthworms, and the A_2 horizon showed a much lower concentration of total carbon but a much higher concentration of 'bomb' carbon. These data suggest that the presence of earthworms had not only helped the mixing of organic matter into the mineral horizons, but also accelerated its decomposition, including the decomposition of the older carbon.

Table 4.3 Distribution of soil organic carbon in Tihoi soil, North Island, New Zealand, below three different exotic pastures. (Sampled 6th September 1978).

	Horizon	Depth (cm)	C (%)	$\delta^{13}C$	$\Delta^{14}C$
Pasture with established earthworm population (748 earthworms m^{-2})	A_1	0–10	5.6	−26.6	+ 152
	A_2	10–20	2.8	−26.0	+ 9
	B_{sh}	20–35	1.3	−25.1	− 47
Pasture recently invaded by earthworms (412 earthworms m^{-2})	A_1	0–10	5.5	−26.3	+ 105
	A_2	10–20	9.5	−25.0	− 10
	B_{sh}	20–35	1.4	−25.4	− 54
Pasture without earthworms after ploughing (no earthworms)	A_1	0–10	6.6	−25.7	+ 96
	A_2	10–20	5.2	−25.6	+ 41
	B_{sh}	20–35	1.5	−24.7	+ 4

Table 4.4 Earthworm populations m^{-2} at Tihoi soil sites, 6 years after introduction of lumbricid species.

	A. caliginosa	A. chlorotica	A. longa	A. rosea	L. rubellus	Total
Site of introduction	467	311	99	4	125	1006
Adjacent site*	130	0	0	0	181	311

* The introduction site and adjacent sites were 50 m apart at NZMS 1, N93, 267548. Populations were sampled on 29th October 1973 by handsorting soil from five pits, each 21.5 × 21.5 cm.

At a third site, where earthworms had not yet invaded the pasture, but where it had been ploughed, there was a more even distribution of 'bomb' radiocarbon through the profile, but the concentration of total carbon showed less reduction than where earthworms were active. This indicates the difference between the two roles of mixing organic matter through the horizons and of decomposition.

4.4.4 Wehenga soil, South Island (Table 4.5)

Samples were taken from a pasture area without introduced lumbricid earthworms and from an adjacent area where *A. caliginosa* was active. The profile without introduced earthworms had a surface mat of plant litter whose very high radiocarbon level indicated its recent origin and the stable isotope ratio showed its very close affinity to that of the growing plant. The mineral soil showed a much lower concentration of radio-carbon, but the stable isotope ratio indicated that considerable de-composition had taken place.

Where earthworms were present, there was no surface mat of litter and the top 18 cm of mineral soil showed enrichment with 'bomb' radio-carbon as well as a higher level of total carbon than in the uninvaded soil. The stable isotope ratio indicates that the organic matter had decomposed significantly. The mean radiocarbon age in the deeper horizons is comparable with that of the Judgeford soil (O'Brien and Stout, 1978), suggesting a similar rate of turnover. Comparison of the two profiles indicates the degree to which incorporation of organic matter into the mineral soil is due simply to root growth or the diffusion of soluble or comminuted material through the profile and how far it is due to the activity of earthworms. In the top 10 cm of the mineral soil without worms the concentration of organic carbon is < 30 mg cm^{-3}, with worms it is > 40 mg cm^{-3}, and the total weight of carbon is 294 and 420 mg cm^{-3} respectively. In the next horizon, there is roughly 60% more total carbon in the horizon with worms (Table 4.5), though there is no difference in the deeper horizons where the radiocarbon age clearly indicates little contamination from 'bomb' carbon. Overall, the mineral horizons to 40 cm depth have a third more carbon in the profile with worms. In part, this is due to the incorporation of the turf mat into the mineral soil by the worms and, in part, to the increased annual input of carbon following the stimulation of plant growth.

4.4.5 Analysis using the diffusion model (by B. J. O'Brien)

It is possible to analyse the stable carbon and radiocarbon data for the Wehenga soil using the diffusion model previously used to study the

Table 4.5 Distribution of soil organic carbon in two profiles of Wehenga soil, South Island, New Zealand.

Soil depth (cm)	C (mg cm^{-3})	C (%)	Total C (mg cm^{-2})	δ^{13}C	Δ^{14}C	Radiocarbon	Modern (%)
Profile without introduced worms							
3–0	3.8	19	114	−29	+400		140
0–10	2.9	3.5	294	−27.5	+89		109
10–17	2.0	2	140	−25.0	−11	70±30	99
17–24	1.2	1.1	85	−26.2	−62	510±40	94
24–40	8	0.7	134	−26.6	−121	1050±50	88
40–56		0.3		−27.6	−250	2360±70	75

Total C (0–40 cm) 653 mg cm^{-2}
Annual input of C 33 mg cm^{-2}

Soil depth (cm)	C (mg cm^{-3})	C (%)	Total C (mg cm^{-2})	δ^{13}C	Δ^{14}C	Radiocarbon	Modern (%)
Profile with introduced worms							
0–10	42	4.9	420	−27.8	+194		120
10–18	29	3	288	−28.0	+102		110
18–25	12	1.1	121	−26.9	−22	170±40	98
25–40	7	0.6	72	−27.0	−127	1100±50	88
40–59		0.3		−27.2	−224	2070±80	78
59–73		0.08		−26.8	−402	4230±90	60

Total C (0–40 cm) 843 mg cm^{-2}
Annual input of C 42 mg cm^{-2}

Judgeford soil (O'Brien and Stout, 1978). In that model, vertical movement of carbon in the soil profile was by diffusion, and respiration was accounted for by first-order kinetics. Under equilibrium conditions this model predicts that the concentration of carbon in the soil profile at depth z, is given by the relation $C(z) = C_0 \exp(-z/z_0)$, where C_0 is the concentration of carbon at the surface. By fitting this expression to the carbon profiles given in Table 4.5, z_0 is estimated to be 17.8 cm. It has about the same value for the profile with earthworms and that without. If F is the flux of carbon into the surface layer of soil, then $F = C_0 \kappa/z_0$ and $\kappa = z_0^2/\tau$, where κ is the diffusivity of carbon in the soil and τ is the mean respiration time (O'Brien and Stout, 1978). Using the $\Delta^{14}C$ data for the 0–10 cm profile of the pasture without worms, and assuming that ^{14}C-enriched carbon was available from 1962 to 1977, F is estimated to be 0.003 g cm^{-2} y^{-1} or 300 kg ha^{-1} y^{-1}. Using the expressions given above, κ is estimated as 1.7 cm^2 y^{-1} and τ as 180 y. In the same way, for the pasture with introduced earthworms, F is estimated as 0.01 g cm^{-2} y^{-1} or 1000 kg ha^{-1} y^{-1}; κ is 4.7 cm^2 y^{-1} and τ is 67 years.

The parameters for the Wehenga soil with introduced worms are fairly similar to those of the Judgeford soil referred to above. The diffusivity of the Wehenga soil is a little lower, the respiration time somewhat shorter and the flux of carbon into the soil, about one half. However, comparison of the parameters for the two Wehenga soil profiles shows that the introduction of earthworms has trebled the diffusivity, the respiration rate and the flux of carbon into the soil. The increase in F and κ can be attributed directly to the activity of earthworms in drawing surface plant material into the soil and in moving organic carbon through the profile. The increased respiration rate undoubtedly results from the increased biological activity caused by the greater flux of new carbonaceous material down the soil profile and probably also by better soil aeration.

4.5 CONCLUSIONS

Knowledge of the carbon isotope ratios and the radiocarbon age of the soil organic matter provides an understanding of the origin, history and decomposition of the soil organic matter. There is a marked contrast between the values for unmodified virgin forest, such as the Hubbard Brook hardwood forest site, and secondary growth forest, such as the Tihoi forest site. This is shown by both the age of the organic accumulation and its distribution which may derive from the growth of roots or the translocation of organic detritus through the profile. The pattern of translocation also reflects soil mineral differences and the nature of the organic detritus. Even though the climate, soil and vegetation at both Hubbard Brook and Tihoi lead to the formation of a podzolized soil, their differences in age, parent material and vegetation provide contrasting patterns of organic accumulation. These patterns must be identified and understood before the role of soil animals can be evaluated.

Earthworms effect two distinct functions in soil: a physical one of cultivation, 'boring, perforating, and loosening the soil, and rendering it pervious to rains and the fibres of plants, by drawing straws and stalks of leaves into it; and, most of all, by throwing up such infinite numbers of lumps of earth called worm-casts', as Gilbert White recognized; and an acceleration of organic decomposition. The former can also be effected by mechanical means, especially by ploughing and harrowing, and the data of the ploughed site of the Tihoi soil illustrate the effectiveness of mechanical ploughing in mixing the organic matter through the topsoil, which may be compared with the progressive mixing of the horizons following earthworm invasion. The earthworm's role in organic decomposition is less easy to evaluate. Its direct contribution to soil metabolism may not appear very large; Phillipson *et al.* (1978) estimate that at the Brogden's Belt site the earthworms contribute about 4–5 % of the total soil metabolism. But indirectly, by providing a well-comminuted pabulum for microbial activity and by constantly redistributing the organic matter throughout the soil profile ensuring the rapid recycling of plant nutrients, they stimulate plant production and affect soil organic matter levels and total soil metabolism. This is shown by both the Tihoi and the Wehenga data. Soil respiratory values give little measure of this effect but carbon isotope values enable them to be much more critically evaluated.

O'Brien and Stout (1978) identified four factors in the movement and turnover of soil organic matter: C input, rate of decomposition, diffusion down the soil profile, and turnover time (Fig. 4.2), and they estimated values for these from the carbon isotope data of the soil. These factors, however, clearly integrate climatic, soil and biological functions, such as productivity, palatability, food chain length, and biological and soil interactions, and such estimates also assume a steady state. One difficulty is the wide difference in physical and chemical mobility of the different organic fractions in soil which is often a function of their age and origin. Jenkinson and Rayner (1977) have analysed the turnover of organic matter in some of the classical Rothamsted experiments and have estimated the differential rates of turnover of the various fractions. The fundamental difficulty, however, is to obtain adequate reliable quantitative data, for, even with only four variables as in the O'Brien and Stout model, because they are interdependent, four equations are required to solve the unknowns. These are provided by a measure of the productivity or input, the stable and radioisotope concentrations, and the volume–weights of the soil carbon. Provided this sort of data is available it is possible to obtain quantitative figures for the four factors and provide a convincing soil organic matter balance sheet. Further, if such data can be obtained for the same soil where or when there has been a major change in the function of any one of the variables, then it becomes possible to provide an estimate for the components of the variable factor and this we have done for the Wehenga site whose history and properties are moderately well documented. The merit of this approach is that it provides an integrated value for organic turnover in the soil, rather than the additive value provided by analysing the various components separately and it facilitates comparison between soils and ecosystems of very different origin and character.

4.6 ACKNOWLEDGEMENTS

The isotope determinations were carried out by H. S. Jansen, R. C. McGill, M. K. Burr and R. G. Currie of the Radiocarbon Laboratory, Institute of Nuclear Sciences, Department of Scientific and Industrial Research, Lower Hutt, New Zealand. The earthworms at the Tihoi site were identified by R. Nielson, formerly of the Ruakura Agricultural Research Centre, Hamilton, New Zealand, who first introduced lumbricid worms into this area. Introductions at the Wehenga site were made by S. M. J. Stockdill.

4.7 REFERENCES

Bormann, F. H. and Likens, G. E. (1979) *Pattern and Process in a Forested Ecosystem.* Springer-Verlag, New York. 253 pp.

Herbert, J. (1980) Structure and growth of dense podocarp forest at Tihoi, central North Island, and the impact of selective logging. *N. Z. J. For.,* **25**, 44–57.

Jenkinson, D. S. and Rayner, J. H. (1977) The turnover of soil organic matter in some of the Rothamsted classical experiments. *Soil Sci.,* **123**, 298–305.

Likens, G., Borman, F. H., Pierce, R. S., Eaton, J. S. and Johnson, N. M. (1977) *Biogeochemistry of a Forested Ecosystem.* Springer-Verlag, New York. 146 pp.

Martin, N. A. (1978) Earthworms in New Zealand agriculture. *Proc. 31st N. Z. Weed and Pest Control Conference,* 176–180.

O'Brien, B. J. and Stout, J. D. (1978) Movement and turnover of soil organic matter as indicated by carbon isotope measurements. *Soil Biol. Biochem.,* **10**, 309–317.

O'Brien, B. J., Stout, J. D. and Goh, K. M. (1981) The use of carbon isotope measurements to examine the movement of labile and refractory carbon in soil. In *Carbon Dioxide Effects Research and Assessment Program, Flux of Organic Carbon by Rivers to the Oceans.* Report of a Workshop, Woods Hole, Massachusetts, 1980. US Department of Energy, CONF-8009140.

Phillipson, J., Putnam, J., Steel, J. and Woodell, S. R. J. (1975) Litter input, litter decomposition and evolution of carbon dioxide in a beech woodland. *Oecologia (Berl)*, **20**, 203–217.

Phillipson, J., Abel, R., Steel, J. and Woodell, S. R. J. (1976) Earthworms and the factors governing their distribution in an English beechwood. *Pedobiologia*, **16**, 258–285.

Phillipson, J., Abel, R., Steel, J. and Woodell, S. R. J. (1978) Earthworm numbers, biomass and respiratory metabolism in a beech woodland – Wytham Woods, Oxford. *Oecologia (Berl)*, **33**, 291–309.

Ragg, J. M. and Miller, R. B. (1978) Soil Survey of part Taieri Uplands Otago, New Zealand. *Soil Bur. Bull. N. Z.*, **39**, 56 pp.

Stockdill, S. M. J. (1982) Effects of introduced earthworms on the productivity of New Zealand pastures. *Pedobiologia*, **24**, 29–36.

Stout, J. D., Goh, K. M. and Rafter, T. A. (1981) Chemistry and turnover of naturally occurring resistant organic compounds in soil. In *Soil Biochemistry* (eds. E. A. Paul and J. N. Ladd), Vol. 5, Marcel Dekker, New York. 1–73.

Chapter 5

Effect of earthworms on the disappearance rate of cattle droppings

P. HOLTER

5.1 INTRODUCTION

It is well known that several earthworm species aggregate under dung pats (Svendsen, 1957; Boyd, 1958; Martin and Charles, 1979). In addition, pot experiments (Guild, 1955; Barley, 1959) have shown that dung is readily eaten, and thereby removed from the surface by earthworms. When sufficiently abundant, worms have therefore been supposed to play an important role in the decay and disappearance of dung pats (e.g. Barley, 1961). There is, however, a lack of studies assessing the actual impact of worms in the field combining (1) quantification of earthworm populations under and in dung pats, and (2) measurements of the disappearance rate of the pats.

This was attempted in two Danish pastures. Some of the material has been published earlier (Holter, 1979) but additional information from another pasture permits a different analysis of the combined data. Further, possible interactions between earthworms and the dung beetle *Aphodius rufipes* (L.) are discussed.

5.2 STUDY AREAS

Both sites are located near Hillerød in northern Zealand, 30–35 km north of Copenhagen. Pasture A (Holter, 1979), belonging to the National Experimental Farm 'Trollesminde', is surrounded by intensively farmed arable land. The soil is a well-drained sandy loam, pH about 6.5, loss on ignition 5 %. Pasture B ('Bøgemose'), at Strødam about 4 km north of A, is surrounded by woodland. The soil is a dark fen soil, pH about 6.0, loss

49

on ignition 34 %; drainage is rather poor. At both sites, an experimental area of 150 m² of pasture was fenced to keep cattle out. The pastures were grazed by heifers and/or calves.

5.3 METHODS

Artificial, standardized dung pats, 22–24 cm diameter, each containing 2 kg of fresh (0–6 h old) cattle dung collected in the pastures and thoroughly mixed before use, were laid out about 1st August on nylon netting (7 mm mesh). The netting permitted later quantitative removal of the partly degraded dung pats and impeded dung removal by dorbeetles (*Geotrupes spiniger*); these large beetles, which occasionally occurred in the pastures, were the only dung animals that were unable to pass through the netting.

Pats, usually four in each sample, were collected after 12–14, 35–37 and sometimes 60–80 days. Their organic matter content (ignition loss at 500°C) was determined (cf. Holter, 1979). This method of recording the remaining dung seemed preferable to dry weight (as used by e.g. Dickinson *et al.*, 1981) which was sometimes strongly affected by soil mineral particles adhering to the dung or brought into pats by earthworms.

Earthworms in the pats were collected by handsorting. Worms in the soil below were expelled by applying 0.3 % formalin solution in three lots of 2.5 litres at 10 min intervals (Satchell, 1971) to a circular area of 0.125 m² concentric with each pat removed. After determination of their total fresh weight, the worms were preserved in 70 % alcohol, identified and counted.

This procedure was followed in 1975 and 1977 on pasture A and in 1979 and 1980 on pasture B. In 1977, two additional treatments were included.

(1) During the first six nights (from about 19:30 h to 07:00 h) some of the pats were covered by gauze tents (Holter, 1977). These tents excluded night-flying insects, notably *Aphodius rufipes* and *A. rufus* which invade pats during the first few nights. Day-flying insects had free access.

(2) Larvae of *A. rufipes* (80), freshly collected in pats about two weeks old, were added to each of eight experimental pats, aged 14 days, that had been covered during the first six nights (Holter, 1979). The other covered pats (treatment 1) were manipulated in the same way but without addition of larvae.

The possibility that dung beetles attract worms was further tested as follows. Five wooden boxes (1.0 m × 0.5 m × 0.3 m) containing a 10-cm layer of sward and topsoil from pasture B were placed outdoors adjacent to the Copenhagen laboratory. A fresh dung pat (1.0 kg, 16 cm diameter) was placed on the turf in each end of the boxes, and a supporting frame for gauze tents (Holter, 1977) was placed around each pat. In each box, about 50 g of earthworms from pasture B were added in a furrow 4–5 cm deep halfway between the two pats. Additionally, 30 adult *A. rufipes* were added to one pat in each box. These five pats were covered by tents for the first five nights to prevent the beetles escaping. The other five pats were covered for the next three nights to prevent invasion by beetles leaving the first set of pats. On day 9, tents and supports were removed from all the pats, now freely exposed to earthworm colonization. After 30 days the earthworms in pats and in the soil of a 0.125 m² area around each pat were collected by handsorting, identified, counted and weighed. Pats with beetles added contained 25 ± 9 (SE) third-instar larvae of *A. rufipes*.

Finally, in 1980 (pasture B), the invasion of worms in twelve dung pats in the field was delayed by removing a 10 cm deep turf (35 cm × 35 cm); lining the hole with a 90 cm × 90 cm piece of nylon net (1 mm mesh) and replacing the turf. The free edges of the net were held by a peg in each corner and a dung pat was placed on the enclosed turf.

5.4 RESULTS AND DISCUSSION

5.4.1 The earthworm fauna attracted by dung pats

The work in pasture A (1977) has been described by Holter (1979) and the data presented here are mainly from pasture B. Figure 5.1 (freely exposed pats, 1980) shows a rapid increase of earthworm biomass to a maximum of 71 g at 36 days. The changes in biomass were accompanied by cor-

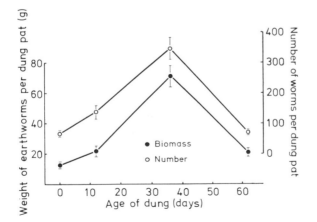

Fig. 5.1 Biomass and number of worms in dung pats in relation to dung age; means ± SE (*n* = 4).

responding changes in the number of worms. The biomass curve seems to indicate temporary aggregation and there is no evidence of substantial changes in the mean weights of the individual earthworms. At 12 and 62 days, worms expelled from the soil contributed almost the entire biomass. At 36 days, worms in the dung contributed about 12 % of the total fresh weight; practically all were *Lumbricus castaneus* (Sav.).

Five or six species occurred in pasture B (Fig. 5.2): *Lumbricus castaneus* (Sav.), *L. festivus* (Sav.), *L. rubellus* Hoffm., *Allolobophora rosea* (Sav.), and *A. caliginosa* (Sav.). From Reynolds (1977), 298 of the latter were identified as *Aporrectodea turgida* and seven as *A. tuberculata* (Eisen). The most abundant species was *L. castaneus* which also contributed most of the *Lumbricus* juveniles. *L. rubellus* and *L. festivus* were much less abundant, slower to arrive and stayed longer. The two *Allolobophora* species contributed little to biomass or numbers; they were apparently not attracted to the dung (cf. Svendsen, 1957; Boyd, 1958).

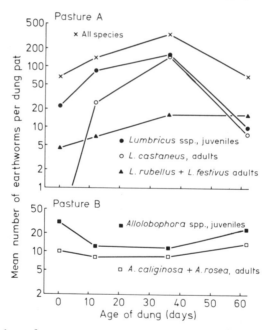

Fig. 5.2 Number of worms per pat in relation to dung age; means ($n = 4$). *Lumbricus* species comprise *L. castaneus*, *L. rubellus* and *L. festivus*. *Allolobophora* species comprise *A. caliginosa* and *A. rosea*.

The fauna of pasture A was quite different: *Lumbricus terrestris* L., *Allolobophora longa* Ude, *A. rosea*, and *A. caliginosa* (again mostly *Aporrectodea turgida*) were the only species present. *L. terrestris* and *A. longa* contributed nearly all the biomass (Fig. 5.3).

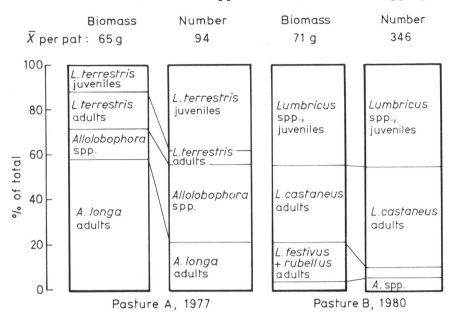

Fig. 5.3 Earthworm populations of dung pats aged 5–6 weeks.

5.4.2 Disappearance of dung in relation to earthworm biomass

Figure 5.4 shows the combined data from pasture A and B. The dung organic matter that disappeared from 12–14 to 35–37 days is plotted against the biomass of earthworms supposed to be active in the dung during that period. This biomass activity is expressed as 'earthworm-g-days', the area under the graph relating earthworm biomass to dung age. The line is fitted by a geometric mean regression (Ricker, 1973).

The individual points resulted from the following field treatments: a–c indicate pats laid out simultaneously in 1977 (pasture A); a were freely exposed; b were covered during the first 6 nights with *Aphodius* larvae added at 14 days; c were covered like b but without addition of larvae. d represents freely exposed pats from 1975 (pasture A); owing to an exceptionally hot and dry late summer earthworms were completely absent. e and f represent freely exposed pats from pasture B (1979–80). g indicates pats laid out simultaneously with f; enclosed in nylon net and hence subject to a considerably delayed invasion of worms; the species that eventually arrived were the same as those in e and f.

The graph exhibits an extremely close ($r = 0.996$) linear relation between rate of dung disappearance and earthworm biomass. Although causal inferences from correlations may be dangerous, Fig. 5.4(A)

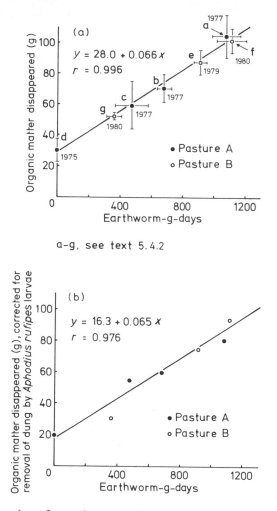

a–g, see text 5.4.2

Fig. 5.4 Regression of organic matter disappearance from dung pats on their earthworm biomass; means $\pm$ SE (A).

strongly supports the idea that earthworms promote the disappearance of dung pats. The close fit seems even more remarkable considering that quite different species occurred in pasture A and B; the net effect was approximately the same in either case: the disappearance of 0.066 g of dung organic matter per 'earthworm-g-day'. As in points a, e and f, 1000 g-days lead to the removal of nearly one-third of the dung initially present.

Besides earthworms, the most important animals in pats aged 13–36 days in August and September were larvae of *Aphodius rufipes*. Their highly variable abundance was determined as in Holter (1979). The amount of dung removed by one larva is roughly known from pot experiments (Holter, 1977) and seems little affected by the presence or absence of earthworms. Hence, a crudely estimated effect of *A. rufipes* larvae could be subtracted from each *y*-value in Fig. 5.4(a), resulting in the corrected relation shown in Fig. 5.4(b). Since the slope is unchanged, one earthworm-g-day still leads to the disappearance of about 0.065 g of dung organic matter.

5.4.3 Interactions between *Aphodius rufipes* and earthworms

Information on this problem was obtained from pasture A. In 1974 and 1977, *Aphodius rufipes* and the less important *A. rufus* were excluded by gauze tents from pats laid out around 1st August; earthworm populations under and in pats were only assessed in 1977. The results suggest that treated pats in 1977 were invaded by worms much more slowly than freely exposed controls (Fig. 5.4(A), a and c). Further, the disappearance rate was strongly affected: in 1974, 75 % disappearance times for treated pats and controls were 54 and 32 days respectively and, in 1977, 64 and 42 days (Holter, 1979). A simple explanation of these findings would be that the activity of *A. rufipes* beetles and larvae promotes the decay of dung indirectly by stimulating the aggregation of earthworms.

This was partly tested by the experiment in which *A. rufipes* larvae were added to pats, aged 14 days, from which the beetles had originally been excluded. The position of point b in Fig. 5.4(A) supports the hypothesis: on average, these pats scored 683 'earthworm-g-days', against 476 g-days for the pats which held neither beetles nor larvae (point c). The covering of pats by gauze tents may have prevented colonization by surface-moving worms, but the tents were only present while the dung was 0–6 days old and probably rather unattractive to worms.

The wooden box experiment (Section 5.3), provides further evidence (Table 5.1). The total quantity of worms, whether expressed as numbers

Table 5.1 Influence of dung beetles (*Aphodius rufipes*) on the number and biomass of earthworms (pasture B species) from 30-day-old dung pats.

Treatment	Quantity	Earthworms per dung pat, mean ± SE (n = 5)		
		All species	*Lumbricus*	*Allolobophora*
No dung insects	Number	36.0 ± 3.2	26.8 ± 2.9	9.2 ± 0.7
	Biomass (g)	19.2 ± 1.8	16.5 ± 1.7	2.8 ± 0.4
Beetles added	Number	58.0 ± 3.3	43.4 ± 3.6	14.6 ± 2.5
	Biomass (g)	28.2 ± 1.1	24.1 ± 1.6	4.0 ± 0.9

or biomass, was significantly higher ($P < 0.02$, t-test for paired comparisons) in pats with dung beetles added. This holds also for *Lumbricus* alone ($P < 0.05$ for numbers and biomass), whereas the difference is nonsignificant for *Allolobophora*. The mean ratio of worm numbers in pats with beetles to worm numbers in pats without beetles was 2.4 for *L. castaneus* and 1.5 for *L. festivus + L. rubellus*. Thus *L. castaneus* particularly preferred pats with beetles. Since *L. castaneus* only made up 22% of the individuals, against 45% in the field (cf. Fig. 5.3), this experiment may underrate the increase of worm aggregation by dung beetles, and perhaps other dung insects, in the field.

In conclusion, the evidence confirms the idea that *A. rufipes*, probably both beetles and larvae, promotes the aggregation of earthworms but the mechanisms responsible are not known. One possibility is that worms may find it difficult to attack the remarkably firm and coherent texture that often develops unless the dung is loosened by beetles and their larvae. Another possibility (Satchell, pers. comm.) is that beetles and their larvae may open up the pat at an early stage, thereby releasing ammonia that might otherwise deter worm colonization. A comparable increased release of ammonia from cattle dung by fly larvae has been reported by Harris *et al.* (1980).

5.5 SUMMARY

The effect of earthworms on dung pat disappearance was assessed by combining quantification of earthworm populations under and in pats, and measurements of the disappearance rate of these pats. Measurements were made in late summer in two Danish pastures (A and B) with quite different earthworm faunas. Pats in A were dominated by *Allolobophora longa* and *Lumbricus terrestris* and pats in B were occupied by *Lumbricus castaneus*, *L. festivus* and *L. rubellus*. Irrespective of species, the same biomass of worms seemed to result in the same disappearance rate of dung. The combined results (seven points) yielded a close linear relation between weight of dung organic matter disappearing during the dung age interval 13–36 days (y), and the supposedly active biomass of worms ('earthworm-g-days') under and in pats (x): $y = 28.0 + 0.066x$ ($r = 0.996$). Hence, 1000 'earthworm-g-days', a normal value in the two pastures, would lead to the disappearance of about 66 g organic matter, nearly one third of that initially present in pats. This result was not affected by a correction of y-values for dung removal by larvae of the dung beetle *Aphodius rufipes*.

The evidence presented demonstrates that *A. rufipes* beetles and larvae stimulate the aggregation of earthworms in pats, thereby indirectly promoting disappearance of the dung.

5.6 ACKNOWLEDGEMENTS

B. Brandt provided skilful technical assistance, including the drawing of figures. I am grateful to J. Satchell for useful suggestions and to C. Overgaard Nielsen for discussions and critical reading of the manuscript.

5.7 REFERENCES

Barley, K. P. (1959) The influence of earthworms on soil fertility. II. Consumption of soil and organic matter by the earthworm *Allolobophora caliginosa* (Savigny). *Aust. J. Agric. Res.*, **10**, 179–185.

Barley, K. P. (1961) The abundance of earthworms in agricultural land and their possible significance in agriculture. *Adv. Agron.*, **13**, 249–268.

Boyd, J. M. (1958) The ecology of earthworms in cattle-grazed machair in Tiree, Argyll. *J. Anim. Ecol.*, **27**, 147–157.

Dickinson, C. H., Underhay, V. S. and Ross, V. (1981) Effect of season, soil fauna and water content on the decomposition of cattle dung pats. *New Phytol.*, **88**, 129–142.

Guild, W. J. McL. (1955) Earthworms and soil structure. In *Soil Zoology* (ed. D. K. McE. Kevan), Butterworths Scientific Publications, London. 83–89.

Harris, R. L., Ilcken, E. H., Blume, R. R. and Oehler, D. D. (1980) The effects of horn fly larvae and dung beetles on ammonia loss from bovine dung. *Southwest. Entomol.*, **5**, 104–106.

Holter, P. (1977) An experiment on dung removal by *Aphodius* larvae (Scarabaeidae) and earthworms. *Oikos*, **28**, 130–136.

Holter, P. (1979) Effects of dung-beetles (*Aphodius* spp.) and earthworms on the disappearance of cattle dung. *Oikos*, **32**, 393–402.

Martin, N. A. and Charles, J. C. (1979) Lumbricid earthworms and cattle dung in New Zealand pastures. In *Proc. 2nd Australasian Conf. Grassland Invertebrate Ecol.* (eds. T. K. Crosby and R. P. Pottinger), Government Printer, Wellington. 52–55.

Reynolds, J. W. (1977) The earthworms (Lumbricidae and Sparganophilidae) of Ontario. *Life Sci. Misc. Pub., R. Ont. Mus., Toronto*. 141 pp.

Ricker, W. E. (1973) Linear regressions in fishery research. *J. Fish Bd. Can.*, **30**, 409–434.

Satchell, J. E. (1971) Earthworms. In *Methods of Study in Quantitative Soil Ecology: Population, Production and Energy Flow* (ed. J. Phillipson). IBP Handbook no. 18, Blackwell, Oxford. 107–127.

Svendsen, J. A. (1957) The behaviour of lumbricids under moorland conditions. *J. Anim. Ecol.*, **26**, 423–439.

Chapter 6

Soil transport as a homeostatic mechanism for stabilizing the earthworm environment

A. KRETZSCHMAR

6.1 EARTHWORMS AS SOIL CONDITIONERS

During the 1880s, many important works on the soil were published and it is of a great interest to put together what Dokuchaev, Darwin and Müller wrote at that time. First, we learn, rewriting slightly Dokuchaev's definition of the soil, that its structural features can be understood as the result of an equilibrium between organizing forces (living organisms) and erosive forces (mainly climate and relief) acting on the parent material. Darwin and Müller described the role of living organisms, dealing respectively with earthworms and with the soil fauna as a whole. Both of them emphasized the effect of the fauna on the transformation of soil into 'good earth' and on its resultant fertility.

Soil transport by earthworms can be considered as a consequence of the behaviour of animals which are obliged to move continuously inside the profile, at least for those which have evolved an endogeic strategy (Satchell, 1980). This constant movement is understandable if it is recognized that diurnal and seasonal variations in temperature, humidity, soil atmosphere and available food produce an environment in which optimal conditions for earthworms are rarely found at the same level at the same time. Conversely, by transporting soil during these movements, the earthworms change their environment: burial of surface litter, fine mixing of organic matter with mineral soil, burrowing at depth and casting at the surface help to minimize the effect on the population of changes in their environment by increasing its stability (Kretzschmar, 1981, 1982a).

6.2 MATERIALS AND METHODS

6.2.1 Measurement of soil parameters

The gut contents of earthworms show great variability: sometimes full of organic matter, sometimes almost solely of mineral soil. In permanent pastures on even slightly leached soils, the colour of some surface casts is extremely different from that of the surface soil, indicating the deep origin of the casts. There is always a little leaching of the clay and therefore an accumulation of clay deeper in the profile. Conversely there is a decreasing proportion of the free and bound organic matter from the surface to the bottom of the profile. It is thus possible to characterize a vertical gradient, decreasing from the surface, by the ratio % organic matter/% clay, as suggested by Monnier (pers. comm.).

In studies on earthworms, free organic matter (f.o.m.) and bound organic matter (b.o.m.) should be discriminated, in order to take into account the variations of the f.o.m. due to litter ingestion. Moreover, we need to determine on the same sample those fractions of very low weight (between 200 and 500 mg). This was done by the following method.

Analyses were made on 500 mg air-dry samples, gently broken up and 500 μm-sieved. Gut contents were individually and wholly analysed. All the operations were done in 50 ml glass tubes adapted to centrifuging. F.o.m. is separated by the densimetric method (bromoform/alcohol mixture, density = 2) (Monnier *et al.*, 1962). The floating material was deposited on carbon-free glass-fibre filters by filtration and the proportion of f.o.m. was determined by carbon measurement on the filter by Anne's method (1945) simplified by Soignet (pers. comm.): 2 h oven-drying at 90°C instead of 5 min boiling.

By successive centrifuging, the densimetric mixture was removed with alcohol. The material remaining with a little alcohol was dried at 50°C and its carbon content determined as before to give percentage b.o.m. The sulphochromic solution was removed by centrifuging and replaced with 5 ml H_2O_2 with approximately 50 ml water; the tube was placed in an oven at 90°C until there were no more gaseous losses. This operation was used to complete the oxidation of bound organic matter and free all the clay. The proportion of clay was measured by counting particles with a Coulter Counter (Walker and Hutka, 1971).

To measure the gut contents, the earthworms were dissected from the ventral side on a cork board with a device for the recovery of the water with which the gut is ringed. The water was removed by alcohol and centrifuging and the gut content was dried at 50°C and weighed. This sample was lightly broken up but not sieved. All the results were transformed to the standard of 500 mg.

For each sample, we obtained three observations: % f.o.m., % b.o.m. and % clay. In the soil profile samples, the ratio % b.o.m./% clay was found to vary more directly with depth than % clay, % f.o.m. or % b.o.m. alone and was therefore calculated for all samples.

6.2.2 Sample characteristics

This method was employed for four types of samples: (1) soil, collected at 5 cm intervals to a depth of 60 cm, two series of samples being analysed, one collected in January and one in June (Fig. 6.1); (2) surface casts

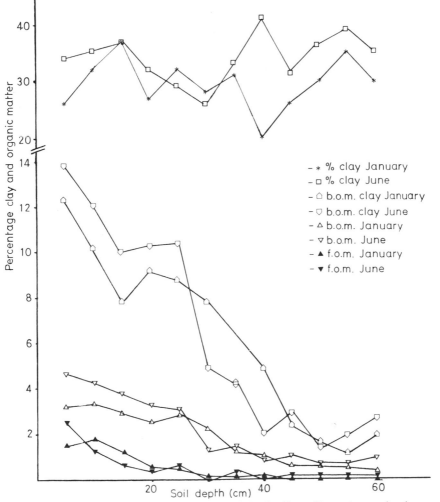

Fig. 6.1 Organic matter and clay content (%) of soil profile to 60 cm depth.

collected weekly during a previous experiment (Beugnot, 1978). Two series of 20 casts were analysed, each sample belonging to a well-identified cast; one series was collected at the time of maximum cast production in autumn and the second at the maximum of spring production; these samples were oven-dried at 105 °C; (3) surface casts collected hourly, between 19:30 h and 00:30 h, one series being collected in November and the other in March, each of 20 samples; (4) gut contents of earthworms (*Allolobophora longa (Nicodrilus longus)*, *A. icterica*) collected at the same time as the casts in March, at the same hours.

All the samples were collected in a permanent pasture near Dijon. The soil is a slightly leached brown earth, pH near 6.5; the vegetation is a *Lolio-cynusoretum*.

6.3 RESULTS

6.3.1 Surface casts collected weekly

There is little difference between the seasons (Table 6.1) except that f.o.m. is higher in autumn when litter production is highest (cf. Kretzschmar, 1977). For the same mean value of the % clay (42.1 in autumn and 40.1 in spring, not significantly different at the 5 % level) the standard deviation is larger in the autumn (43.1 against 15.7). Moreover, the % b.o.m. and the % clay are higher in the casts than in the soil, although their ratio would be equal, on average, to that of the upper level of the soil.

6.3.2 Surface casts collected hourly

The differences observed in the former samples between casts and profile are more noticeable here (Table 6.1); % f.o.m. and % b.o.m. are higher and % clay lower than in soil. The ratio % b.o.m./ % clay is higher than in the weekly collected casts but there are no differences between hours or seasons.

6.3.3 Gut contents

At 21:30 h the gut contents have a high proportion of b.o.m. (Table 6.2) (at the time of this experiment the earthworms were already feeding, Kretzschmar, 1977); the % clay is on average lower than in the casts of the same hour, and the % b.o.m. is less. At 00:30 h, % f.o.m. and b.o.m. are higher and % clay lower than in the corresponding casts. The two species show very marked differences: the mean values of % b.o.m./ % clay for *A. longa* and *A. icterica* are respectively 12.1 and 3.7 at 21:30 h and 51.4 and 24.2 at 00:30 h.

Table 6.1 Organic matter and clay content (%) of casts.

	f.o.m.			b.o.m.			Clay			$\frac{\text{b.o.m.}}{\text{Clay}} \times 100$		
	(n)	Mean	S.E.	(n)	Mean	S.E.	(n)	Mean	S.E.	(n)	Mean	S.E.
Casts collected weekly												
Autumn	(20)	2.645	0.128	(20)	3.640	0.100	(19)	42.111	1.549	(19)	8.942	0.577
Spring	(20)	1.815	0.415	(20)	3.900	0.083	(20)	40.11	0.908	(20)	9.83	0.330
Casts collected hourly												
Autumn												
19:30h	(6)	4.417	0.804	(5)	5.060	0.129	(6)	25.0	1.519	(5)	19.5	0.849
23:30h	(6)	3.233	0.362	(7)	4.386	0.235	(7)	27.786	1.282	(7)	15.728	0.821
00:30h	(7)	4.014	0.660	(7)	4.329	0.201	(5)	26.74	2.136	(5)	16.02	1.143
Spring												
19:30h	(4)	3.8	0.915	(4)	4.325	0.865	(4)	27.675	2.848	(4)	15.875	3.120
20:30h	(5)	5.27	0.663	(5)	5.1	0.270	(5)	26.1	1.896	(5)	20.2	2.416
23:30h	(5)	3.52	0.292	(5)	4.96	0.160	(5)	26.2	1.135	(5)	19.16	1.342
00:30h	(6)	2.583	0.258	(6)	5.033	0.182	(6)	27.433	1.229	(5)	18.467	0.797

Table 6.2 Organic matter and clay content (%) of gut content of two species and casts collected at night.

	f.o.m.			b.o.m.			Clay			$\dfrac{\text{b.o.m.}}{\text{Clay}} \times 100$		
	(n)	Mean	S.E.	(n)	Mean	S.E.	(n)	Mean	S.E.	(n)	Mean	S.E.
A. longa												
21:30h	(8)	4.188	0.406	(7)	2.514	0.451	(8)	20.55	2.674	(7)	12.143	1.905
00:30h	(8)	6.3	0.378	(8)	3.888	0.754	(8)	16.775	3.790	(8)	51.35	20.024
A. icterica												
21:30h	(2)	2.7	0.2	(2)	0.7	0.5	(2)	21.15	14.955	(2)	3.65	2.85
00:30h	(2)	3.15	0.25	(2)	4.4	1.7	(2)	17.55	2.85	(2)	24.15	5.75
Casts												
21:30h	(10)	3.89	0.378	(9)	2.111	0.443	(10)	20.67	2.148	(9)	10.256	1.975
00:30h	(10)	5.67	1.793	(10)	3.99	0.650	(10)	16.93	3.022	(10)	45.91	16.229

6.4 DISCUSSION

The ratio % b.o.m./% clay decreases in the order: gut content > hourly collected casts > weekly collected casts > soil. It is therefore necessary to take account of the marked transformation caused by transit through the gut as well as its rate of transport.

6.4.1 Transformation

The analysis of the gut content of *A. longa* at 21:30 h suggests that both litter and surface soil were inside the gut but unmixed. Three hours later, at 00:30 h, when the animals had been ingesting litter continuously, mixing had begun inside the gut and the % b.o.m. increased. The casts at the same hour (00:30 h) are clearly higher in clay and lower in f.o.m. Hourly and weekly collected casts at the same season were not significantly different.

6.4.2 Transport

The high standard deviation of % clay in the aged casts can be interpreted in terms of burrowing activity. In autumn, when the burrows are made at all levels down to 1 m, casts may come from any level; in spring, the burrow system being established, the worms are probably feeding mainly in the upper levels of the soil (Kretzschmar, 1982b).

The two species show clear differences. *A. icterica* at 21:30 h, has the gut full of deep soil whereas at 00:30 h the gut content is similar to the surface soil enriched with organic matter but without ingested litter. The gut transit duration is known to be rather short in such endogeic species (Fernandez, 1974; Lavelle, 1975; Bolton and Phillipson, 1976), and the deep soil contained in the gut is probably cast when the worms are feeding near the surface.

6.4.3 Effects of soil transportation on the stability of the earthworm environment

The main consequences of soil transport by earthworms are to increase the soil b.o.m. during passage through the gut and to introduce the organic-rich upper soil deeper into the profile. The structural stability of soil is maintained mainly by the products of the early stages of soil organic matter decomposition (Monnier, 1965) but is improved by the fine mixing of organic and mineral matter by earthworms as demonstrated by the high stability of earthworm casts. By binding organic and mineral fractions, earthworms thus create conditions of porosity, aeration and drainage favourable to their own survival. Moreover, by transferring surface soil down the profile, they create a habitat they can occupy below the surface at depths where humidity is more constant. By these means earthworms increase the stability of their own environment.

6.5 CONCLUSION

The analysis of these three pedological characteristics indicates the complex interaction of the movements of animals under the pressure of their environment and the simultaneous transformation by them of the soil in which they dwell. The movement, transport and transformation of soil and organic matter can be interpreted as components of a strategy of adjusting and stabilizing the environmental conditions close to their optimal value.

6.6 REFERENCES

Anne, P. (1945) Dosage du carbone organique du sol. *Ann. Agron.*, Serie A, 165.

Beugnot, M. (1978) *Recherches sur la Dynamique de Production des Turricules de Vers de Terre d'une Prairie Permanente.* D.E.S., Fac. Sci. Dijon, 55 pp.

Bolton, P. J. and Phillipson, J. (1976) Burrowing, feeding, egestion and energy budgets of *Allolobophora rosea* (Savigny) (Lumbricidae). *Oecologia (Berl)*, **23**, 225–245.

Fernandez, A. (1974) *Introduction à l'Étude des Vers de Terre.* D.E.A., E.N.S., Paris, 14 pp.

Kretzschmar, A. (1977) Étude du transit intestinal lombriciens anéciques. II. Résultats et interprétation écologique. In *Soil Organisms as Components of Ecosystems* (eds. U. Lohm and T. Persson), *Ecol. Bull. (Stockholm)*, **25**, 210–221.

Kretzschmar, A. (1981) Problèmes théoriques liés à la fonction biologique dans les sols. *C. R. Séminaire de l'Ecole de Biologie Théorique*, Paris, E.N.S.T.A.T., 523–537.

Kretzschmar, A. (1982a) Eléments de l'activité saisonnière des vers de terre en prairie permanente. *Rev. Ecol. Biol. Sol.*, **19**, 2, 193–204.

Kretzschmar, A. (1982b) Description des galeries de vers de terre, et variation saisonnière des résaux. (Observations en conditions naturelles). *Rev. Ecol. Biol. Sol.*, **19**, 4, 579–591.

Lavelle, P. (1975) Consommation annuelle de terre par une population naturelle de vers de terre *Millsonia anomala* Omodeo (Acanthodrilidae, Oligochètes) dans la savane de Lamto (Côte d'Ivoire). *Rev. Ecol. Biol. Sol.*, **12**, 11–24.

Monnier, G. (1965) *Action des matières organiques sur la stabilité structurale des sols.* Thèse, Fac. Sci. Paris, I.N.R.A., 140 pp.

Monnier, G., Turc, L. and Jeanson-Luusinang, V. (1962) Une méthode de fractionnement densimétrique par centrifugation des matières organiques du sol. *Ann. Agron.*, **13**, 55–63.

Satchell, J. E. (1980) r worms and K worms: a basis for classifying lumbricid earthworm strategies. In *Soil Biology as Related to Land Use Practices* (ed. D. L. Dindal). *Proc. VII Int. Soil Zool. Coll.*, E.P.A., Washington D.C., 848–854.

Walker, P. H. and Hutka, J. (1971) Use of the Coulter Counter (Model B) for particle-size analysis of soil. Division of soil, technical paper no. 1. Commonwealth Sci. Ind. Res. Organ., Melbourne, 1–40.

Chapter 7

Earthworm ecology in grassland soils

J. K. SYERS and J. A. SPRINGETT

7.1 INTRODUCTION

The environmental requirements of earthworm species in grassland ecosystems have been reviewed by Satchell (1958, 1967), Barley (1961), Gerard (1967), Bouché (1972), Lofty (1974) and Edwards and Lofty (1977). This chapter considers the physical, chemical and biotic effects of earthworms on managed temperate grassland soils and their effects on pasture yields.

7.2 PHYSICAL EFFECTS

7.2.1 Incorporation and mixing

The important role of earthworms in the incorporation of plant residues and dung in pasture ecosystems is well established. Earthworms mechanically break up organic matter and mix it with the mineral soil, thereby:

(i) Redistributing nutrients, bringing them closer to roots, a factor of great importance when soil surface layers dry out, and reducing loss of nutrients through surface run-off;
(ii) Blocking phosphate (P)-sorbing sites on soil components, effectively increasing plant available P;
(iii) Increasing the degree of mixing with soil microflora and thereby facilitating further enzyme activity. This increases the overall decomposition of the organic matter and the rate of release of nutrients.

Using ^{32}P labelling of litter and casts, Mansell *et al.* (1981) showed that by incorporating litter into casts, earthworms increased the short-term plant-availability of P derived from litter by a factor of approximately three. Similarly, Sharpley *et al.* (1979) showed that by removing pasture litter from the soil surface, earthworms may substantially reduce the amounts of P and nitrogen (N) in surface runoff.

Stockdill and Cossens (1966) showed that *Allolobophora caliginosa* (Savigny) had a pronounced effect on the vertical movement of DDT in New Zealand pasture soils and that this had considerable implications for the control of grass-grub (*Costelytra zealandica* (White)). DDT top-dressed on soils without *A. caliginosa* remained concentrated in the immediate surface and the control of the deeper feeding grass-grub larvae was poor. *A. caliginosa* increased the downward movement of DDT and its effectiveness against grass-grub.

The same authors studied the effect of feeding and burrowing by *A. caliginosa* on mixing lime into soils. In the absence of this earthworm, surface-applied lime (up to 5 t ha^{-1}) remained in the surface 2.5 cm 4 to 5 years after application. In the presence of *A. caliginosa*, lime was mixed throughout the top 20 cm of soil.

Our recent work on lime incorporation into soil by earthworms has shown that different species operate in different ways. *A. caliginosa*, and particularly *A. longa* Ude, mixed surface-applied lime into soil contained in boxes in the laboratory. *Octolasion cyaneum* (Savigny) did not mix lime into the soil, as shown by the similar pH distribution of the control (Fig. 7.1). In the field, *A. longa* was added to a soil already containing *A. caliginosa* and *Lumbricus rubellus* Hoffmeister and soil pH was measured in May, 1981 some 10 months after applying the earthworms and the lime. The results indicate that *A. longa* had mixed the added lime to a much greater depth than the resident species (Table 7.1).

Recent work with phosphate rock (Mackay *et al.*, 1982) shows that when earthworms mix materials of low water solubility into soil, plant growth is increased. In a glasshouse study with perennial ryegrass, a mixed population of *L. rubellus* and *A. caliginosa* increased the availability of P from a pelleted phosphate rock material by 15 to 30 %, relative to the no-earthworm treatment (Fig. 7.2). The amounts of water-extractable P in the soil at three depths (Fig. 7.3) show that, in the presence of earthworms, most of the surface-applied pelleted phosphate rock material had been mixed through the upper 6 cm. In contrast, water-extractable P levels indicate that the phosphate rock was restricted to the 0–2 cm layer in the absence of earthworms (control).

Unpublished data obtained by Mackay and co-workers suggest that there is little direct incorporation of surface-applied phosphate rock material into the soil by earthworms. Rock particles can be washed into the soil when burrows are open to the surface, but this does not explain all the increase in yield. The greatest effect of the earthworm on the agronomic performance of phosphate rock appears to result from an

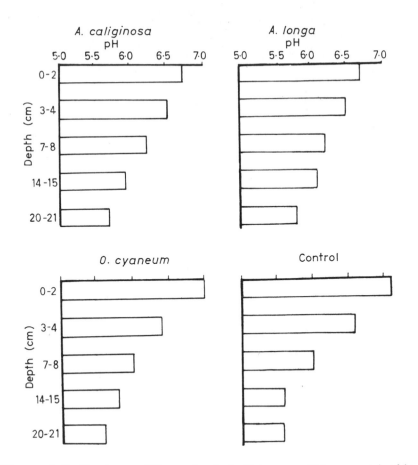

Fig. 7.1 Soil pH values at different depths in Tokomaru silt loam contained in boxes to which lime (5 tons ha⁻¹) was surface applied and three species of earthworms were separately added.

Table 7.1 Soil pH values at different depths in Tokomaru silt loam (already containing *A. caliginosa* and *L. rubellus*) 10 months after surface-applying *A. longa* and lime (5 t ha⁻¹).

Depth (cm)	*A. longa* + resident earthworms	Resident earthworms only	Control (no lime)
0–2	6.8	7.2	5.7
2–4	6.5	6.4	5.5
4–6	6.4	5.6	5.3
6–8	6.0	5.4	5.3
8–10	5.6	5.4	5.1
10–12	5.6	5.4	5.2
12–16	5.5	5.2	5.2

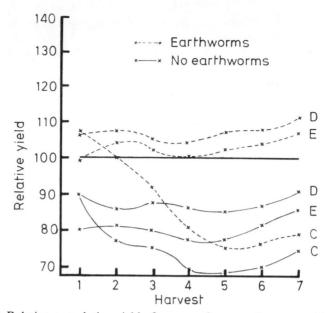

Fig. 7.2 Relative cumulative yield of ryegrass for seven harvests with pelleted phosphate rock in the presence and absence of earthworms. Surface-applied superphosphate in the absence of earthworms = 100. D, pelleted phosphate rock; E, pelleted phosphate rock incorporated; C, Control. (Reproduced with permission from Mackay *et al.* (1982).)

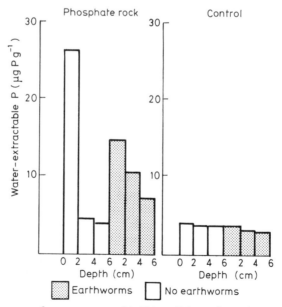

Fig. 7.3 Amounts of water-extractable P in soil from three depths as influenced by earthworms and the surface application of phosphate rock (100 kg of P ha^{-1}). (Redrawn from Mackay *et al.* (1982).)

increase in the degree of contact between phosphate rock particles and soil surfaces during passage through the earthworm, and also from coatings of cast material on phosphate rock particles in soils. This increased contact would accelerate dissolution and explain the 32% increase in chemical extractability of P obtained when earthworms were present.

7.2.2 Soil physical characteristics

The usefulness of the earthworm effect, in terms of increased plant growth, depends on what is limiting pasture production and, as physical properties are interrelated, the cause of changes in physical properties is often difficult to analyse.

Most of the effect of earthworms on soil physical properties can be interpreted in terms of changes in pore size distribution (Fig. 7.4). The induced pores can vary from large channels (2–11 mm) created by burrowing, to the medium pores resulting from casting and mixing. In addition, the effectiveness of a change in pore size distribution in a particular soil must depend on the stability of the induced pores. As soils

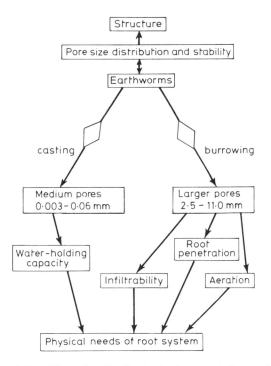

Fig. 7.4 Interrelationships of soil physical characteristics as influenced by earthworm activities.

vary widely in their pore size distributions and physical limitations to plant growth, earthworms may be expected to have a varying effect on the soil physical characteristics relevant to plant growth. The larger pores created by burrowing influence infiltrability, aeration and root penetration, and the medium pores created by casting influence water-holding capacity. All these characteristics are important to the functioning of the root system.

Soil structure is commonly discussed but poorly understood. It has two aspects: pore size distribution and the stability of that distribution to applied stresses e.g. raindrops and mechanical compaction. Earthworms influence both by burrowing and by ingesting and mixing organic matter with soil material but the mechanisms are far from clear. The water stability of aggregates also increases in the presence of earthworms but the relative significance of the secretion of cementing substances, the burial of plant residues, the enrichment of burrow walls with mucus and urine, which stimulates micro-organisms, and the effect of fungal hyphae, is not known. It would be expected to vary with earthworm species, agricultural practice and perhaps with soil type but the subject has advanced little since the studies of Swaby (1949) and Parle (1963).

The effect of earthworms on infiltrability in pasture soils is better established. Stockdill and Cossens (1966) report a substantial increase in infiltrability resulting from earthworm introduction (Table 7.2). Conversely, infiltrability was reduced threefold when earthworms were eliminated from a pasture soil (Sharpley et al., 1979), resulting in a twofold increase in the volume of surface run-off. Channeling by

Table 7.2 Effect of earthworm introduction on soil physical characteristics (Stockdill and Cossens, 1966).

Characteristic	Depth (cm)	Earthworms	
		Absent	Present
Gravimetric water content at field capacity (%)	0–10	42	52
	0–30	37	43
15 bar retention (%)	0–10	16	16
	0–30	14	14
Bulk density (g ml^{-1})	0–10	0.68	0.86
	0–30	0.96	0.96
Available moisture (mm)	0–10	18	31
	0–30	66	84
Macroporosity (%)	0–10	35	22
	0–30	29	23
Infiltrability (mm h^{-1})		14	26

earthworms (Ehlers, 1975) and the opening of burrows to the surface appear to be primarily responsible for the increased infiltrability. Unless the large channels created by burrowing vent to the surface and are in contact with free water, they remain air-filled and have little effect on infiltration. Earthworm species vary widely in the extent to which they create burrows open to the surface and so have different effects on infiltrability.

Infiltrability depends strongly on the saturated hydraulic conductivity which, in turn, is determined by pore size distribution. As the flow through a channel is proportional to radius to the power four, large pores or channels are the most important for water movement, assuming laminar flow is maintained. Ehlers (1975) showed that the channels of *L. terrestris* (L.) commonly extended down to 80 cm in a non-tilled soil.

Stockdill and Cossens (1966) suggest that the available water-holding capacity increases when earthworms are introduced into pasture (Table 7.2), particularly in the top 10 cm of soil. Structural rearrangement would probably result in fewer macropores but more fine pores and this is consistent with the higher values for field capacity and available moisture. Pores which are important for storing available water range from approximately 0.003 to 0.06 mm in diameter and if earthworms increase the proportion of these pores, water-holding capacity will increase. The reported increase in available moisture under field conditions in the presence of earthworms (Stockdill and Cossens, 1966) may be explained in terms of the change in pore size distribution because of casting and organic matter incorporation, rather than any change in bulk density. Stockdill and Cossens (1966) found that total porosity was the same in the presence and absence of earthworms which supports the above suggestion.

There is an apparent contradiction in Table 7.2. Earthworms have lowered the macroporosity (pores > 0.06 mm ϕ), but have increased the infiltrability. This occurs because a few earthworm burrows open to the surface would increase infiltrability but have only a minor effect on macroporosity.

An increase in available water-holding capacity of the top 10 cm of soil could have important effects on plant nutrient uptake. Because roots extract water and nutrients from the upper 10 cm of soil for most of the growing season, and as available plant nutrients tend to be concentrated in the same horizon, improved water-holding capacity is particularly important in soils with low available water capacities in regions subject to water stress.

The effect of earthworms on aeration is not clear. Both Barley (1959) and Satchell (1967) consider that the effect is minor. Rogaar and Boswinkel (1978) have recently demonstrated a marked increase in the air permeability of the surface horizon of a grass-mulched orchard sited on reclaimed polder soils. The increase in air permeability would only be effective if wind-induced viscous flow contributes significantly to aeration. However, Kimball and Lemon (1971) indicate that this is unlikely and that molecular diffusion, which is not restricted by pore diameter, is the dominant mechanism responsible for oxygen transport.

7.3 CHEMICAL EFFECTS

Earthworms need the usual major and minor nutrients required for cell development. Carbon compounds are required in an energy-rich form but the extent to which earthworms make direct use of plant and fungal structural polysaccharides is uncertain. It is generally assumed that more complex molecules are not broken down within the earthworm (Nielson, 1962), although Laverack (1963) has reported the presence of several enzymes, including cellulase, in extracts of earthworm alimentary canal tissue. Nitrogen, sulphur and phosphorus are also required in organic form as proteins or amino acids.

Earthworm casts contain more plant nutrients than the soil matrix but less than plant litter (Parle, 1963; Graff, 1970; Vimmerstedt and Finney, 1973). Earthworms cannot increase the total amount of nutrients in the soil but can make them more available (Barley and Jennings, 1959; Sharpley and Syers, 1977) and they may increase the rate of nutrient cycling, thereby increasing the quantity available at any one time.

It is assumed that the source of nutrients affected by earthworms is soil organic matter, which includes mainly dead material at the soil surface and also roots and organic matter in the soil with its complement of microflora and microfauna. The direct effect of earthworms on the release of nutrients by weathering soil minerals may be regarded as insignificant over the short term. Earthworms chemically influence nutrients in the soil:

(i) By direct enzyme action on organic matter in the intestine;
(ii) by turning plant organic matter with a high C/N ratio into tissue with a lower C/N ratio, the excess C being lost by respiration and the tissue protein decomposing rapidly after the death of the earthworms;
(iii) By metabolizing organic material and releasing metabolic products into the soil, in particular N.

The complexity of the interaction amongst these factors has been demonstrated by Sharpley and Syers (1976, 1977) and by Mansell et al. (1981). Sharpley and Syers (1977) showed that the amounts of loosely bound inorganic P and organic P changed significantly through the season (Fig. 7.5). Maximum release to solution of inorganic P coincided with maximum surface-casting activity of *A. caliginosa* and *L. rubellus*. This change may be related to the quality of the plant material the earthworms are ingesting. Early in the season, fresh litter is available at the soil surface but by mid-July most of it has disappeared from these pastures. Syers *et al.* (1979) showed that the oxidizable carbon in surface casts declined during the season.

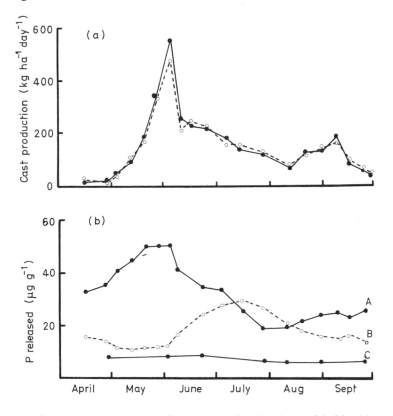

Fig. 7.5 Seasonal variation in surface cast production (oven-dried basis) on 6° (——) and 13° (----) slopes (a), and release to 0.1 M-NaCl of inorganic P from casts (A) and underlying 0–10 cm soil (C), and of organic P from casts (B) during 1 h at a solution/solid ratio of 400:1(b). (Reproduced with permission from Sharpley *et al.* (1977).)

A second factor may be the effect of low temperature late in the season on microbial activity and therefore on phosphatase enzyme activity as measured by the release of *p*-nitrophenol (Fig. 7.6, Sharpley and Syers, 1977). Sharpley *et al.* (1979) also showed that surface litter released more dissolved inorganic P to surface run-off than did earthworm casts. It might be expected then that plant litter would be a better source of plant-

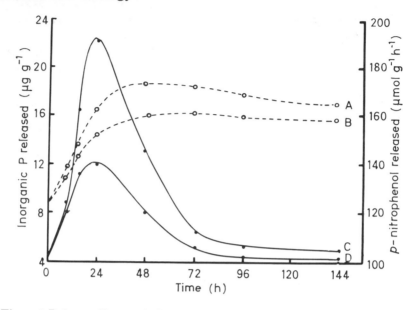

Fig. 7.6 Release of inorganic P to 0.1 M-NaCl from casts incubated at 16°C (A) and 4°C (B), and release of p-nitrophenol from casts incubated at 16°C (C) and 4°C (D) as a function of time of incubation. (Reproduced with permission from Sharpley *et al.* (1977).)

available P than casts but Mansell *et al.* (1981) found that passage through earthworms increased the short-term availability of P in litter by a factor of two or three. The increase in availability had a greater effect on the yield of ryegrass on a low P status soil than on a higher P status soil.

Mackay *et al.* (1982) have shown that the effect that earthworms have of increasing the availability of P is not limited to their effects on organic matter. Whereas earthworms would appear to stimulate P uptake from organic matter by redistribution and by increasing phosphatase activity, the increased availability of P from phosphate rock may result from intimate mixing of the fertilizer particles with soil.

Similarly, earthworms can affect the cycling of N. Barley and Jennings (1959) showed that 6.4% of the non-available N ingested by growing *A. caliginosa* was excreted in casts in plant-available form. Needham (1957) stated that up to half the N excreted by earthworms is in the form of protein, probably mucus, most of the remainder being ammonia and urea.

Several workers (Russell, 1910; Satchell, 1967) have suggested that decomposing earthworm tissue is the probable source of the increased plant-available N in pot experiments. Mackay *et al.* (1982) have shown that, although earthworm biomass in pots fell by 63% in 9 months, mainly by loss in weight of individuals rather than by reduction in numbers, this loss in biomass contributed only 17 μg N g^{-1} to the N pool in the soil, or 3.4% of the total inorganic N added as ammonium nitrate.

Earthworm casts have been shown to have increased microbial populations (Atlavinyte, 1975) and enhanced nitrification (Parle, 1963). Syers et al. (1979) reported that urease activity of casts was higher than that of soil, was temperature-dependent, and reached a maximum one day after deposition of the casts (Fig. 7.7). The latter workers concluded that the annual turnover of inorganic N in surface casts is agronomically insignificant in productive grass–clover pastures but this takes no account of the redistribution of the organic matter in the soil profile or the instantaneous rate of NH_4–N and NO_3–N production throughout the year in relation to root distribution and nutrient uptake.

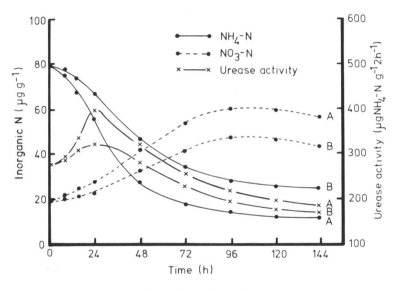

Fig. 7.7 Amounts of NH_4–N and NO_3–N produced in casts incubated at 16°C (A) and 4°C (B) and urease activity in casts incubated at the same temperature. (Reproduced with permission from Syers et al. (1979).)

Earthworms feed on organic matter and fleetingly increase mineralization rates, at least of N and P. It is likely that similar processes operate on other nutrients in plant litter, but data are lacking. Unless a peat or turf mat is developing, almost all the organic material is recycled during a year and the importance of earthworms appears to be not in the total turnover of the nutrient pool, but in producing frequent small increases in enzyme activity and nutrient availability.

Earthworms have been shown to prevent the formation of a turf mat and to remove all deposited surface litter. According to Barley (1961), earthworms play only a minor direct metabolic role in the decomposition process. Estimates of the earthworm component of the decomposition of organic matter vary from 0.4 % on sheep pasture (Nowak, 1975) to 8 % in a *Fraxinus-Quercus* wood (Satchell, 1967). Figure 7.8 has been derived using field data for earthworm biomass and an initial crop of litter with no further input, and published data on nutrient contents (Crampton and Harris, 1969) and ingestion rates (Lofty, 1974). For simplicity, a linear increase in biomass of 0.8 kg ha^{-1} day^{-1} over 140 days is assumed. Although the earthworms apparently use very little of the energy of the plant litter they are quite efficient at extracting nutrients from it. Assuming that plant litter is the only food source, by July (114 days), 10 % of the N content of the plant litter has been transferred to earthworm tissue; after 140 days, 17 % of the N has been transferred. The derived curve, showing the total amount of food ingested over the season, suggests that earthworms need to ingest the equivalent of the standing crop of litter almost three times to assimilate this amount of N. Because of differences

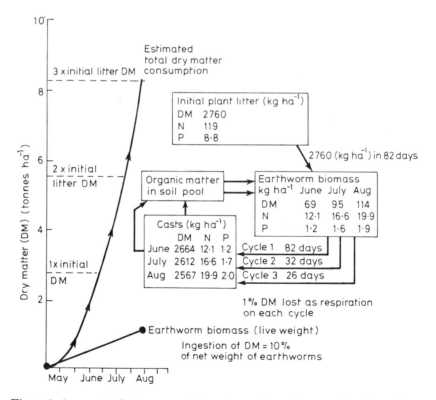

Fig. 7.8 Amounts of nitrogen and phosphorus in earthworm biomass and cast material in relation to cumulative litter consumption.

in pasture, species, management and climate, the rates and quantities of dry matter and nutrients would vary appreciably in space and time. Nevertheless, Fig. 7.8 does suggest that the relatively high mineral nutrient requirements of earthworms and their low energy requirement may be the reason why earthworm populations often appear to be food limited in the midst of an abundant supply of organic material.

7.4 BIOTIC EFFECTS

Earthworm activity appears to affect plant growth in ways that cannot be explained by either changes in soil physical characteristics or nutrient availability. Nielsen (1965) isolated indole compounds, which acted as plant growth stimulators, from both the bodies and casts of earthworms. Springett and Syers (1979) showed that earthworm casts could affect plant growth in an environment with adequate nutrients and that root morphology was also affected. Graff and Makeschin (1980) have recently shown that leachates from pots with earthworms increase the growth of ryegrass although the origin of this effect was not established.

The concept of growth stimulators (McNaughton, 1979) or growth inhibitors (Wood et al., 1970) passing between members of a food web is not new. Dyer (1980) has demonstrated the presence of a polypeptide plant growth stimulator in mouse submaxillary gland extract. Odum (1971) has discussed mechanisms of circular stimulation, and has indicated that the maintenance of a loop reinforcement requires a low-energy, high-information feedback in any system. There is no reason to suppose that decomposer systems differ from plant–herbivore systems. However, one could expect the identification of low-energy, high-information messenger substances from soil to be inordinately difficult.

7.5 EFFECTS ON PASTURE YIELD

The effects of earthworms on pasture growth have been measured in three situations. Firstly, in field areas where there have been no earthworms present and a well-defined turf mat has developed, introduced worms, usually A. caliginosa, have broken down and buried the turf mat. The release of nutrients into the system can bring about an initial increase in yield of about 70 %. In the New Zealand pastures studied by Stockdill and Cossens (1966), the initial large increase in yield was not sustained and over the long term, the real increase in dry matter was of the order of 25–30 % (Lacy, 1977).

Pot experiments in which plants are sown with earthworms present, or after the soil has been worked by earthworms which have then been removed, have given variable results. van Rhee (1965) reported that clover growth was increased 10 times but Waters (1955) reported no effect on clover growth. Growth may be stimulated primarily by N released from the bodies of dead earthworms in the pots but the data of Mackay *et al.* (1982) suggest that this is not the only explanation.

Studies in which an existing population of earthworms is modified by the addition of another species of earthworm or by increasing the population density of the earthworms present are closer to the 'real world' but little experimental information is available. In New Zealand there are examples of pasture productivity increasing after a new species has invaded an area already occupied by *A. caliginosa*. In a recent experiment designed to quantify this effect, *A. longa* was introduced at $150 \, m^{-2}$ with a biomass liveweight of $83 \, g \, m^{-2}$ into a pasture already containing populations of *A. caliginosa* at $950 \, m^{-2}$ with a biomass of $142 \, g \, m^{-2}$, *L. rubellus* at $250 \, m^{-2}$ with a biomass of $75 \, g \, m^{-2}$, and *O. cyaneum* at $25 \, m^{-2}$ with a biomass of $25 \, g \, m^{-2}$. The experiment has been in progress for only 10 months. Over the summer period there is very little growth of grass and the earthworms are not active but the first autumn growth on the plot with *A. longa* exceeded that of the control by 20%.

So far these data indicate that earthworms, in particular *A. caliginosa*, can increase grass growth if there has been a block in the decomposer system and a breakdown in the recycling processes. Earthworms appear to increase the availability of nutrients, giving a surge in production which

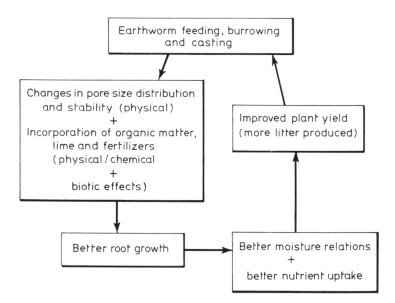

Fig. 7.9 Interrelationships between earthworm activities, soil properties, and plant growth.

settles back again to a higher constant level of production. The mechanisms by which the earthworms do this are likely to be complex and one or several mechanisms may operate at any one time. Low-P-status soils can benefit from earthworm activity by the increase in P availability, but a high-P-status soil with poor structure can also benefit from earthworm burrowing and tunnelling.

Interrelationships between certain activities of earthworms, soil properties, and the growth of plants are summarized in Fig. 7.9. These relationships are interactive, cyclical and complex. Because of this complexity, any substantial progress in understanding earthworm ecology in grassland soils will depend on interdisciplinary research.

7.6 ACKNOWLEDGEMENTS

We would like to thank Mr A. D. Mackay for allowing us to use his unpublished work and Dr D. R. Scotter for helpful discussion.

7.7 REFERENCES

Atlavinyte, O. (1975) *Ecology of Earthworms and their Effect on the Fertility of Soils in Lithuanian SSR.* Mokslas Publications, Vilnius.

Barley, K. P. (1959) Earthworms and soil fertility. IV. The influence of earthworms on the physical properties of a red brown earth. *Aust. J. Agric. Res.*, **10**, 371–376.

Barley, K. P. (1961) The abundance of earthworms in agricultural land and their possible significance in agriculture. *Adv. Agron.*, **13**, 249–268.

Barley, K. P. and Jennings, A. C. (1959) Earthworms and soil fertility. III. The influence of earthworms on the availability of nitrogen. *Aust. J. Agric. Res.*, **10**, 364–370.

Bouché, M. B. (1972) *Lombriciens de France, Ecologie et systématique.* Institut National de la Recherche Agronomique, Paris. 671 pp.

Crampton, E. W. and Harris, L. E. (1969) *Applied Animal Nutrition; the Use of Feedstuffs in the Formulation of Livestock Rations*, 2nd edn. W. H. Freeman. 753 pp.

Dyer, M. I. (1980) Mammalian epidermal growth factor promotes plant growth. *Proc. Natl. Acad. Sci. U.S.A.*, **77**, 4836–4837.

Edwards, C. A. and Lofty, J. R. (1977) *Biology of Earthworms.* 2nd edn. Chapman and Hall, London. 333 pp.

Ehlers, W. (1975) Observations on the earthworm channels and infiltration in tilled and untilled loess soil. *Soil Sci.*, **119**, 242–249.

Gerard, B. M. (1967) Factors affecting earthworms in pastures. *J. Anim. Ecol.*, **36**, 235–252.

Graff, O. (1970) The phosphorus content of earthworm casts. *LandbForsch-Völkenrode*, **20**, 33–36.

Graff, O. and Makeschin, F. (1980) Crop yield of ryegrass influenced by the excretions of three earthworm species. *Pedobiologia*, **20**, 176–180.

Kimball, B. A. and Lemon, E. R. (1971) Air turbulence effects upon soil exchange. *Proc. Soil Sci. Soc. Am.*, **35**, 16–21.

Lacy, H. (1977) Putting new life in wormless soil. *N. Z. Farmer*, **98**, 20–22.

Laverack, M. S. (1963) *The Physiology of Earthworms*. Pergamon Press, London. 206 pp.

Lofty, J. R. (1974) Oligochaetes. In *Biology of Plant Litter Decomposition* (eds. C. H. Dickinson and G. J. F. Pugh), Academic Press, London. **2**, 467–488.

Mansell, G. P., Syers, J. K. and Gregg, P. E. H. (1981) Plant availability of phosphorus in dead herbage ingested by surface casting earthworms. *Soil Biol. Biochem.*, **13**, 163–167.

Mackay, A. D., Syers, J. K., Springett, J. A. and Gregg, P. E. H. (1982) Plant availability of phosphorus in superphosphate and a phosphate rock as influenced by earthworms. *Soil Biol. Biochem.*, **14**, 281–287.

McNaughton, S. J. (1979) Grazing as an optimization process: Grass–ungulate relationships in the Serengeti. *Am. Nat.*, **113**, 691–703.

Needham, A. E. (1957) Components of nitrogenous excreta in the earthworms *L. terrestris* and *E. foetida*. *J. Exp. Biol.*, **34**, 425–446.

Nielsen, C. O. (1962) Carbohydrases in soil and litter invertebrates. *Oikos*, **13**, 200–215.

Nielsen, R. L. (1965) Presence of plant growth substances in earthworms demonstrated by paper chromatography and the Went pea test. *Nature (London)*, **208**, 113–114.

Nowak, E. (1975) Population density of earthworms and some elements of their production in several grassland environments. *Ekol. Pol.*, **23**, 459–491.

Odum, H. T. (1971) *Environment, Power, and Society*. John Wiley & Sons Inc., New York. 331 pp.

Parle, J. N. (1963) A microbiological study of earthworm casts. *J. Gen. Microbiol.*, **31**, 13–22.

Rogaar, H. and Boswinkel, J. A. (1978) Some soil morphological effects of earthworm activity; field data and X-ray radiography. *Neth. J. Agric. Sci.*, **26**, 145–160.

Russell, E. J. (1910) Soil conditions and plant growth. *J. Agric. Sci. Camb.*, **3**, 246–257.

Satchell, J. E. (1958) Earthworm biology and soil fertility. *Soils Fertil.*, **21**, 209–219.

Satchell, J. E. (1967) Lumbricidae. In *Soil Biology* (eds. A. Burges and F. Raw), Academic Press, London. 259–322.

Sharpley, A. N. and Syers, J. K. (1976) Potential role of earthworm casts for the phosphorus enrichment of runoff waters. *Soil Biol. Biochem.*, **8**, 341–346.

Sharpley, A. N. and Syers, J. K. (1977) Seasonal variation in casting activity and in the amounts and release to solution of phosphorus forms in earthworm casts. *Soil Biol. Biochem.*, **9**, 227–231.

Sharpley, A. N., Syers, J. K. and Springett, J. A. (1979) Effect of surface-casting earthworms on the transport of phosphorus and nitrogen in surface runoff from pasture. *Soil Biol. Biochem.*, **11**, 459–462.

Springett, J. A. and Syers, J. K. (1979) Effect of earthworm casts on ryegrass seedlings. *Proc. 2nd Australasian Conf. Grassld. Invertebr. Ecol.*, 47–49.

Stockdill, S. M. J. and Cossens, G. G. (1966) The role of earthworms in pasture production and moisture conservation. *Proc. N.Z. Grassld. Assoc.*, 168–183.

Swaby, R. J. (1949) The influence of earthworms on soil aggregation. *J. Soil Sci.*, **1**, 195–197.

Syers, J. K., Sharpley, A. N. and Keeney, D. R. (1979) Cycling of nitrogen by surface-casting earthworms in a pasture ecosystem. *Soil Biol. Biochem.*, **11**, 181–185.

van Rhee, J. A. (1965) Earthworm activity and growth in artificial cultures. *Pl. Soil*, **22**, 45–48.

Vimmerstedt, J. P. and Finney, J. H. (1973) Impact of earthworm introduction on litter burial and nutrient distribution in Ohio strip-mine spoil banks. *Proc. Soil Sci. Soc. Am.*, **37**, 388–391.

Waters, R. A. S. (1955) Numbers and weights of earthworms under highly productive pasture. *N.Z. J. Sci. Technol.*, **36**, 516–525.

Wood, D. L., Silverstern, R. M. and Nakajima, M. (eds.) (1970) *Symp. on Control of Insect Behaviour by Natural Products.* Academic Press, New York. 345 pp.

Chapter 8

Effect of earthworms on grassland on recently reclaimed polder soils in the Netherlands

M. HOOGERKAMP, H. ROGAAR and
H. J. P. EIJSACKERS

8.1 INTRODUCTION

The number and biomass of earthworms in grassland may vary widely in space as well as temporally. In old grassland in the Netherlands, numbers normally range from 300 to 900 m^{-2} with an average population density of 500 m^{-2} and biomass of 2500 kg ha^{-1}. *Allolobophora caliginosa* is the most common species in grassland soil and other common species are *A. rosea, A. chlorotica, A. cupulifera, A. longa, Lumbricus terrestris, L. rubellus, L. castaneus* and *Octolasium cyaneum* (van Rhee, 1970). Nevertheless in some grassland soils earthworms may be scarce or absent because of unfavourable climatic or edaphic conditions, application of certain pesticides or fertilizers, or flooding; in grassland seeded on arable land the number is low initially.

In young polder soils reclaimed recently from Lake Ijssel, earthworms are virtually absent, although after draining, the ecological conditions are favourable. In intensively used pastures the following problems connected with this lack of earthworms may occur:

Detachment of the sward by grazing cattle and sheep, especially in early spring and late summer;
Development of weeds soon after seeding;
Compaction of top soil by treading;
Damage to sward by urine scorch.

These problems can cause a drop in production a few years after seeding and to overcome them many pastures have been ploughed up and reseeded, sometimes with an intermediate potato or maize crop.

This report presents the first results of a research project on the influence of earthworms on soil conditions and grassland production in these polders and the increase and spread of earthworm populations, inoculated at several places some ten years ago. Earlier work on this subject has been done in the Netherlands by van Rhee (1977).

Recent summaries of the effects of earthworms on grassland are given by Syers and Springett (Chapter 7) and Stockdill (1982).

8.2 MATERIALS AND METHODS

8.2.1 Sites

The experimental sites are located in pastures near Biddinghuizen, Swifterbant and Lelystad, Eastern Flevoland. Earthworms were inoculated at about 50 points shortly after sowing in 1971 and 1972. In the Biddinghuizen field about 3000 worms of the species *A. caliginosa* (2790) and *L. terrestris* (244) were inoculated in an area of a few dm². In the Lelystad and Swifterbant fields, mixtures of *A. caliginosa*, *A. longa*, *L. rubellus* and *L. terrestris* were inoculated by lifting the sod at a number of points in areas of 0.25–1 ha. The density of inoculation amounted to 10 m⁻². Most research was done in the fields near Biddinghuizen (5° 38′ 8″/52° 33′ 2″) and Swifterbant (15° 47′ 6″/52° 27′ 6″).

The pastures, predominantly of *Lolium perenne*, are mostly intensively used for dairy farming with a yearly dressing of 300–550 kg fertilizer-N ha⁻¹ and a stocking rate of 3–4 cattle units ha⁻¹. Some pastures are also grazed by sheep. The climate is temperate, humid and maritime with an annual rainfall of about 700 mm fairly evenly spread over the year and an annual mean temperature of about 9.2 °C.

The soils are developing in stratified, calcareous silt loam to silty clay deposits of lacustrine origin and in reclamation since 1957. The range of carbonate contents of the upper 50 cm is 7.7–11.1 %, the organic matter content by loss-on-ignition 2.1–4.6 %, the pH (KCl) 7.1–7.5. The upper 20–30 cm has been mixed by ploughing and is of silt loam texture on the Biddinghuizen field and of silty clay loam texture on the Swifterbant fields. The soils are artificially drained by ditches and tile drains. Groundwater levels are within 100 cm depth at Biddinghuizen and 160 cm at Swifterbant. Soil formation is as yet restricted to the formation of a system of cracks and root channels down to the groundwater level and the development of a mat layer at the surface. An A_1 horizon is developing where earthworms are active. The soils are classified as aeric Fluvaquents (Soil Conservation Service, 1975) or calcaric Fluvisol (FAO/UNESCO, 1974).

8.2.2 Methods

Earthworm populations were estimated by digging and handsorting soil samples of $0.5 \times 0.5 \times 0.5$ m^3. At Biddinghuizen the earthworms were sampled at regular intervals along two axes crossing at right angles at the inoculation point. Sampling was extended until one worm or none were found in two consecutive samples, the last sample being considered as the dispersal front. Samples were taken in 1972, 1973 and 1979. For each year, total numbers of worms were plotted and a mean distribution line was drawn by eye. The total population number was estimated from a dispersal model (Appendix A).

In Swifterbant, samples were taken along each radius, at two points, at one-third of the radius length where the worm population was well developed and there was no litter layer, and at two thirds of the radius in the transition zone where the worm population was developing and the litter layer was starting to disappear. Eight samples were taken at each site.

Spade and auger were used to determine the general soil features and to survey the areas. Root and worm channels were counted at horizontal and vertical faces, at every 10 cm depth in the soil pits. C was determined after Kurmies (wet oxidation) and N after Kjeldahl. Total organic matter was determined by loss-on-ignition. Infiltration capacities were measured in the field in 30 cm ϕ soil columns *in situ* (Bouma, 1977) and the total water- and air-filled pore volumes in 50 cm^3 ring samples 5 cm high were measured in the laboratory. For water tension a desorption procedure was chosen, starting at saturation. Penetration resistance was measured in the field with a penetrometer (conus: 1 or 5 cm^2) and in the laboratory at fixed water tension values in 50 cm^3 ring samples of 5 cm ϕ and with a conus of 1 mm ϕ. Shear strength was measured in the field by a CL-600 Torvane Torsional vane shear device with standard vane.

Repeated thermographical detection of the worm plots in the Swifterbant fields was done by Infrared Line Scanning (IRLS, 8–14 μm wavelength) from an altitude above ground level of 1000 m (1979 and 1980) and 300 m (1981) respectively; during the spring flight of 1981, a series of colour infrared photographs (wavelength $\pm$ 0.4–0.9 μm) was taken from the aircraft used. By recording the output of the infrared detector as digital data on a magnetic tape and by transforming the digital data with a colour graphic system, grey images and colour images could be produced at choice.

Root quantities were estimated in the field by counting on fracture planes of core samples of 8 cm ϕ at several depths, as well as by rinsing and weighing in the laboratory (Böhm, 1979).

The botanical composition of the sward was analysed in terms of the number of tillers 0.25 dm^{-2} (about 50 samples), species frequency 0.25 dm^{-2} (about 100 samples) or the dry weight proportions of the species (about 50 samples) (de Vries, 1949). In some instances the coverage of species and bare spots was estimated. Yields of fresh herbage and dry matter were determined by cutting strips protected from grazing by cages (four replicates per area).

The worm plots, mostly situated in the centres of the fields, were compared with the surrounding worm-free areas within the same pastures.

8.3 RESULTS

8.3.1 Increase and dispersal of the earthworm populations

The distribution curves along the axes for *A. caliginosa* demonstrate that dispersal in different directions was unequal. Figure 8.1 shows an increased dispersal along axes 2 and 4 presumably due to a cow path along them. However, the numbers along the two halves of the other axis differed too, possibly due to slight soil differences. The total numbers along the dispersal radii consequently have large standard deviations but it is nevertheless possible (see Appendix A) to calculate the population density. Both population development and dispersal show a lag phase (Fig. 8.2) after which there is linear increase of the dispersal distance and a log-linear increase of earthworm numbers. The specific growth rate r (for unlimited growth $N_t = N_0 e^{rt}$) was calculated as 0.07, slightly less than calculated by van Rhee (1969). The dispersal rates indicate mean yearly

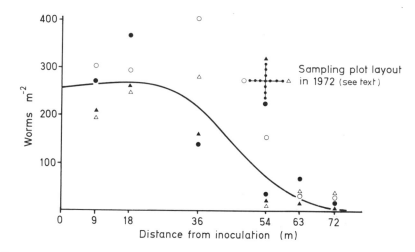

Fig. 8.1 Distribution of *A. caliginosa* at Biddinghuizen along axes intersecting at the point of inoculation.

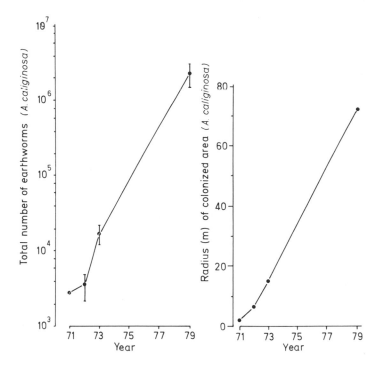

Fig. 8.2 Population development of *A. caliginosa*.

dispersal distances of 9 m for *A. caliginosa* and 4.5 m for *L. terrestris*. The age distribution along the axes shifts towards a higher proportion of mature worms at the periphery (Table 8.1).

In Swifterbant, the peripheral zone contained *A. caliginosa*, *A. longa*, *A. rosea* and *L. rubellus* indicating that these species must have about the same dispersal rate (Table 8.2). The ratio of mature/immature worms in this table shows again that the pioneers are predominantly adult,

Table 8.1 Mature and immature worms at different distances from inoculation point; Biddinghuizen.

Distance from inoculation point (m)	Number of *A. caliginosa* per soil sample (0.125 m³)		
	Mature	Immature	Mature/immature
9	16	229	0.07
18	33	259	0.13
36	21	226	0.09
54	8	55	0.14
63	13	35	0.37

occupying the peripheral zone and producing cocoons from which immature worms develop. In Biddinghuizen and Swifterbant the density in a well-established area is about 200 worms m^{-2}, whereas in the peripheral zone, still with a thin mat, the density is 90 worms m^{-2}. In this paper the duration of the earthworm activity at a certain distance from the inoculation area is expressed in years from colonization by the fastest species, mainly *A. caliginosa*, and estimated from the model (Fig. 8.1). At the inoculation points the period of inoculation is taken as starting point. From the annual increase in the size of the area occupied, the time taken for the litter layer to disappear, and the population development rate one can calculate with a few rough assumptions that about 2000 earthworms have to be inoculated at 50 m intervals to eliminate the mat from a pasture within 5 years.

Table 8.2 Composition and size of the worm population in the centre (C) and periphery (P) of a worm plot; inoculated winter 1971–1972, sampled 18/12/1980, Swifterbant.

	Number per soil sample (0.125 m^3)					
	Mature		Immature		Mature/immature	
Species	C	P	C	P	C	P
A. caliginosa	29	17	184	24	0.16	0.71
A. longa	6	7	25	7	0.24	1.00
A. rosea	3	0	1	0	3.00	—
L. rubellus	2	3	20	8	0.10	0.37
L. terrestris	0	0	2	0	—	—
Total	40	27	232	39	0.17	0.69

8.3.2 Soil characteristics

Without earthworms, dead and decaying organic matter accumulates on the surface of the mineral soil forming a mat layer or O horizon. It is composed of dark reddish-brown fibrous material (roots, leaves, manure), living roots and fine dark-coloured moder humus. At one Swifterbant site its thickness reached 0.5 to 2.5 cm in ten years, with an average fresh weight of 195 × 10^3 kg ha^{-1}, a dry matter content of 43 % and an organic matter content of 41 %. A considerable amount of mineral soil is incorporated by mice, treading etc. The transition to the mineral soil is abrupt and the connection is formed by dead and living grass roots. In the mineral material, soil formation is restricted to a system of cracks and fine grass-root channels down to ground-water level with gley mottling along these voids.

When earthworms invade an area, the mat is ingested and incorporated into the soil in about three years and a dark-coloured A_1 horizon starts to develop. Initially this horizon has a mixed character and is composed of very dark organic worm casts and lighter-coloured unchanged mineral soil. Only at the surface a very dark-coloured, 1–2 cm thick layer of a homogeneous character is present. This layer gradually increases in thickness to 5–8 cm in 8 or 9 years, but with a decreasing colour intensity (Fig. 8.3).

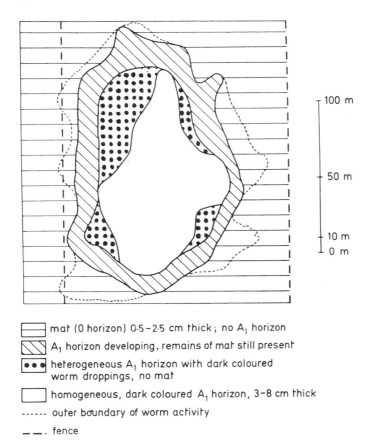

100 m

50 m

10 m
0 m

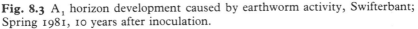 mat (0 horizon) 0.5–2.5 cm thick ; no A_1 horizon

A_1 horizon developing, remains of mat still present

heterogeneous A_1 horizon with dark coloured worm droppings, no mat

homogeneous, dark coloured A_1 horizon, 3–8 cm thick

------ outer boundary of worm activity

— —. fence

Fig. 8.3 A_1 horizon development caused by earthworm activity, Swifterbant; Spring 1981, 10 years after inoculation.

Because of the working of the soil and casting by the earthworms, the weak plate-like or moderate block-like structure of the upper decimetre becomes granular. The root channels formed initially are disturbed and replaced by a system dominated by 1–4 mm wide burrows and wider

Table 8.3 Soil C and N contents and ratios in relation to estimated duration of earthworm activity, Biddinghuizen.

	C (%)					N (%)					C/N	
Years of earthworm activity:	0	2	4	6	8	0	2	4	6	8	0	8
Depth												
2–0 cm (mat)	16.5	—	—	—	—	0.67	—	—	—	—	24.6	—
0–2 cm	1.8	5.9	3.4	3.8	3.3	0.18	0.49	0.28	0.29	0.25	10.0	13.2
2–5 cm	1.2	1.9	2.4	2.9	2.9	0.10	0.19	0.20	0.24	0.22	12.0	13.2
5–10 cm	1.1	1.3	1.4	1.6	1.8	0.10	0.11	0.10	0.13	0.19	11.0	12.8
10–20 cm	1.1	1.2	1.1	1.1	1.2	0.09	0.10	0.09	0.09	0.09	12.2	13.3

chambers of *A. caliginosa*, the commonest species (Rogaar and Boswinkel, 1978). The lower boundary of this system gradually extends to reach 50 cm depth after 8 years as observed at Biddinghuizen, although the complete working of the soil was restricted to the upper 8 cm. Where adults of *L. terrestris* or other larger species are active, a few wider vertical burrows are found somewhat below the soil surface and down to the ground-water level. At all depths the filled-in burrows far outnumber the open ones especially in the uppermost soil.

Together with the removal and incorporation of the mat, the soil organic matter content, represented by the C and N contents, of the mineral soil increases. In consecutive years the total amounts obviously do not change anymore, but a redistribution occurs, extending the depth of augmentation together with the development of the A_1 horizon. The C/N ratio tends to increase (Tables 8.3 and 8.4). The data of Table 8.4 are calculated, assuming base levels of C $= 1,1\%$ and N $= 0,1\%$ (Table 8.3).

The working of the soil and the formation of a channel structure enhances the infiltration capacity of the soils considerably (Table 8.5). Methylene Blue added to the infiltration water showed that the voids left by the granular structure are also important in water transport, especially in the surface layer where open burrows are virtually absent.

Conductivity for water at saturation, conductivity for air at pF-2 and the oxygen diffusion at pF-2 are higher in the upper 20 cm of the soil (Table 8.6). In deeper layers, no distinct changes in conductivity caused by earthworm activity were found, the number of channels and degree of reworking being still very limited at these depths after 8 to 10 years. The

Table 8.4 Increase in C and N contents in the upper 20 cm of mineral soil in relation to estimated duration of earthworm activity, Biddinghuizen (see text).

Years of earthworm activity	0	2	4	6	8
kg C $\times 10^3$ ha^{-1}	1.78	16.6	16.6	16.9	16.9
kg N $\times 10^3$ ha^{-1}	0.204	1.34	0.84	1.21	1.10

Table 8.5 Infiltration capacity for water of plots with and without earthworms.

	Infiltration capacity (m 24 h^{-1})	
	Without worms	8–10 years worm activity
Biddinghuizen	0.039	4.6
Swifterbant	0.047	6.4

Table 8.6 Conductivity for water at saturation and conductivity for air and oxygen diffusion at pF2 of soil without worms and after 8 years of earthworm activity, Biddinghuizen; means and ranges ($n = 8$).

	Conductivity for water (m 24 h^{-1})			
	− Worms		+ Worms	
Depth (cm)	Mean	Range	Mean	Range
0–10	0.95	0.46– 1.50	22.2	1.0–88.4
10–20	0.42	0.11– 0.64	6.5	0.8–28.6
20–30	14.00	1.15–19.20	4.2	0.4–25.6

	Conductivity for air (cm^{-2} × 10^{-8})			
	− Worms		+ Worms	
Depth (cm)	Mean	Range	Mean	Range
0–10	8.40	4.7– 12.5	54.3	12.9–114.8
10–20	7.10	2.9– 13.7	36.0	20.5– 57.4
20–30	92.20	49.2–143.5	19.6	8.4– 68.9

	O$_2$ diffusion (cm^2 s^{-1} × 10^{-2})			
	− Worms		+ Worms	
Depth (cm)	Mean	Range	Mean	Range
0–10	0.27	0.15–0.37	0.68	0.33–1.06
10–20	0.28	0.18–0.38	0.57	0.39–0.81
20–30	0.38	0.71–0.93	0.33	0.28–0.37

cracking pattern of the soil is the dominant factor. The wide variation in values is caused partly by the method applied but also by the irregular distribution of open worm channels. In the Swifterbant soil, worm activity produced a higher pore volume, and higher water and air contents in the top soil (Table 8.7) as well as more available moisture at field capacity. In the Biddinghuizen soil no marked changes could be established at the prevailing pore volume of about 50%.

In the upper decimetre, the penetration resistance of the soil is reduced by worm activity (Table 8.8, Fig. 8.4). In deeper layers the differences were not significant or were determined by other factors such as soil texture. The shear strength at the surface was clearly greater on the worm

Table 8.7 Mean water and air contents at different moisture tensions of soil with and without earthworms, Swifterbant; ($n = 8$).

| pF value | Depth 0–10 cm | | | | Depth 10–20 cm | | | |
| | − Worms | | + Worms | | − Worms | | + Worms | |
	Water	Air	Water	Air	Water	Air	Water	Air
Saturated	44.9	—	48.9	—	48.1	—	49.0	—
1.0	43.6	1.3	46.1	2.8	45.1	3.0	45.5	3.5
1.5	43.1	1.8	45.6	3.3	—	—	—	—
2.0	42.4	2.5	44.3	4.3	43.0	5.1	43.1	5.9
2.3	40.7	4.2	42.4	6.5	40.4	7.7	40.4	8.6
2.7	38.7	6.2	40.2	8.7	38.2	9.9	37.6	11.4

Table 8.8 Mean penetration resistance ($kg\,cm^{-2}$, $n = 8$) of soil with and without earthworms, measured at different moisture tensions in ring samples; conus 1 mm ϕ; Swifterbant.

| Depth (cm) | pF-2.0 | | | | pF-2.7 | | | |
| | − Worms | | + Worms | | − Worms | | + Worms | |
	$\bar{x}$	SD	$\bar{x}$	SD	$\bar{x}$	SD	$\bar{x}$	SD
0–10	24.4	3.3	13.9	2.6	33.4	3.7	21.0	8.3
10–20	14.3	3.7	17.2	4.3	22.5	1.8	19.9	2.8
20–30	21.6	5.6	12.2	2.5	29.3	3.3	18.1	4.1

plots, the mat being easily stripped off the worm-free plots (Table 8.9). Despite its looser character, the mineral topsoil of the worm plot is more cohesive than on the plot without worms, evidently an effect of the higher number of roots binding the soil.

The working of the topsoil by earthworms, resulting in better aeration, higher infiltration capacity and water conductivity, reduces water stagnation, resulting in less reducing conditions in the top soil in wet periods. This may be deduced from the Swifterbant soil which was less mottled where worms were present than where they were absent (Table 8.10).

In standard colour aerial photographs the worm-plots were not visible. Infrared line scanning (aerial photography) showed that the inoculated places were warmer by night and cooler by day than the surrounding worm-free areas. The effects are particularly outstanding in early spring, when the grass is still short, the differences corresponding to the presence

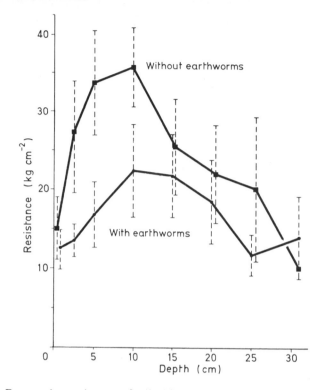

Fig. 8.4 Penetration resistance of soil with and without earthworms, mean values and standard deviations; Biddinghuizen, April 1976; conus 5 cm².

Table 8.9 Mean shear strengths (kg cm⁻², $n = 8$) of soil with and without earthworms, Swifterbant.

	− Worms		+ Worms	
	$\bar{x}$	SD	$\bar{x}$	SD
Grass surface	0.25	0.02	0.41	0.03
Transition mat layer/mineral soil	0.19	0.03	—	—
Mineral soil, 3 cm depth	—	—	0.32	0.04
Mineral soil, 5 cm depth	0.45	0.03	0.41	0.03

and absence of the mat layer. Without this layer the heat exchange between soil and air is better, diminishing the daily fluctuations at the surface (Fig. 8.5). The pictures produced by transforming the digital data with a colour graphic system also show differences in temperature within the worm-plots; the reason for these differences are not yet understood.

Table 8.10 Degree of gley mottling of 246 samples of the surface layer of soil with and without earthworm activity; Swifterbant; spring 1981.

Proportions of exposed surface	− Worms		+ Worms	
	n	%	n	%
< 2%	29	16	45	70
> 2%	153	84	19	30
Total	182	100	64	100

On false colour photographs (wavelength band $\pm$ 0.4–0.9 μm) the worm-plots have a more intense red colour than the surrounding grassland; presumably the biomass of the grass is the determining factor here.

8.3.3 Grassland aspects

In the years after sowing, many grasslands deteriorate, i.e. sown grasses, mainly *Lolium perenne*, are replaced by inferior ones such as *Poa annua* and by dicotyledonous weeds (e.g. *Taraxacum officinale*) and by the development of bare spots. In Table 8.11, areas within pastures, with and without earthworms, are compared about 10 years after sowing. Considerable differences in species abundance are evident, the worm-plots showing generally a better botanical composition. Variations were considerable, however, both spatial and temporal. The botanical decline in the sward may be caused by drought, frost and urine scorch. It was of a temporary character, a clear recovery occurring on the observation plots in favourable periods. It may, however, be irreversible if damage is very severe or conditions for recovery less favourable.

On four fields at Swifterbant yield determinations were started (Table 8.12) in 1981. Yields per cut and per field varied, but those of the worm areas were highest in most instances. The mean increase, taking all the plots and cuts together, was about 10%.

Determination of penetration resistance and shear strength (Tables 8.8 and 8.9, Fig. 8.4) indicated a lower bearing capacity, but a greater resistance to slip of the silty loamy soils, which were invaded by earthworms. Nevertheless, penetration resistance remained above the critical bearing capacity of 7.5 kgf cm^{-2} (measured by a 5 cm^2 conus after Haans, 1979). Some damage by treading and soiling of grass was observed on the worm-plots in wet periods, indicating a loss of production and damage to the sward. In fields without earthworms with a well-developed mat, poaching and soiling of the grass are almost precluded,

Fig. 8.5 Differences in radiant temperature of soil with and without earthworms. Worm-plots in field centres 9 years after inoculation. Swifterbant (courtesy Begeleidingscomm. Remote Sensing-BCRS).

Table 8.11 Botanical composition of grassland with and without earthworms.

	Cover (%)		Tillers dm^{-2}				Dry wt. (%)			
	I*		II*		III*		IV*		V*	
	−Worms	+Worms	−Worms	+Worms	−Worms	+Worms	−Worms	+Worms	−Worms	+Worms
Lolium perenne	57	74	63	85	122	125	47	77	72	88
Poa trivialis and P. pratensis	2	1	5	2	1	1	0	2	0	3
P. annua	14	12	58	58	9	10	9	7	8	4
Phleum pratense	5	3	2	+	1	+	11	3	0	0
Trifolium repens	1	7	< 0.5	1	3	4	3	8	0	0
Elytrigia repens	0	0	+	0	0	0	2	0	15	3
Dicot. weeds	10	0	3	1	1	+	28	3	5	2
Bare ground	11	3								

*Sites: I–IV Swifterbant (two fields), V Biddinghuizen.
Sampling dates: I, April 1981; II, III, May 1981; IV, May 1980; V, August 1979.

Table 8.12 Dry matter yields (kg ha^{-1}) in 1981 (six cuts) of pastures at Swifterbant with and without earthworms.

| | Field | | | | | | | | $\bar{x}$ | $\bar{x}$ | Mean increase |
| | 23.9 | | 24.9 | | 26.2/23.2 | | 24.4 | | | | |
Cut	−worms	+worms	−worms	+worms	−worms	+worms	−worms	+worms	−worms	+worms	(%)
1	3 232	4 047	3 502	3 775	3 039	3 964	2 505	2 924	3 070	3 678	19.8
2	4 086	4 088	2 596	2 921	3 853	4 287	4 175	4 432	3 678	3 931	6.9
3	1 754	1 989	2 408	2 140	1 639	2 037	1 696	1 854	1 874	2 005	7.0
4	3 254	3 074	3 152	3 147	2 886	3 351	3 075	3 221	3 092	3 198	3.4
5	2 147	2 001	1 964	1 837	1 985	2 191	1 304	1 669	1 850	1 925	4.1
6	1 432	1 514	809	1 381	1 043	1 312	1 377	1 452	1 165	1 415	21.5
Total	15 905	16 713	14 431	15 201	14 445	17 142	14 132	15 552	14 729	16 152	9.7

because of the high penetration resistance and the protective carpet. This is considered an advantage.

On the worm-plots, the grass roots grow directly into the mineral soil with a concentration in the upper few centimetres. In field and laboratory tests (Table 8.13), the quantity of roots in the upper decimetres of the mineral soil seemed higher in the worm-plots, but because of the abundance of dead roots in the worm-free plots this was difficult to establish. The absolute depth reached by the roots is the lowest ground-water level and is not different for areas with worms and without. The shape of the roots in the top soil is more elongated and less branched on the worm-plots than on the worm-free plots with relatively high soil strength (Table 8.8).

Table 8.13 Root contents of the soil, 0–15 cm depth in relation to estimated duration of earthworm activity, Biddinghuizen.

Worm activity (years)	Root content ($g\ 1000\ cm^{-3}$ dry weight)		
	n	$\bar{x}$	SD
0	4	0.38	0.17
1	4	0.38	0.13
4	2	0.50	0.21
6	4	1.25	0.45
8	4	1.31	0.64

Particularly in warm weather and with moist soil the sward is susceptible to damage by urine of the grazing cattle (urine scorching) (Keuning, 1981). The grass is killed and a yellow patch of a few square decimetres develops, which is colonized later by weeds and *Poa annua*. In worm-free areas these patches are more common than on the worm-plots (Table 8.14), probably because the conductivities for water and air are lower and moisture contents at field capacity are higher in the worm-free soils.

Table 8.14 Incidence of urine scorch on grassland with and without earthworms.

		Killed patches per $100\ m^2$	
		− Worms	+ Worms
Biddinghuizen	Nov. 1976	1.7	0.8
Swifterbant	Oct. 1979	18.5	2.8
Swifterbant	Oct. 1980	8.3	2.5

On the worm-free plots the mat layer virtually forms the turf. Its attachment to the mineral subsoil is weak and tufts of grass are easily detached by grazing cattle and sheep (Table 8.15). This occurs especially in late summer, autumn and early spring when the surface may be covered by little tufts of dead grass.

Table 8.15 Frequency of tufts of grass detached from the sward; Lelystad, March 1980.

	− Worms	+ Worms
Tufts (100 m^{-2})	316	18
Fresh weight (g 100 m^{-2})	7090	370

A rather unexpected effect of the earthworm inoculations was an increased interest in these areas of several bird species such as lapwing and gull. In July 1980 when wet weather forced the worms to the surface, birds damaged the sward, although the visible effects lasted only a couple of days. Mole activity also damages the sward. Previously moles hardly occurred in the new polders, but when they migrated into the field at Biddinghuizen they were particularly active in the marginal zones of the worm-plots where the earthworms were living in the mat layer and upper mineral soil. In this, moles formed extensive superficial tunnel systems, doing much damage to the sward. At the centre of the worm-plot, the situation seemed to be more stabilized with scattered tunnels and hills as in old grassland.

8.4 CONCLUSIONS

The first results indicate a variable increase in grass production caused by earthworm activity. This agrees with the results of Stockdill (1966) and other workers in New Zealand, although they found higher yield increases. The sward is less susceptible to deterioration in botanical composition, is better attached to the soil and is less susceptible to urine scorch where earthworms are active. Conditions for root growth also are improved by increased infiltration capacity, permeability and aeration of the upper soil layers and a considerable decrease in soil compaction. Worms also remove the organic surface layer and incorporate and redistribute it in the soil, promoting the formation of an A_1 horizon which increases in depth with time. All these effects positively influence root growth and root distribution in the soil. Adverse effects of earthworm activity were some decrease in the bearing capacity of the soil, increasing the chance of poaching, attracting moles that damage the sward and increased muddiness of the surface, soiling the grass. The consequences of a change in heat exchange have still to be worked out. The economies of worm inoculations require further study. The inoculation effort is considerable for the slow rate of dispersal and the benefits are not realized for a number of years.

Teledetection of the worm-plots was possible in the thermal infrared wavelength band (8–14 μm) and in the near-infrared wavelength band (0.7–0.9 μm).

8.5 ACKNOWLEDGEMENTS

We are much indebted to Mr J. Bodt, Mr A. Hoogerbrugge,ing. G. W. J. C. Peek, Mr H. Schlepers, Mr R. Bosch, Mr J. P. M. Noordman, Mr H. Oldenziel, Mr P. Vaandrager and Mr A. Maris technical assistants of the institutes and students of the Agricultural University of Wageningen for their work and enthusiasm. The thermographic pictures were prepared and interpreted under the guidance of ir. H. J. Buiten from the Department of Landsurveying and Photogrammetry of the Agricultural University of Wageningen. Part of the analytical work was done by the Institute of Soil Fertility at Haren. Mrs A. H. van Rossem revised the text. Finally we acknowledge the co-operation of Mr W. P. Boer, farmer at Biddinghuizen and dr. ir. E. G. Kloosterman and his staff of the 'Ir. A. P. Minderhoudhoeve', experimental farm at Swifterbant.

8.6 REFERENCES

Böhm, W. (1979) Methods of studying root systems. *Ecological Studies*, **33**, Springer-Verlag, Berlin. 188 pp.

Bouma, J. (1977) Soil survey of the study of water in the unsaturated soil. *Soil Survey Papers*, **13**, Stiboka, Wageningen.

de Vries, D. M. (1949) Survey of methods of botanical analysis of grassland. *5th Int. Grassld. Congr.; Noordwijk.* 143–148.

FAO/UNESCO (1974) *Soil Map of the World* (1:5.000.000) vol. 1 (legend), Paris.

Haans, J. C. F. M. (ed.) (1979) *De Interpretatie van Bodemkaarten; rapport 1463*, Stiboka, Wageningen.

Keuning, J. A. (1981) Urinebrandplekken in grasland. *Bedrijfsontwikkeling*, **12**, 453–458.

Rogaar, H. and Boswinkel, J. A. (1978) Some soil morphological effects of earthworm activity; field data and X-ray radiography. *Neth. J. Agric. Sci.*, **26**, 145–160.

Soil Conservation Service (1975), *Soil Taxonomy*, AH-436, U.S.D.A., Washington D.C.

Stockdill, S. M. J. (1966) The effect of earthworms on pastures. *Proc. N.Z. Ecol. Soc.*, **13**, 68–75.

Stockdill, S. M. J. (1982) Effect of introduced earthworms on the productivity of New Zealand pastures. *Pedobiologia* (in press).

van Rhee, J. A. (1969) Development of earthworm populations in polder soils. *Pedobiologia*, **9**, 133–140.

van Rhee, J. A. (1970) De regenwormen (*Lumbricidae*) van Nederland. *Wet. Meded. K.N.N.V.*, **84**, 23.

van Rhee, J. A. (1977) A study of the effect of earthworms on orchard productivity. *Pedobiologia*, **17**, 107–114.

APPENDIX A: ESTIMATION OF THE EARTHWORM POPULATION DENSITY AT BIDDINGHUIZEN

J. H. OUDE VOSHAAR and H. J. P. EIJSACKERS

The total number of earthworms was estimated by two methods. Both assume that earthworms disperse equally in all directions so that the dispersal pattern may be described by circles centred at the inoculation point. The data showed no evidence against this assumption. In the first method, the simplest one, the earthworm numbers are plotted against the radius r (Fig. 8.1) and the function $g(r)$ is obtained by fitting a curve by eye. This function relates the density of earthworms to the distance from the centre. The total number of worms is given by the formula $2\pi \int_0^\infty r g(r)\, dr$ which is computed by numerical integration (trapezoidal rule, Isaacson and Keller, 1966). The second method involves division of the experimental area into concentric rings. Sampling sites are located halfway between the inner and outer borders of each ring. As we sampled along four radii, each ring contains generally four sampling sites (0.25 m^2 each). We assume that the samples are taken at random and are representative for the population within this ring. Moreover, we assume that numbers in samples within a ring and between rings are independent.

If there are k sampling sites within ring i, the total number of worms in the ith ring (which is denoted by n_i) will be estimated as follows:

$$\hat{N}_i = \frac{4}{k} \times o_i \times \sum_{j=1}^{k} X_{ij}$$

where o_i is the area of the ring i and X_{ij} is the number of earthworms in the jth sample in the ith ring (random variables are written in capitals and the symbol $\hat{\ }$ is used to distinguish an estimate from its corresponding parameter). The variance of this estimator can be estimated by:

$$\text{vâr}\, \hat{N}_i = \frac{16}{k} \times o_i^2 \times \frac{1}{k-1} \left[\sum_j X_{ij}^2 - \frac{1}{k} \left(\sum_j X_{ij} \right)^2 \right]$$

The total number of worms (denoted by n) is estimated by $\hat{N} = \sum_i \hat{N}_i$. Its variance is estimated by

$$\text{vâr}\, \hat{N} = \sum_i \text{vâr}\, \hat{N}_i.$$

The 95% confidence interval of the total worm number can be approximated by:

$$\hat{N} \pm 2\,(\text{vâr}\, \hat{N})^{\frac{1}{2}}$$

REFERENCE

Isaacson, E. and Keller, H. B. (1966) *Analysis of Numerical Methods*. Wiley, New York.

Chapter 9

The activities of earthworms and the fates of seeds

J. D. GRANT

9.1 INTRODUCTION

Viable seeds which fall on to the soil surface may germinate, die, be lost or pass into the seed bank (Sagar, 1970). Seeds at the soil surface are more vulnerable to predation by birds, rodents and insects and to germination in unfavourable conditions (Roberts, 1970). Seed burial is a potent factor in prolonging the survival of seeds (Harper, 1957). The studies of Brenchley (1918), Chippendale and Milton (1934) and Rabotnov (1969) have shown that vast numbers of viable seeds lie buried in soil.

The mechanisms by which seeds are buried have not been extensively studied. Darwin (1881) reported downward movement of seeds by earthworms and Kropač (1966) and others suggested that earthworm activity is likely to contribute to seed burial. McRill (1974) and Corral (1978) have demonstrated seed burial by *Lumbricus terrestris* L. in wormeries. In this study the ability of *L. terrestris* and *Allolobophora longa* Ude to ingest seeds selectively and the effect of passage through the gut on subsequent seed germination were investigated.

9.2 MATERIALS AND METHODS

9.2.1 Seed ingestion, recovery and germination

Mature individuals of *L. terrestris* (min. fresh wt. 2.5 g) and *A. longa* (min. fresh wt. 1.3 g) obtained from grassland soil and seeds from eight species of grass from commercial sources were used. After 24 h, to allow egestion of the gut contents, each earthworm was placed on filter paper moistened with 6 ml of deionized water in a Petri dish containing either 20 seeds of a single species or ten seeds each of two species. The treatments, replicated

20 times, were randomized and kept in the dark for 18 h at 15°C. The earthworms were then removed and the numbers of seeds ingested were recorded.

After removal, the earthworms were left for 48 h on moist filter paper in fresh Petri dishes to egest the gut contents. The numbers of seeds recovered were recorded. Sets of control seeds which had not been offered to earthworms, of egested seeds and of seeds offered but not ingested were sown on moist filter paper in separate Petri dishes and kept in the dark at 15°C. Germination was recorded for 21 days.

9.2.2 Removal of seeds from the soil surface

(a) Seeds sown under laboratory conditions

In two separate trials, 45 plastic tubs (18 × 18 × 20 cm deep) were filled to 15 cm with John Innes soil and two mature *L. terrestris* (min. fresh wt. 2.5 g), three mature *A. longa* (min. fresh wt. 1.3 g) or no earthworms were added to each tub. Each treatment was replicated 15 times and the tubs were kept moist in a refrigerator at 11°C.

In the first trial 25 seeds of *Lolium perenne* L. and 25 of *Festuca rubra* L. were sown alternately 1.5 cm apart on the soil surface in a 10 × 5 grid. Seed losses from the surface were recorded for 12 days. In the second trial, *Dactylis glomerata* L. and *Holcus lanatus* L. seeds were used and seed losses from the surface were recorded for 85 days. After 5 months the tubs were transferred to a glasshouse (min. temp. 20°C) for 4 months where further seedling emergence was monitored. Seeds still on the soil surface were removed and kept moist on filter paper in Petri dishes and their germination was recorded. The soil from the tubs was emptied into seed trays and kept moist and dark for 4 weeks before being transferred to the glasshouse for a further 4 weeks. Emergence of seedlings was recorded. Seed losses were analysed using *t*-tests, and χ^2 analyses were used to compare numbers of seedlings emerging.

(b) Seeds sown under field conditions

Seeds of *L. perenne* (25) and of *F. rubra* (25) were sown on the surface of grassland soil cores (10 cm ϕ × 20 cm). The cores were placed in nylon mesh bags, inserted into plastic tubing (11 cm ϕ × 21 cm) and returned to their original positions. Two mature *L. terrestris*, three mature *A. longa* or no earthworms were added to each core with 15 replicates of each treatment. The number of seeds lost from the soil surface was recorded and analysed using the Kruskal–Wallis one-way analysis of variance (Siegel, 1956).

9.2.3 Seeds and seedlings in wormcast material

(a) Seeds present in wormcast material

Surface wormcasts (210) were collected from grassland at Treborth Botanic Garden and Pen y Ffridd Experimental Station, Bangor. These and the egested gut contents of mature *L. terrestris* and *A. longa* collected in late autumn from the grassland at Treborth were kept moist on John Innes No. 2 soil in a glasshouse (min. temp. 20°C). Emergent seedlings were identified where possible and recorded.

(b) Seedling emergence from grassland wormcasts

Forty quadrats (10 × 10 cm) in the grassland at Treborth with similar numbers of surface wormcasts were assigned one of four treatments: A (control), wormcasts left *in situ;* B, wormcasts removed; C, wormcasts left *in situ* and wormcasts from treatment D added; D, wormcasts removed but replaced by casts from treatment B. Transferred worm-casts were placed in positions corresponding to those from which they had been removed. The number and position of wormcasts and dicotyledon-ous seedlings emerging were recorded. The data were analysed using the Kruskal–Wallis test.

(c) Earthworm activity and seedling emergence

To each of 45 soil cores two mature *L. terrestris*, three *A. longa* or no earthworms were added. The surface vegetation was removed and surface wormcast production and the number of seedlings, identified where possible, were recorded for 6 months. The results were analysed as for Section (c).

9.3 RESULTS

9.3.1 Seed ingestion, recovery and germination

L. terrestris ingested significantly more seeds than *A. longa* of *Poa pratensis, Cynosurus cristatus* and *D. glomerata* (Fig. 9.1). The percentage of seeds ingested varied considerably and the ability of both earthworms to distinguish between species was shown when seed mixtures were offered (Table 9.1). *Poa trivialis* and *Agrostis tenuis* seeds in particular were taken more frequently than the seeds offered with them. In most species the proportion of seeds ingested from mixtures was similar to that of the same species offered alone and for *A. longa* none of the differences

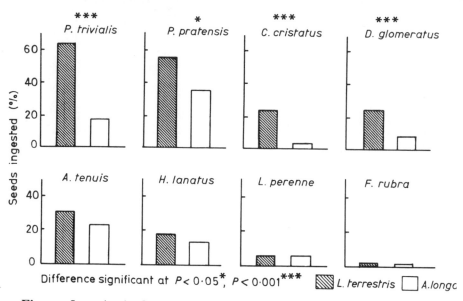

Fig. 9.1 Ingestion by *L. terrestris* and *A. longa* of seeds offered in Petri dishes. Difference significant at $P < 0.005*$, $P < 0.001***$.

Table 9.1 Ingestion (%) of seeds offered as single species or two species mixtures.

	L. terrestris			A. longa	
Species pair	1 sp.	2 spp.	Species pair	1 sp.	2 spp.
P. trivialis	66.5	46.0*	P. trivialis	16.8	12.5
D. glomerata	23.8	16.0	C. cristatus	0.5	1.0
P. pratensis	55.5	33.0*	P. pratensis	32.8	27.5
H. lanatus	18.3	33.5*	D. glomerata	6.0	9.5
A. tenuis	32.3	32.5	A. tenuis	22.3	23.0
F. rubra	1.3	0.5	L. perenne	2.8	3.0
C. cristatus	24.8	29.5	H. lanatus	14.0	11.5
L. perenne	2.5	7.5	F. rubra	2.3	1.5

* Significantly different ($P < 0.05$) when offered alone.

were significant. *L. terrestris*, however, ingested significantly less *P. pratensis* and *P. trivialis* seeds when they were offered in mixture with *H. lanatus* and *D. glomerata* respectively than when offered alone (Table 9.1). For most species 75–90 % of seeds ingested by *L. terrestris* and *A. longa* were recovered (Table 9.2). The worms egested 70–100 % of the

Table 9.2 Recovery of ingested seeds after removal of worms from seed source.

	Lumbricus terrestris					*Allolobophora longa*				
	Number of seeds ingested	Numbers of seeds egested			Egested/ ingested (%)	Number of seeds ingested	Numbers of seeds egested			Egested/ ingested (%)
		0–24 h	24–28 h	Total			0–24 h	24–48 h	Total	
A. tenuis	129	95	16	111	86	89	60	11	71	80
C. cristatus	99	80	7	87	88	4	2	0	2	50
D. glomerata	95	69	5	74	78	24	7	3	10	42
F. rubra	5	4	0	4	80	9	7	1	8	89
H. lanatus	73	50	6	56	77	56	48	4	52	92
L. perenne	10	10	0	10	100	11	6	1	7	64
P. pratensis	222	169	17	186	84	131	75	28	103	79
P. trivialis	261	181	14	195	77	67	45	5	50	75

total number of seeds finally recovered (Table 9.2) in the first 24 h after removal from the seed source.

Ingestion by *A. longa* had no significant effect on the number of seeds subsequently germinating. However, the germination of *D. glomerata*, *P. trivialis* and *P. pratensis* seeds egested by *L. terrestris* was significantly lower than those of equivalent controls (Fig. 9.2). The germination peaks of *A. tenuis*, *P. trivialis*, *H. lanatus* and *P. pratensis* seeds after ingestion by *L. terrestris* or *A. longa* were 24–48 h later than those of the controls (Fig. 9.3). The mean time to 50 % germination of egested seeds was significantly greater than that of control seeds (Fig. 9.4). Seeds of other species showed a similar trend but were too few to allow satisfactory analysis.

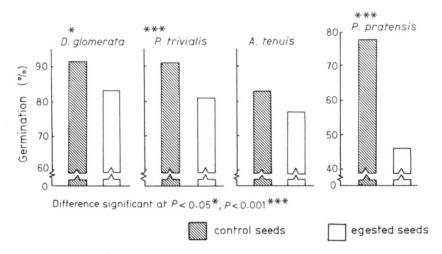

Fig. 9.2 Effect of passage through the gut of *L. terrestris* on seed germination after 21 days. Difference significant at $P < 0.05*$, $P < 0.001***$.

9.3.2 Removal of seeds from the soil surface

(a) Seeds sown under laboratory conditions

In the first trial, significantly ($P < 0.001$) more seeds were lost in the presence of *L. terrestris* (44 %) than of *A. longa* (12 %). Significantly more seeds of *L. perenne* were lost than of *F. rubra* with both earthworms (Fig. 9.5). There were no losses in the controls. In the first 3 weeks of the second trial, significantly ($P < 0.001$) more *D. glomerata* and *H. lanatus* seeds were lost in the presence of *L. terrestris* (52 %) than of *A. longa* (27 %). Losses of *D. glomerata* were similar to those of *H. lanatus* (Fig. 9.6).

Significantly more *D. glomerata* and *H. lanatus* seeds were found on the soil surface after 7 months in the controls than in either of the earthworm

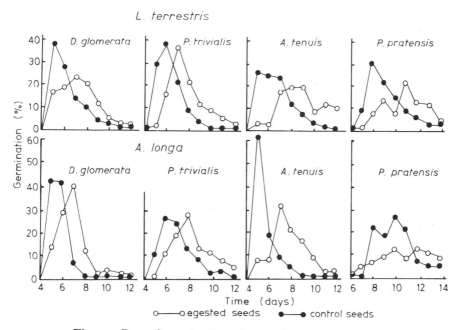

Fig. 9.3 Rate of germination of egested and control seeds.

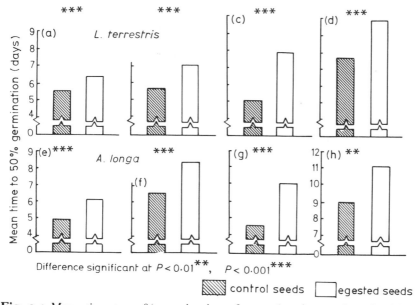

Fig. 9.4 Mean time to 50% germination of control and egested seeds. (a, e) *A. tenuis*; (b, f) *P. trivialis*; (c, g) *H. lanatus*; (d, h) *P. pratensis*. Difference between *L. terrestris* and *A. longa* significant at $P < 0.001$**, $P < 0.0001$***.

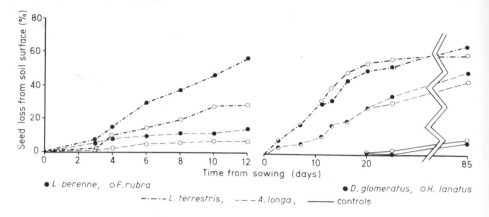

Fig. 9.5 Effect of earthworm activity on loss of surface-sown seeds in laboratory conditions.

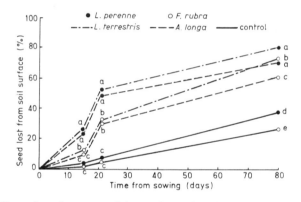

Fig. 9.6 Effect of earthworm activity on loss of surface-sown seeds under field conditions. Points sharing a common letter at a given date do not differ at $P < 0.05$.

treatments (Table 9.3, $P < 0.05$). When the soil was placed in trays, seedling emergence (which only occurred under glasshouse conditions) was greater in the earthworm treatments than from the control. There were significantly more seedlings of *H. lanatus* than of *D. glomerata*. More of the buried *H. lanatus* seeds germinated than of those that had remained on the surface (Table 9.3, $P < 0.05$).

(b) Seeds sown under field conditions

Losses of both *L. perenne* and *F. rubra* from the surface of the soil cores which contained earthworms were greatest in the first 3 weeks after

Table 9.3 Effect of earthworm activity on seed germination after seven months.

	Seeds still on soil surface	Germination (%)	Seeds lost from soil surface	Germination (%)
Control				
D. glomerata	199	2.0	34	2.9
H. lanatus	194	1.6	39	10.3
Allolobophora longa				
D. glomerata	71	1.4	156	3.2
H. lanatus	79	3.8	155	11.6
Lumbricus terrestris				
D. glomerata	51	2.0	198	3.0
H. lanatus	43	0.0	186	9.7
Total				
D. glomerata	321	1.9	388	3.1
H. lanatus	316	1.9	380	10.5

sowing (Fig. 9.6). After 2 weeks, significantly more *L. perenne* than *F. rubra* seeds were lost in both earthworm treatments. Seeds disappeared from the control cores throughout the experiment at a relatively constant rate which was similar to the rate of loss from the earthworm treatments after the first 3 weeks. After 80 days, losses of *L. perenne* and *F. rubra* seeds from the *L. terrestris* cores were significantly higher than those from the *A. longa* cores. *L. perenne* losses were greater than those of *F. rubra* in both earthworm treatments and the controls (Fig. 9.6).

9.3.3 Seeds and seedlings in wormcast material

(a) Seeds present in gut contents and wormcasts

Seedlings emerged from the gut contents of 23 % of the *A. longa* and 30 % of the *L. terrestris* collected (Table 9.4). The commonest species were *Prunella vulgaris*, *Anagallis arvensis* and *H. lanatus* (Table 9.5). The wormcasts from Pen y Ffridd produced more seedlings than those from Treborth (Table 9.4). Many grass seedlings, especially *H. lanatus* and *P. pratensis*, emerged from the Pen y Ffridd wormcasts. The commonest seedlings from the Treborth casts were *P. vulgaris*, *A. arvensis* and *Bellis perennis* (Table 9.5).

(b) Seedling emergence from surface wormcasts in grassland

Wormcast production decreased from late April (day 175) and ceased by June (day 215). Casting began again in early September (day 280) and

Table 9.4 Seedling emergence from earthworm gut contents and casts.

	No. of worms or casts	% gut contents or casts producing seedlings	Total seedlings	Mean no. seedlings per gut or cast
Gut contents				
A. longa	61	23	25	0.4
L. terrestris	142	30	57	0.4
Wormcasts				
Pen y Ffridd	210	28	197	0.9
Treborth	210	30	90	0.4

Table 9.5 Numbers of seedlings emerging in earthworm casts and gut contents.

	Wormcasts		Gut contents	
	Pen y Ffridd	Treborth	L. terrestris	A. longa
Agrostis tenuis Sibth.	10	3	4	1
A. stolonifera L.	5	2	2	—
Anagallis arvensis L.	—	10	8	6
Arenaria serpyllifolia L.	—	1	—	—
Bellis perennis L.	—	8	2	—
Capsella bursa-pastoris (L.) Medic.	1	1	—	—
Cerastium holosteoides Fr.	4	2	2	—
Dactylis glomerata L.	—	1	—	—
Holcus lanatus L.	56	4	5	2
Juncus bufonius L.	22	2	—	—
Lolium perenne L.	2	1	—	—
Myosotis arvensis (L.) Hill.	—	2	—	—
Plantago major L.	—	4	—	—
Poa annua L.	1	—	—	—
P. pratensis L.	35	—	—	—
P. trivialis L.	7	4	4	1
Prunella vulgaris L.	—	25	20	11
Ranunculus repens L.	8	2	—	—
Sagina procumbens L.	2	5	1	1
Spergula arvensis L.	2	—	—	—
Stellaria media (L.) Vill.	12	1	—	—
Taraxacum officinale Weber.	—	1	—	—
Trifolium repens L.	5	—	—	—
Urtica dioica L.	2	1	—	—
Veronica sp.	—	4	1	—
Unidentified dicotyledons	23	6	8	3
Total	197	90	57	25

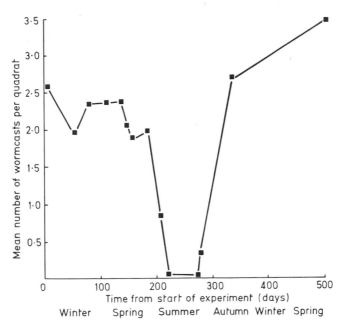

Fig. 9.7 Surface wormcast production in 18 months in grassland quadrats.

casts were abundant by mid-autumn (day 320, Fig. 9.7). The mean surface area of the wormcasts before treatment was 24–28 % of the surface area of the quadrats (Table 9.6).

Seedling emergence in all treatments was low in the first 10 weeks but in March and April (days 108–158) there was a rapid increase in seedling numbers in the three wormcast treatments (Fig. 9.8). After mid-April (day 158) the numbers of seedlings in these treatments were significantly higher than in those without wormcasts. In September and October (days

Table 9.6 Percentage of quadrat area covered by wormcasts and numbers of seedlings emerging.

| | Treatment | | | |
	A, control	B, wormcasts removed	C, added wormcasts	D, replaced wormcasts
Before treatment	28.2	26.1	24.1	23.5
After treatment	28.2	0.0	39.9	26.5
Total number of seedlings	110	39	109	99
Percentage of seedlings on wormcasts or wormcast sites	69.1	66.7	66.1	69.7

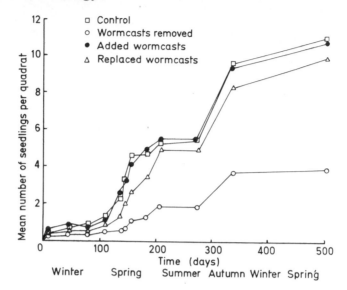

Fig. 9.8 Effect of wormcasts on seedling emergence in grassland.

276–340) more seedlings again emerged from the three wormcast treatments than from the quadrats with wormcasts removed (Fig. 9.8).

(c) Earthworm activity and seedling emergence

The cores containing *A. longa* produced most surface wormcasts, those containing *L. terrestris* had less and the controls least. Very few seedlings emerged in any treatment during the first month but by April, significantly more seedlings had emerged from the *A. longa* cores than from those with *L. terrestris* or the controls (Fig. 9.9). More seedlings and more seedling species were produced by the *A. longa* cores than by the *L. terrestris* or control cores (Table 9.7).

9.4 DISCUSSION

The effects of earthworms on seeds are clearly important in plant population dynamics. Both *L. terrestris* and *A. longa* ingested seeds offered in Petri dishes. *L. terrestris* ingested more seeds than *A. longa* and this is consistent with Perel's (1977) description of *L. terrestris* as the more active species responding more readily to various stimuli. The seeds of *P. pratensis* and *P. trivialis* were preferred to those of *L. perenne* and *F. rubra*.

Table 9.7 Effect of earthworm activity on number of seedlings emerging from soil cores.

	A. longa	L. terrestris	Control
Anagallis arvensis	66	16	19
Bellis perennis	8	2	1
Capsella bursa-pastoris	1	—	—
Cerastium holosteoides	6	2	5
Cirsium sp.	1	1	—
Crepis sp.	7	—	1
Leontodon taraxacoides (Vill.)			
Merat.	2	1	1
Lotus corniculatus L.	6	1	8
Medicago lupulina L.	17	6	14
Plantago lanceolata L.	2	1	1
P. major	2	—	3
Prunella vulgaris	9	3	3
Ranunculus sp.	3	3	1
Senecio jacobaea L.	1	—	—
Sonchus sp.	1	1	—
Taraxacum officinale	1	—	—
Trifolium sp.	1	1	—
Veronica agrestis L.	2	1	3
V. chamaedrys L.	2	1	2
Veronica sp.	4	—	2
Unidentified dicotyledons	45	21	26
Total	187	61	90

Some physical and chemical properties of seeds appear to influence their ingestion by earthworms (McRill, 1974; Grant, 1979) and the same mechanisms as have been described for leaf litter selection by L. terrestris (Satchell and Lowe, 1967) may be involved.

Some 75–90% of the seeds ingested by L. terrestris and A. longa were recovered confirming previous reports of high rates of recovery of viable seeds from earthworm casts (Beccari, 1886–1890; Jones, 1973; Corral, 1978).

As 70–100% of seeds recovered were egested in the first 24 h after removal from the seed source, the time it takes for material to pass through the gut of L. terrestris (Parle, 1963), it seems likely that most of the 'lost' seeds had been destroyed by the earthworm through gizzard contraction and enzyme activity. Seed losses of 30% may not have very great effects on vegetation dynamics compared with the effects of the much larger losses from other causes reported by Sagar and Mortimer (1976).

Seed viability was only slightly affected by ingestion; the reduced germination of seeds of P. pratensis, P. trivialis and D. glomerata egested

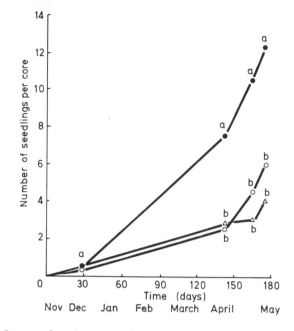

Fig. 9.9 Influence of earthworm activity on seedling emergence from soil cores. a, b, Treatments sharing the same letter on a given date are not significantly different at $P < 0.05$.

by *L. terrestris* may have been due to scarification and enzyme activity or to delayed germination. Ridley (1930) reported that germination was slower in seeds egested by birds, although the proportion finally germinating was the same as in the control. Germination of seeds egested by *L. terrestris* and *A. longa* was delayed 24–48 h. Earthworm faeces are poorly aerated and rich in ammonia (Satchell, 1967), conditions known to induce seed dormancy or delay germination (Kidd, 1914; Bibbey, 1948; Crocker and Barton, 1953; Mullverstedt, 1963).

Seeds of many species survive better buried than when left on the surface (Chepil, 1946) and for some burial is essential (Chew and Chew, 1970). There appear to be few natural mechanisms to explain how seeds are buried and earthworm activity may be essential in the formation of seed banks. Seed selection by earthworms may help to explain the frequently recorded differences between the species composition of buried seeds and the standing vegetation.

Conditions which promote the germination of seeds occur at or near the soil surface and any disturbance which brings buried seeds near to the surface will therefore increase their chances of germination (Roberts and Dawkins, 1967). Soil is moved upwards by burrowing rabbits, badgers and moles but surface-casting earthworms are more widespread and are probably far more important in bringing buried seeds to the surface. The presence in wormcasts of seeds of species not present in the standing vegetation indicates ingestion in the soil rather than on its surface. Since *L. terrestris* feeds mainly at the surface (Darwin, 1881; van Rhee, 1963) and *A. longa* feeds on sub-surface soil (Gerard, 1963) they may act on different populations of seeds from different depths. *L. terrestris* appears to be much the more important species in seed burial.

Removal of wormcasts from the soil surface significantly reduced the number of seedlings which appeared subsequently and, in plots where casts were not removed, about 70 % of the seedlings that emerged were on the sites of wormcasts. Earthworm activity may therefore be an important factor in plant population dynamics, floristic composition and weed control.

9.5 ACKNOWLEDGEMENTS

The author would like to thank Professor G. R. Sagar for his help and encouragement and Professor J. L. Harper for providing facilities and the stimulating atmosphere in which the work was carried out. Financial support was provided by the Ministry of Agriculture, Fisheries and Food.

9.6 REFERENCES

Beccari, O. (1886) Compendium of Botanical Observations on the Plants of the Indo-Malaysian and Papuan Archipelagos. *Malesia*, **3**, 324–329.

Bibbey, R. O. (1948) Physiological studies of weed seed germination. *Plant Physiol.*, **23**, 467–484.

Brenchley, W. E. (1918) Buried weed seeds. *J. Agric. Sci.*, **9**, 1–31.

Chepil, W. S. (1946) Germination of weed seeds. I. Longevity, periodicity of germination and vitality of seeds in cultivated soil. *Sci. Agric.*, **26**, 307–346.

Chew, R. M. and Chew, A. E. (1970) Energy relationships of the mammals of a desert shrub (*Larrea tridentata*) community. *Ecol. Monogr.*, **40**, 1–21.

Chippendale, H. G. and Milton, W. E. J. (1934) On the viable seeds present in the soil beneath pastures. *J. Ecol.*, **22**, 508–531.

Corral, G. (1978) Some effect of *Lumbricus terrestris* L. on seed movement, seedling emergence, establishment and growth. M.Sc. dissertation, University of Wales.

Crocker, W. and Barton, L. V. (1953) Physiology of seeds. *Chronica Botanica*, Waltham, Mass. 267 pp.

Darwin, C. (1881) *The Formation of Vegetable Mould Through the Action of Worms, with Observations on their Habits*. John Murray, London. 326 pp.

Gerard, B. M. (1963) The activities of some species of Lumbricidae in pasture-land. In *Soil Organisms* (eds. J. Doeksen and J. Van Der Drift), North Holland Publ. Co., Amsterdam. 49–54.

Grant, J. D. (1979) The influence of earthworm activity on the fates of seeds. Ph.D. Thesis, University of Wales.

Harper, J. L. (1957) The ecological significance of dormancy and its importance in weed control. *Proc. 4th Int. Congr. Crop Prot. Hamburg*, 415–420.

Jones, D. G. (1973) An investigation of the ingestion of seeds by the earthworm *Allolobophora longa* Ude B.Ed. Dissertation, University of Wales.

Kidd, F. (1914) The controlling influence of carbon dioxide in the maturation, dormancy and germination of seeds. *Proc. R. Soc. London Ser. B*, **87**, 408–421 and 609–625.

Kropač, Z. (1966) Estimation of weed seeds in arable soil. *Pedobiologia*, **6**, 105–128.

McRill, M. (1974) Some botanical aspects of earthworm activity. Ph.D. Thesis, University of Wales.

Mullverstedt, R. (1963) Untersuchungen uber die Keimung von Unkrantsamen in Abhangigkeit vom Sauerstoffpartialdruck. *Weed Res.*, **3**, 154–163.

Parle, J. N. (1963) Micro-organisms in the intestines of earthworms. *J. Gen. Microbiol.*, **31**, 1–11.

Perel, T. S. (1977) Differences in lumbricid organisation connected with ecological properties. In *Soil Organisms as Components of Ecosystems* (eds. U. Lohm and T. Persson), *Proc. 6th Int. Soil Zool. Coll. Ecol. Bull. (Stockholm)*, **25**, 56–63.

Rabotnov, R. A. (1969) Plant regeneration from seed in meadows of the U.S.S.R. *Herb. Abstr.*, **39**, 269–277.

Ridley, H. N. (1930) *The Dispersal of Plants Throughout the World*. L. Reeve & Co., Ashford, Kent. 744 pp.

Roberts, H. A. (1970) Viable weed seeds in cultivated soils. *Rep. Natl. Veg. Res. Stn. for 1969*, 25–38.

Roberts, H. A. and Dawkins, P. A. (1967) Effect of cultivation on the numbers of viable weed seeds in soil. *Weed Res.*, **7**, 290–301.

Sagar, G. R. (1970) Factors controlling the size of plant populations. *Proc. 10th Br. Weed Control Conf.*, 965–979.

Sagar, G. R. and Mortimer, A. M. (1976) An approach to the study of the population dynamics of plants with special reference of weeds. In *Applied Biology* (ed. T. H. Coaker), Vol. 1. Academic Press, London. 1–47.

Satchell, J. E. (1967) Lumbricidae. In *Soil Biology* (eds. A. Burgess and F. Raw), Academic Press, London. 259–322.

Satchell, J. E. and Lowe, D. G. (1967) Selection of leaf litter by *Lumbricus terrestris*. In *Progress in Soil Biology* (eds. O. Graff and J. E. Satchell), North Holland Publ. Co., Amsterdam. 102–119.

Siegel, S. (1956) *Non-parametric Statistics*. McGraw Hill, London. 312 pp.

van Rhee, J. A. (1963) Earthworm activities and the breakdown of organic matter in agricultural soils. In *Soil Organisms* (eds. J. Doeksen and J. van der Drift), North Holland Publ. Co., Amsterdam. 55–59.

Chapter 10

Earthworm ecology in cultivated soils

C. A. EDWARDS

10.1 INTRODUCTION

Most cultivated soils in temperate regions support earthworms. Population densities vary widely and the species present differ in relation to climatic conditions, soil type and cropping. Cultivation by mechanical disturbance completely changes the environment in which earthworms live, destroying the habitat and changing the soil temperature and moisture and the availability of food. All these factors influence the size of earthworm populations, their species diversity, dominance and vertical distribution (Fig. 10.1).

There are few analytical studies of the changes in earthworm ecology brought about by arable cropping and virtually no experimental studies. Even less is known about earthworms in cultivated soils in the tropics although there is little doubt that the impact of cultivation is greater there than in temperate regions (Perfect *et al.*, 1979).

10.2 SPECIES OF EARTHWORMS IN CULTIVATED SOILS

The numbers of species of earthworms supported by agricultural soils depends mainly upon the kind and extent of plant cover and to some extent its permanence. Thus, permanent pastures usually have the greatest species diversity of earthworms with, in the British Isles, up to twelve species present (Table 10.1). *Bimastos eiseni* (Levinsen) and *Dendrobaena octaedra* (Savigny) are sometimes found in rough pasture, particularly on acid mor soils.

When pasture is ploughed and arable crops grown, the species diversity gradually decreases. As the amount of soil organic matter decreases with

123

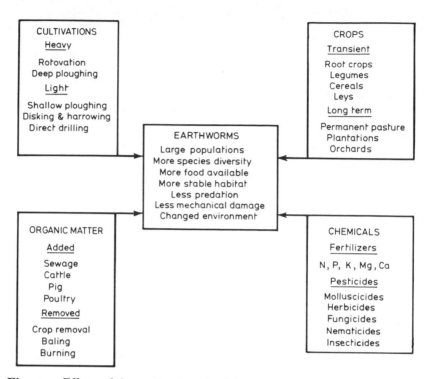

Fig. 10.1 Effects of the production of cultivated crops on earthworm ecology.

Table 10.1 Lumbricidae common in cultivated soils in Britain.

	Orchard (Grass)	Pasture	Arable
Dominant	A. caliginosa L. terrestris	A. caliginosa A. longa A. nocturna	A. chlorotica A. caliginosa
Common	A. longa L. castaneus A. chlorotica A. rosea	L. terrestris A. rosea A. chlorotica O. cyaneum L. castaneus	A. longa L. terrestris A. rosea
Often present	O. cyaneum D. subrubicunda A. icterica L. rubellus	A. limicola L. festivus D. octaedra B. eiseni D. rubida A. tuberculata L. rubellus	L. castaneus L. rubellus

continual arable cropping, the species most dependent upon it, e.g. *Lumbricus terrestris* and *Allolobophora longa*, become less common. Usually *A. chlorotica* becomes dominant, *A. caliginosa* remains numerous and *A. rosea*, *Octolasion cyaneum* and occasionally *L. castaneus* persist. In cultivated orchards where there is a considerable addition of organic matter as leaf fall, *L. terrestris*, *A. longa* and *A. caliginosa* are usually very common, with *L. terrestris* and to a lesser extent *A. longa* dominant in biomass and numbers. From a survey of arable farms in Kent there is recent evidence of competition between species occupying the same ecological niche (Edwards and Lofty, 1982a). In seven fields that had been direct drilled for several years, there were strong inverse correlations between numbers of *A. longa* and *L. terrestris* and of *A. chlorotica* and *A. caliginosa* (Fig. 10.2).

The species of earthworms present in tropical soils is reviewed by Lee (Chapter 15). In Nigeria, *Hyperodrilus africanus* Beddard is dominant in the uncultivated bush and *Eudrilus eugeniae* (Kinberg) is scarce. When the bush is cleared, cultivated and cropped, *E. eugeniae* becomes dominant and *H. africanus* virtually disappears (Madge, 1969).

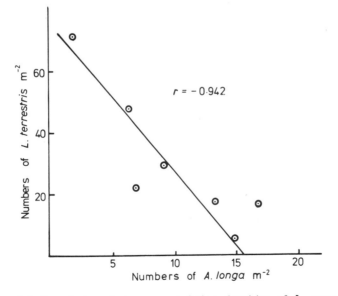

Fig. 10.2(a) Correlations between population densities of *L. terrestris* and *A. longa*.

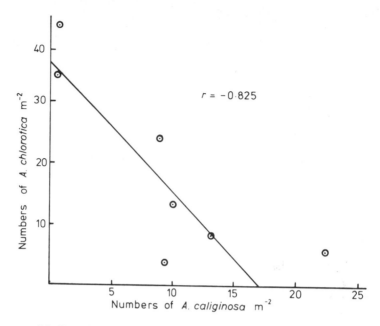

Fig. 10.2(b) Correlations between population densities of *A. chlorotica* and *A. caliginosa*.

10.3 EARTHWORM POPULATIONS IN CULTIVATED SOILS

Earthworm populations reported in the literature are about three times as large in orchards with grass cover as they are in arable soils (Table 10.2). Ploughing usually decreases the earthworm populations of permanent pasture (Low, 1972; Russell, 1973). However, one or two cultivations can actually increase populations (Edwards and Lofty, 1978), although repeated arable cropping undoubtedly reduces population densities. This was attributed to mechanical damage during cultivation by Zicsi (1958), but in later work he showed that the use of disk cultivators actually led to increased numbers of some species (Zicsi, 1969). There is little experimental data to support the view that the fall in population density following cultivation is due to loss of the insulating layer of vegetation, but the gradual decrease of organic matter with repeated arable cropping (Evans and Guild, 1948; Edwards and Lofty, 1978) is most important. Arable soils to which organic matter is added can support large earthworm populations despite repeated cultivation (Edwards and Lofty, 1982b). Cultivation may favour some species such as *A. chlorotica* by mixing organic matter into the upper strata. There is a tendency for

Table 10.2 Earthworm populations in cultivated soils.

Crop	No. m^{-2}	Wt. m^{-2} (g)	Site	Extraction method	Reference
Orchard (grass)	848	230	Cambs., UK	Formalin and Handsorting	Raw (1959)
	300–500	75–112	Holland	Handsorting	Van Rhee and Nathans (1961)
	542	182	Cambs.	Formalin	Raw and Lofty (unpub.)
	420	238	Cambs.	Formalin	Raw and Lofty (unpub.)
	254–344	63–86	USSR	Handsorting	Dzangaliev and Belousova (1969)
	292	146	Kent	Formalin	Lofty (unpub.)
	250	213	Cambs.	Formalin	Raw and Lofty (unpub.)
Orchard (arable)	218	193	Cambs.	Formalin	Raw and Lofty (unpub.)
	140	153	Cambs.	Formalin	Raw and Lofty (unpub.)
Arable					
Cereal, sugar beet, grass	101–453	13–40	Denmark	Formalin	Andersen (Chapter 11)
	287	76	Bardsey Island	Handsorting	Reynoldson (1955)
Barley	22–280	7–95	Rothamsted	Formalin	Edwards and Lofty (1982b)
	220	48	Germany	Wet sieving	Kruger (1952)
	146	50	N. Wales	Handsorting	Reynoldson (1955)
	56–143	—	Scotland	Formalin	Gerard and Hay (1979)
	80–109	42–59	Scotland	Formalin	Gerard and Hay (1979)
Wheat	100	—	Rothamsted	Handsorting	Morris (1922a)
	74–98	36–45	Woburn, Herts.	Formalin	Edwards and Lofty (1982a)
	6–92	3–70	Broadbalk Rothamsted	Formalin	Edwards and Lofty (1982b)
	27–72	14–33	Boxworth Herts	Formalin	Edwards and Lofty (1982a)
	15–64	9–35	Rothamsted	Formalin	Edwards and Lofty (1982a)
Root crops	11–76	7–46	Barnfield Rothamsted	Formalin	Edwards and Lofty (1982b)

shallow-working species to penetrate deeper when the soil is regularly cultivated.

10.4 EFFECTS OF CULTIVATION

Uncultivated soils are insulated against climatic changes by the litter layer in orchards and woodlands and by the grass and root mat in pastures. Under cultivation, this insulation disappears as the organic matter breaks down, temperature and moisture changes are more pronounced (Hay et al., 1978) and freezing is more frequent. In many cultivated soils there is little organic matter near the surface to attract deep-burrowing species to feed there. In many undisturbed soils, wetting and drying causes a network of cracks to develop around root channels and earthworm burrows. After cultivation, these cracks disappear from the upper 20 cm and the soil becomes more open, favouring earthworm activity. Cultivated soils tend to drain more rapidly than undisturbed soils and dry out more quickly.

Regular cultivation destroys the upper parts of the relatively permanent burrows of *L. terrestris*, *A. longa* and *A. nocturna*. Although they can withstand this for one or two seasons and re-form their burrows (Edwards and Lofty, 1978), when it is repeated for many years numbers of these species are depressed and species of the upper soil horizons with no permanent burrows become dominant. Cultivation destroys the worms' environment, damages them mechanically and brings them to the surface where they are preyed on by birds (Cuendet, Chapter 36).

When grass is destroyed with a broad-spectrum herbicide and a crop is drilled directly into the uncultivated soil (direct drilling), large earthworm populations can be maintained. Populations in direct-drilled soil are almost always much larger than in the same soil when it is ploughed and cropped in the traditional way (Schwerdtle, 1969; Graff, 1969; Edwards and Lofty, 1972, 1982a; Gerard and Hay, 1979; Barnes and Ellis, 1982). These differences are enhanced when straw is left on the surface of the uncultivated soil rather than being burnt or baled and removed.

The deep-burrowing species *L. terrestris*, *A. longa* and *A. nocturna* benefit most from minimum cultivation (Fig. 10.3). Chisel ploughing, shallow tining, disking and harrowing usually result in earthworm populations intermediate between those in direct-drilled and ploughed soils (Edwards and Lofty, 1982a) (Fig. 10.4), although in one study the effects on populations of *A. longa* and *A. caliginosa* of tined cultivation were more adverse than ploughing (Gerard and Hay, 1979).

The difference between earthworm populations under direct-drilled crops and crops grown after ploughing can become very large. Edwards

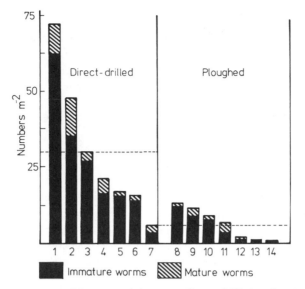

Fig. 10.3 Populations of *L. terrestris* in seven direct-drilled and seven ploughed. fields in Sussex.

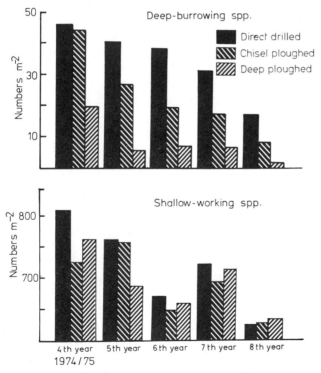

Fig. 10.4 Populations of earthworms in plots that had been direct-drilled, chisel-ploughed or deep-ploughed for 8 years at Boxworth.

and Lofty (1982a) reported a thirty-fold difference after eight years of direct drilling. However, even when crops have been grown for many years with no cultivation, earthworm populations tend to decrease as the soil organic matter becomes limiting (Fig. 10.4). In uncultivated soils growing cereals earthworm burrows provide channels for root growth lined with more available mineral nutrients than the surrounding soil (Edwards and Lofty, 1978, 1980).

10.5 EFFECTS OF CROPPING

Soil under orchards or fruit bushes contains more organic matter in the form of dead leaves and can support more earthworms than that under annual crops. Evidence as to how different annual crops and crop rotations affect earthworm activity is presented by Lofs-Holmin (Chapter 12). Continuous cereal growing tends to build up earthworm populations as illustrated by the experimental field Broadbalk at Rothamsted. Plots that had grown cereals continuously from 1843 to 1979 had many more earthworms than those with root crops or fallow (Fig. 10.5), probably because the root system of wheat, which forms 50% of its biomass, remains in the soil after harvest and together with straw provides a food source. Similar data were obtained from a field in Kent sampled in autumn. Mean numbers of $L.$ $terrestris$ were $180 \, m^{-2}$ under wheat, $60 \, m^{-2}$ under legumes and $35 \, m^{-2}$ under sugar beet. Much of the biomass of legume crops is returned to the soil and supports large populations. Most of the organic matter of root crops is harvested and they sustain small populations.

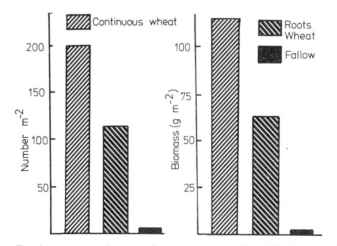

Fig. 10.5 Earthworm populations after 136 years on Broadbalk under different cropping regimes.

10.6 EFFECTS OF PESTICIDES

The effects of pesticides on earthworm populations have been reviewed by Davey (1963), Edwards and Thompson (1973), Edwards and Lofty (1977) and Edwards (1980). Most pesticides are not acutely toxic to earthworms, although there is little information on long-term chronic toxicity.

Most harmful are the fumigant nematicides such as D-D and metham sodium, chloropicrin and methyl bromide, but these are not extensively used in agricultural soils. The nematicide, aldicarb, when applied to dry soils is only moderately toxic and soon breaks down to a non-toxic form but when wet soils are treated, the chemical, which is very soluble, is taken up in large quantities into earthworm tissues. This irritates the worms so that they come to the soil surface where they may be eaten by birds.

The organochlorine insecticides, with the exception of chlordane and endrin, are not very toxic to earthworms but can accumulate in their tissues. They are now little used in developed countries. Several organophosphates, e.g. phorate, parathion, are moderately toxic to worms but their use is restricted in Europe and USA. Carbaryl paralyses earthworms and produces ulcers on their bodies; they may survive for a considerable time in a paralysed state but eventually die.

The greatest current danger to earthworms in arable land is the systematic fungicide carbendazim and the related compounds, benomyl and thiophanate methyl. These are used increasingly on agricultural soils, 30% of British cereal fields being treated annually. They are persistent and very toxic to earthworms whether the worms are exposed to them in the soil or through feeding on decaying plant material containing residues.

Earthworm species differ little in susceptibility to pesticides but some are more at risk because of their habits. *L. terrestris* is most exposed to pesticide residues because of its activity on the soil surface where residues from foliar sprays are concentrated and because it pulls contaminated leaves and straw into its burrows and consumes them. There is evidence that pesticides in ingested soil may be transported through the soil profile. Earthworm populations are generally not greatly at risk from pesticides except in orchards which are regularly treated with a wide range of toxic chemicals.

10.7 EFFECTS OF FERTILIZERS

Nearly all arable soils are treated with fertilizers. In the first half of this century, fertilizer application was predominantly as farmyard manure. With the intensification of pig and cattle breeding in isolated units this was gradually superseded by inorganic fertilizers containing mainly nitrogen, phosphorus and potassium, with smaller amounts of other elements such as sodium and magnesium. However, with increasing costs, these are being partly replaced by pig and cattle solids, sludges and slurries and 46 % of human sewage is now applied as activated sludge or sewage cake to arable soils. In 1978, 29 % of all agricultural land in the UK was treated with organic materials.

10.7.1 Organic fertilizers

The effects of organic fertilizers on earthworms was reviewed by Marshall (1977). In arable crops these have been studied by Morris (1922, 1927), Doeksen (1959) Atlavinyte (1975), Edwards and Lofty (1982b) and Lofs-Holmin (Chapter 12), and in grassland sites by Jefferson (1955, 1956), Zajonc (1975), Curry (1976), Edwards and Lofty (1975), Cotton and Curry (1980a,b).

Morris (1922b, 1927) reported 2–2.5 times as many earthworms in arable plots receiving farmyard manure as in those to which no fertilizer was applied and more recent work has confirmed that farmyard manure encourages the build-up of earthworm populations (Satchell, 1955; Marshall, 1977; Edwards and Lofty, 1975, 1977; Edwards, 1980). The effect is much greater in arable land than in grassland probably because the deficiency of decaying organic matter is greater there than under grass.

In a two-year study, Andersen (Chapter 11) compared the effects on earthworms of 250, 500, 1000 and 2000 kg N ha^{-1}, applied as farmyard manure and animal slurry to a field cropped with a rotation of cereals, sugar beet and grass. Both forms of organic matter increased numbers of *L. terrestris* and *A. caliginosa* but *A. longa* increased only in response to farmyard manure. The highest rates of application tended to decrease populations of most earthworm species. *L. terrestris* increased most because it feeds directly on surface organic matter but the influence of organic matter on all species seems to be through increasing their food supply, whether they feed on it directly or on the micro-organisms it supports.

In an experiment at Rothamsted, the effect on earthworms of farmyard manure was compared with that of sewage cake. Both increased population densities but sewage cake favoured *A. chlorotica*, and *L. terrestris* was influenced more by farmyard manure. Liquid slurries, sometimes applied to arable land, may have an initial detrimental effect on earthworm populations, dead worms appearing on the surface after treatment. Populations may then increase if applications are not too frequent.

10.7.2 Inorganic fertilizers

Effects of nitrogenous inorganic fertilizers on earthworm populations in arable land are described by Andersen (Chapter 11), Barnes and Ellis (1982), Edwards and Lofty (1975, 1982b), Gerard and Hay (1979) and Zajonc (1975). Zajonc concluded that high doses of nitrogenous fertilizers favoured earthworms, particularly species that live close to the surface. This was confirmed by Edwards and Lofty (1982b) who reported that annual N applications to cereals for more than 130 years increased earthworm populations in proportion to the amount of N applied. Regressions of earthworm numbers on the amounts of N applied annually and the organic C content of the soil (Figs. 10.6 and 10.7) suggest that higher levels of N influence earthworms by increasing the plant biomass and, when this dies, the amount of decomposing organic matter. For equal rates of N application, organic N had a greater effect than inorganic N.

Gerard and Hay (1979) also studied the effect of inorganic N on earthworms in arable land. They assessed the effects of 0, 50, 100 and 150 kg of inorganic N ha^{-1} applied annually, on populations in a barley field two years and five years after treatments began, and reported the largest numbers of worms in plots treated with 100 kg N ha^{-1}. Differences between treatments of 100 and 150 kg N ha^{-1} were not significant.

The form of the inorganic N can also affect earthworm populations. In a root crop experiment at Rothamsted, farmyard manure plus ammonium

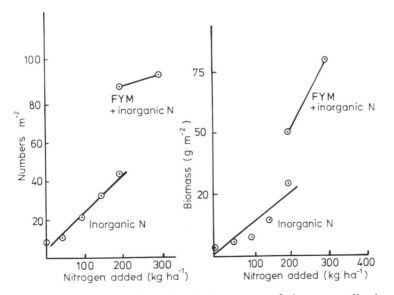

Fig. 10.6 Earthworm populations in relation to rates of nitrogen application.

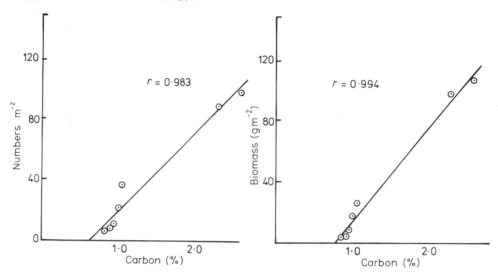

Fig. 10.7 Earthworm populations in relation to soil carbon content.

sulphate increased populations more than farmyard manure plus sodium nitrate. This is surprising because several workers have found ammonium sulphate harmful to earthworms in grassland (Escritt and Arthur, 1948; Rodale, 1948; Jefferson, 1955). The deleterious effect of ammonium sulphate may have arisen from increased soil acidity as observed by Satchell (1955).

Evidence that fertilizer calcium favours earthworms (Spannagel, 1960; Franz, 1959; Kleinschmit, 1962; Edwards and Lofty, 1975) was reviewed by Marshall (1977). The mechanism is not clear but earthworms require calcium and many species do not tolerate very acid conditions. Superphosphates also tend to favour earthworm populations (Doerell, 1950; Sears and Evans, 1953).

10.8 CONCLUSIONS

With the gradual decrease in cultivation occurring in many parts of the world, earthworms are becoming increasingly important in arable soils. There is good evidence that, as in grassland, they contribute to soil fertility by maintaining soil structure, aeration and drainage and by breaking down organic matter and incorporating it into the soil. Many earthworm species can withstand a considerable amount of soil disturbance and the greatest influence on populations in arable soils seems to be the availability of organic matter. If the soil organic matter content is maintained and the use of harmful chemicals is avoided, earthworm populations can be maintained at high densities. A better understanding of the ecology of earthworms could enable their activities to be manipulated to improve soil fertility.

10.9 REFERENCES

Atlavinyte, O. (1975) Effect of chemical substances on the activity of Lumbricidae in the process of straw disintegration. In *Progress in Soil Zoology* (ed. J. Vanek), *Proc. Vth Int. Soil Zool. Coll.*, Academia, Prague, 375–387.

Barnes, B. T. and Ellis, F. B. (1982) The effects of different methods of cultivation and direct drilling, and of contrasting methods of straw disposal on populations of earthworms. *Soil Sci.*, (in press).

Cotton, D. C. F. and Curry, J. P. (1980a) The effects of cattle and pig slurry fertilizers on earthworms (Oligochaeta, Lumbricidae) in grassland managed for silage production. *Pedobiologia*, **20**, 181–188.

Cotton, D. C. F. and Curry, J. P. (1980b) The response of earthworm populations (Oligochaeta, Lumbricidae) to high applications of pig slurry. *Pedobiologia*, **20**, 189–196.

Curry, J. P. (1976) Some effects of animal manures on earthworms in grassland. *Pedobiologia*, **16**, 425–438.

Davey, S. P. (1963) Effects of chemicals on earthworms. A review of the literature. *Special Scientific Report, Wildlife No. 74*. USA Fish and Wildlife Service.

Doeksen, J. (1959) Earthworms, manuring and tillage. *Stikstof* (May), 17–24.

Doerel, E. C. (1950) Was sagen die Regenwurmer zur Mineraldungung. *Dtsch. Landwirt. Presse*, **4**, 19.

Dzangaliev, A. D. and Belousova, N. K. (1969) Earthworm populations in irrigated orchards under various soil treatments. [in Russian] *Pedobiologia*, **9**, 103–105.

Edwards, C. A. (1980) Interactions between agricultural practice and earthworms. In *Soil Biology as Related to Land Use Practices* (ed. D. L. Dindal), *Proc. VIIth Int. Coll. Soil Zool., E.P.A. Washington DC*, 3–12.

Edwards, C. A. and Lofty, J. R. (1972) Effects of pesticides on soil invertebrates. *Rep. Rothamsted Exp. Stn. for 1971*, 210–212.

Edwards, C. A. and Lofty, J. R. (1975) Effects of direct drilling on the soil fauna. *Outlook Agric.*, **8**, 243–244.

Edwards, C. A. and Lofty, J. R. (1977) *Biology of Earthworms*. 2nd edn. Chapman and Hall, London. 333 pp.

Edwards, C. A. and Lofty, J. R. (1978) The influence of arthropods and earthworms upon the root growth of cereals after five years of direct drilling. *J. Appl. Ecol.*, **15**, 789–795.

Edwards, C. A. and Lofty, J. R. (1980) Effects of earthworm inoculation upon root growth of direct drilled cereals. *J. Appl. Ecol.*, **17**, 533–543.

Edwards, C. A. and Lofty, J. R. (1982a) The effect of direct drilling and minimal cultivation on earthworm populations. *J. Appl. Ecol.*, **19**, 723–724.

Edwards, C. A. and Lofty, J. R. (1982b) Nitrogenous fertilizers and earthworm populations in agricultural soils. *Soil Biol. Biochem.*, **14**, 515–521.

Edwards, C. A. and Thompson, A. R. (1973) Pesticides and the soil fauna. *Resid. Rev.*, **45**, 1–79.

Escritt, J. R. and Arthur, J. H. (1948) Earthworm control. *J. Bd. Greenkeep. Res.*, **7**, 162–172.

Evans, A. C. and Guild, W. J. McL. (1948) Studies on the relationships between earthworms and soil fertility. V. Field populations. *Ann. Appl. Biol.*, **35**, 485-493.

Franz, H. (1959) Das biologische Geschen im Waldboden und seine Beeinflussung durch Kalkdüngung. *Allg. Forstztg.*, **70**, 178-181.

Gerard, B. M. and Hay, R. K. M. (1979) The effect on earthworms of ploughing, tined cultivation, direct drilling and nitrogen in a barley monoculture system. *J. Agric. Sci. Camb.*, **93**, 147-155.

Graff, O. (1969) Regenwurmtätigkeit in Ackerboden unter verschiedenem Bedeckungsmaterial, gemessen an der Losungsablage. *Pedobiologia*, **9**, 120-128.

Hay, R. K. M., Holmes, J. C. and Hunter, E. A. (1978) The effects of tillage, direct drilling and nitrogen fertilizer on soil temperature under a barley crop. *J. Soil Sci.*, **19**, 174-183.

Jefferson, P. (1955) Studies on the earthworms of turf. A. The earthworms of experimental turf plots. *J. Sports Turf Res. Inst.*, **9**, 6-27.

Jefferson, P. (1956) Studies on the earthworms of turf. *J. Sports Turf Res Inst.*, **9**, 166-179.

Kleinschmit, J. (1962) Untersuchung von Kalkdungungsversuchsflachen im Lehrforstamt Escherode. *Forstarchiv*, **33**, 25-29.

Kruger, W. (1952) Influence of soil cultivation on animal communities of fields. [in German] *Z. AckerPflanzenbau*, **95**, 269.

Low, A. J. (1972) The effect of cultivation on the structure and other physical characteristics of grassland and arable soils (1945-1970). *J. Soil Sci.*, 363-380.

Madge, D. S. (1969) Field and laboratory studies on the activities of two species of tropical earthworms. *Pedobiologia*, **9**, 188-214.

Marshall, V. G. (1977) *Effects of Manures and Fertilizers on Soil Fauna: A Review.* Commonwealth Bureau of Soils. 79 pp.

Morris, H. M. (1922a) On a method of separating insects and other arthropods from the soil. *Bull. Ent. Res.*, **13**, 197.

Morris, H. M. (1922b) Insect and other invertebrate fauna of arable land at Rothamsted. *Ann. Appl. Biol.*, **9**, 282-305.

Morris, H. M. (1927) The insect and other invertebrate fauna of arable land at Rothamsted. Part II. *Ann. Appl. Biol.*, **14**, 442-463.

Perfect, T. J., Cook, A. G., Critchley, B. R., Critchley, U., Davies, A. L., Swift, M. J., Russel-Smith, A. and Yeadon, R. (1979) The effect of DDT contamination on the productivity of a cultivated forest soil in the subhumid tropics. *J. Appl. Ecol.*, **16**, 705-720.

Raw, F. (1959) Estimating earthworm populations by using formalin. *Nature (London)*, **184**, 1661-1662.

Reynoldson, T. B. (1955) Observations on the earthworms of North Wales. *N.W. Naturalist*, 291-304.

Rodale, J. I. (1948) Do chemical fertilizers kill earthworms? *Organic Gardening*, **12**, 12-17.

Russell, E. J. (1973) *Soil Conditions and Plant Growth.* 10th edn. Longmans, London. 849 pp.

Satchell, J. E. (1955) Some aspects of earthworm ecology. In *Soil Zoology* (ed. D. K. Mc. E. Kevan), Butterworths, London. 180-201.

Schwerdtle, F. (1969) Untersuchungen zur populationsdichte von Regenwurmen bei herkommlicher Bodenarbeitung und bei 'Direktsaat'. *Z. PflKrank. PflPath. PflSchutz.*, **76**, 635-641.

Sears, P. D. and Evans, L. T. (1953) Pasture growth and soil fertility. III. The influence of red and white clovers, super phosphate lime and dung and urine on soil composition and on earthworms and grass grub populations. *N. Z. J. Sci. Technol.*, **35A**, *Suppl.* **1.**, 45–52.

Spannagel, G. (1960) Humusbildung unter dem Einfluss von Kalk in Verbindung mit der Entwicklung einer reichen Bodenfauna. *Trans. 7th Int. Congr. Soil Sci.*, 695–701.

van Rhee, J. A. and Nathans, S. (1961). Observations on earthworm populations in orchard soils. *Neth. J. Ag. Sci.*, **9**, 94–100.

Zajonc, I. (1975) Variations in meadow associations of earthworms caused by the influence of nitrogen fertilizers and liquid manure irrigation. In *Progress in Soil Zoology* (ed. J. Vanek), Dr W. V. Junk, The Hague. 497–503.

Zicsi, A. (1958) Einfluss der Trockenheit und der Bodenarbeitung auf das Leben der Regenwürmer in Ackerböden. *Acta Agron. Hung.*, **8**, 67–75.

Zicsi, A. (1969) Uber die Auswirkung der Nachfrucht und Bodenarbeitung auf die Aktivität der Regenwurmer. *Pedobiologia*, **9**, 141–146.

Chapter 11

Nitrogen turnover by earthworms in arable plots treated with farmyard manure and slurry

N. C. ANDERSEN

11.1 INTRODUCTION

Important studies on the effects of earthworms on nitrogen turnover have been made by Satchell (1967) in woodland and Syers *et al.* (1979) in grassland but data are sparse for arable soils. Most arable crops are able to utilize about 70% of the available inorganic N and the balance of their growth requirements must be met from the pool of organic N after mineralization. This study attempts to assess the contribution of earthworms to the turnover of organic N on plots receiving farmyard manure (FYM) or slurry (SLU) in relation to the crops' N demand. It assumes that the output of mineralizable tissue N from dead worms, or eventually from their predators, is equivalent to the amount produced in new tissue, disregarding cocoon production. Estimates are calculated on an annual basis recognizing that if the system is not in balance, pulses of input and output may not coincide temporally. N output in earthworm faeces, urine and mucus undergoes microbial immobilization and may become partially adsorbed on clay colloids or stable organic matter complexes. Inputs and outputs of these N fractions similarly will vary with, *inter alia*, different agricultural operations, climate and changes in earthworm population density. To render their analysis less intractable they are nevertheless treated here as being in steady state.

11.2 PRODUCTIVITY

A growth curve (Fig. 11.1) was obtained from measurements made on *Aporrectodea caliginosa* maintained in cultures in the dark at 12°C and fed

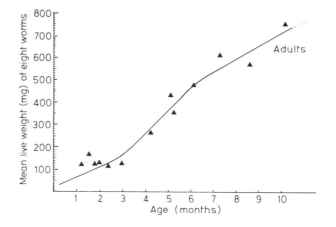

Fig. 11.1 Growth of *A. caliginosa* on pulverized straw at 12°C.

on pulverized straw at the rate of 10–400 g of soil. Faster growth is obtained in cultures fed on FYM (Lofs-Holmin, 1980) but under field conditions, the growth of *A. caliginosa* in manured plots probably corresponds well to growth in cultures with added straw powder.

Time production was estimated for each of the three size classes from the samples collected in October 1976, May 1977, October 1977 and October 1978. Estimates were made in four ways, three using weighted biomass days $(\overline{B}T)$ and the fourth using weighted population density days $(\overline{N}T)$. To simplify calculations the following generalizations were made:

(1) Because *A. caliginosa* was overwhelmingly dominant, all the material was treated as being of this species;
(2) Because of low activity during adverse winter and summer conditions, new tissue was assumed to be produced, under Danish conditions, only during 47% of the year;
(3) Dry weight of tissue was assessed as 17% of the preserved weight of worms (Lakhani and Satchell, 1970) and its energy content as 4.92 k cal g^{-1} dry wt. (Phillipson *et al.*, 1978).

Productivity was estimated on the basis of biomass days calculated from the general expression (11.1) below derived by Petrusewicz and MacFadyen (1970).

$$\sum_{s_1}^{s_3} \overline{B}\,T = \frac{B_{I_s} + B_{II_s}}{2} t_1 + \frac{B_{II_s} + B_{III_s}}{2} t_2 + \frac{B_{III_s} + B_{IV_s}}{2} t_3 \qquad (11.1)$$

B_I-B_{IV} = biomass on successive samplings
$\quad t_1-t_3$ = time between successive samplings
$\quad s_1-s_3$ = developmental stages

Production may be estimated from this formula by multiplication by:
(1) the relative growth rate V', the increase in biomass g wet wt. g^{-1} worm
wet wt. day^{-1}; (2) the calorific equivalent of oxygen consumption using
the respiration equivalent of McNeil and Lawton (1970); (3) population
respiration calculated in (2) and by respiration/production ratio. The first
method uses only data from the cultures, the second literature data, and
the third culture and literature data. Production can be expressed in
calorific equivalents or as g $N m^{-2} g^{-1}$.

11.2.1 Method 1

For the estimate based on relative specific growth rates (V'), values for the
different size classes were taken from the growth curve obtained for
A. caliginosa (Fig. 11.1). Expressed as $g g^{-1}$ (wet wt.) $month^{-1}$, these were
0.103 for adults, 0.262 for the medium-sized class and 0.665 for the small-
size class. $B T$ was multiplied by these values and the products seemed to
give an estimate of production over the two year period. This was
corrected to one year and reduced by 47% to allow for inactivity.

11.2.2 Method 2

The estimate based on oxygen consumption used the relationship
between population respiration and production given for relatively long-
lived poikilotherms by McNeill and Lawton (1970) as:

$$\log P = 0.8233 \log R - 0.2367 \qquad (11.2)$$

where P and R are expressed in calories.

Oxygen consumption of the different size classes was estimated from
Byzova's (1965) data on *A. caliginosa* and from Phillipson and Bolton's
(1976) data on three size classes of *A. rosea*. Taking the calorific equivalent
of oxygen as 1 ml O_2 = 4.775 cal (Heilbrunn, 1947), the following were
obtained:

Adults: 50.1 ml $O_2 g^{-1}$ (wet wt.) $month^{-1}$ = 239 cal
Medium sized: 87.1 ml $O_2 g^{-1}$ (wet wt.) $month^{-1}$ = 416 cal
Small sized: 66.8 ml $O_2 g^{-1}$ (wet wt.) $month^{-1}$ = 319 cal

These values were inserted in expression (11.1) and the estimates of
population respiration so obtained were inserted in (11.2).

11.2.3 Method 3

An estimate of the respiration/production ratio was obtained from literature data on oxygen consumption and the relative specific growth rate measured in cultures. The R/P values obtained were: adults 2.70, medium sized 1.92, small sized 0.57. These ratios were applied to the estimates obtained for respiration of the three size classes to give, after correction by 47%, another estimate of production.

11.2.4 Method 4

An estimate was obtained from equation (11.1) and the absolute growth rate V obtained for the three size classes from worms grown in culture. The number of individuals m^{-2} (N) is used instead of biomass.

11.3 NITROGEN EXCRETION

Nitrogen other than faecal N was estimated from Needham's (1957) data for total N excreted by *A. caliginosa*. His value of 87.5 μg N g^{-1} wet wt. of worms day^{-1} at 23°C when corrected to 12°C by Krogh's (1914) curve ($\times 0.36$) yields an estimate of 31.5 μg which was then expressed as N excreted g^{-1} biomass m^{-2} y^{-1} and reduced by 47% as before.

Faecal nitrogen production was estimated from measurements of faecal production and faecal N concentration. Faeces production was recorded from cultures of *A. caliginosa* kept between glass plates, 15 × 20 × 0.3 cm (Bolton and Phillipson, 1976) in darkness at 12°C. Observations were continued for 10 days, but the data for the first four days were discarded as unrepresentative. Faeces were collected from the cages by three people who were found to have different criteria for distinguishing faeces from soil. To avoid overestimating faeces production, only the data obtained by the operator who recorded the lowest weights of faeces were used. Faeces production was estimated also at Roskilde Experimental Station in a field experiment where the population consists almost exclusively of *A. tuberculata* (Eisen), a species very similar to *A. caliginosa* except for being slightly larger. Faeces were sampled weekly from the surface of four 0.5 m^2 quadrats from July to December. At each sampling, population size and vertical distribution of the worms were determined. One week in which the maximum weight of faeces was produced was selected as a basis for estimating faecal production. During this week, 26/8 to 2/9/80, 80% of the population was active and 71.5% of the active worms were in the upper 10 cm of the soil. It was assumed that 50% of the faeces produced by the active worms was deposited on the surface, corresponding to 33% of the total faecal production of population. The estimate from the field population was the same as the lowest estimate from the cultures between glass plates.

Concentrations of faecal N were estimated for NH_4 and $NO_3–N$ values obtained from *L. terrestris* faeces collected in a grass field adjacent to the Askov study site. Needham estimated the ratio between N output in *L. terrestris* and *A. caliginosa* excreta as 0.325. This was used to obtain an estimate of N production in *A. caliginosa* faeces which was related to mean biomass and again corrected by 47% to allow for inactivity. The N content of earthworm tissue was as 8.75% of the dry weight on the basis of the N content of protein.

11.4 RESULTS

The earthworm population densities at Askov varied widely from year to year in response to climatic conditions (Fig. 11.2). The data obtained in 1979 and the spring of 1980 were considered abnormal and estimation of population metabolism was therefore based on the samples taken in 1976–1978. The population density of all species combined was in general greater on the manured than the control plots and numbers in the FYM and SLU plots were not consistently different throughout the five years of observations. Larger populations were found on plots receiving $100\,t\,ha^{-1}$ than $50\,t\,ha^{-1}$ for both FYM and SLU treatments. Numbers and biomass

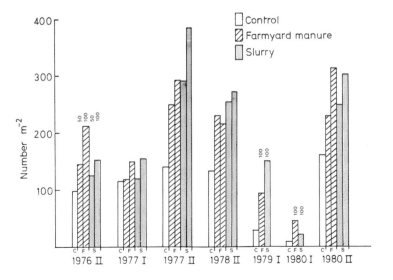

Fig. 11.2 Earthworm population densities at Askov. I, spring; II, autumn; Control ($80\,kg\,N\,ha^{-1}$ as NPK fertilizer). For the manure, values for 50t and $100\,t\,ha^{-1}\,y^{-1}$ are given except for 1979 and 1980 spring.

Table 11.1 Population metabolism estimates, Askov experimental plots.

Application (ha⁻¹)	Control (NPK) 120 kg of N	FYM				SLU			
		50 t	100 t	200 t	400 t	50 t	100 t	200 t	400 t
Numbers (m^{-2})	122.8	184.6	219.3	226.5	223.1	196.5	236.9	230.8	181.5
Biomass (g wet wt. m^{-2})	23.6	23.0	29.4	30.8	26.0	23.6	26.8	20.8	21.4
Production									
Method 1 (g wet wt. m^{-2}y^{-1})	36.6	37.1	45.2	48.0	39.3	35.6	39.1	33.9	27.9
Method 2 (g wet wt. m^{-2}y^{-1})	28.8	30.3	36.7	40.0	32.0	30.8	34.3	28.1	26.1
Method 3 (g wet wt. m^{-2}y^{-1})	30.5	37.2	45.4	47.4	39.6	35.7	40.1	34.0	28.1
Method 3 (kcal m^{-2}y^{-1})	25.5	31.1	37.9	39.6	33.1	29.8	33.5	28.4	23.5
Method 4 (g wet wt. m^{-2}y^{-1})	51.2	73.4	87.3	94.5	85.1	83.4	99.0	90.4	71.0
Respiration (litres O$_2$ m^{-2}y^{-1})	19.3	20.6	25.9	28.5	22.0	21.0	24.0	18.8	17.2
Respiration (kcal m^{-2}y^{-1})	92.4	98.2	123.0	136.3	61.1	100.3	114.4	89.6	82.1
Assimilation (kcal m^{-2}y^{-1})	117.9	129.3	161.8	175.9	94.2	130.1	147.9	118.0	105.6
Consumption (t ha^{-1}y^{-1} (dry wt.))	2.95	3.22	4.04	4.40	2.35	3.25	3.70	2.95	2.64
Consumption (kcal m^{-2}y^{-1})	1179	1293	1618	1759	942	1301	1479	1180	1056

values are given in Table 11.1. The FYM plots were dominated by
A. caliginosa and *A. longa* and the SLU plots by *A. caliginosa* alone.
Further details of species composition and treatment effects are given in
Andersen (1980).

The estimates of biomass production obtained by Methods 1–3 are
similar though the somewhat lower values obtained by Method 2 suggest
that McNeil and Lawton's estimate of R/P for long-lived poikilotherms
may not fit lumbricid populations particularly well. The fourth method
using numbers m^{-2} yielded estimates of productivity almost twice those
based on biomass. This undoubtedly arose from an overestimation of the
productivity of small and newly hatched individuals, as the variations in
individual size in this size class are too great.

The respiration rates in Table 11.1, derived from the biomass
estimates, and respiration rates, taken from the literature and expressed in
calories, were used in combination with the Method 3 production
estimates in calories to estimate assimilation. From this a rough estimate
of consumption was obtained from an assimilation/consumption efficiency of 10% derived from Crossley *et al.* (1971). Taking the energy
content of organic matter consumed as 4000 cal g^{-1}, the weights of organic
matter consumed shown in Table 11.1 were then calculated.

Faecal production by *A. caliginosa* was estimated as 1.52 g (dry wt.) g^{-1}
worm (wet wt.) day^{-1}. The *L. terrestris* data on which faecal nitrogen
production estimates are based are shown in Table 11.2. The ratio
between N output in *L. terrestris* and *A. caliginosa* excreta of 0.325
obtained by Needham for non-faecal material was applied to the
L. terrestris data to give an estimate of 50.1 μg N (NH$_4$ and NH$_3$–N) g^{-1}
faeces dry wt. day^{-1}.

Table 11.2 Nitrogen and carbon analyses of *L. terrestris* faeces.

	NH$_4$N (mg kg^{-1})	NO$_3$N (mg kg^{-1})	Total N (%)	Total C (%)	C/N
Fresh faeces	221.7	7.2	0.37	4.80	13.0
Aged faeces	155.4	7.2	0.39	4.00	10.4
Soil (0–5 cm)	18.3	4.9	0.34	4.00	11.6

The amount of N returned to soil annually in the tissues of dead worms
or dead worm-predators, is assumed, disregarding cocoon production, to
be approximately equivalent to the amount contained in the annual
production of worm tissues. By combining this with the amounts
contained in the annual production of faecal and non-faecal excreta,
estimates of the total N turnover by earthworm metabolism were obtained
(Table 11.3).

The estimates of the total amount of nitrogen metabolized annually by the earthworms are derived from a biomass estimate based on formalin-preserved samples. Phillipson *et al.* (1978) considered that biomass determinations made on preserved material may be 25% too low. However, from a more extensive set of data, Satchell (1969) calculated the weight loss attributable to preservation as 10%. This figure has been used in the present estimates (Tables 11.3 and 11.4).

Table 11.3 Annual nitrogen turnover by earthworms in the Askov experimental plots.

	Control	FYM				SLU			
Application (ha^{-1}):	120 kg N	50 t	100 t	200 t	400 t	50 t	100 t	200 t	400 t
Tissue N (kg ha^{-1})	4.54	5.51	6.73	7.13	5.92	5.35	6.00	5.02	4.21
Faecal N (kg ha^{-1})	3.04	2.97	3.00	3.98	3.36	3.05	3.46	2.68	2.77
Non-faecal excreted N (kg ha^{-1})	1.28	1.24	1.59	1.66	1.40	1.28	1.45	1.12	1.16
Total N turnover (kg ha^{-1})	9.47	10.69	13.33	14.05	11.75	10.65	12.00	9.70	9.00

In Table 11.4, the relations between barley production, application of fertilizer and manures, and N turnover by earthworms are given. Because of the form of the rotation experiment, it has only been possible to give these relationships for 50 and 100 tons of manure applied, for 200 tons applied in the 2nd year and for 400 tons applied in the 2nd and 4th years.

In the estimates of uptake of organic N contained in the manures, it is assumed that 70% of the inorganic N content is taken up by the crop in the first growing season after application. In the successive years, i.e. for 200 and 400 t, it is assumed that all nitrogen taken up by the crop originates from mineralization of organic bound N given in the manures.

Slurry contains 6.5% dry matter and farmyard manure 25%, however, FYM contains approximately 10% sand. After correction for this it was calculated that both SLU and FYM organic matter contains 2.8% N. Plant residues which are recirculated contain 0.52% N.

Crop production for 200 t SLU in the second year after application was 60% of that of 100 t annually but production in the second year after application of 400 t was 83% of that from 100 t SLU applied annually. Estimates for FYM are 69 and 90% respectively. This difference is attributed to the higher amount of organic matter given in FYM compared with the same weight of SLU. Estimates for the fourth year after application of 400 t SLU and FYM were 43 and 63% respectively.

Table 11.4 Nitrogen uptake in barley and turnover of organic matter in the Askov experimental plots.

	Control	SLU					FYM				
	NPK annually	50 tons annually	100 tons annually	200 tons 2nd year	400 tons 2nd year	400 tons 4th year	50 tons annually	100 tons annually	200 tons 2nd year	400 tons 2nd year	400 tons 4th year
Inorganic N input $(g\,N\,m^{-2}\,y^{-1})$	12.0	13.3	26.6	0	0	0	7.3	14.7	0	0	0
Total plant uptake $(g\,N\,m^{-2}\,y^{-1})$	18.9	19.8	23.0	13.9	19.2	9.8	18.6	21.1	14.5	18.9	13.2
NH_3–N taken up $(g\,N\,m^{-2}\,y^{-1})$	8.4	9.3	18.6	—	—	—	5.1	10.3	—	—	—
Uptake from org. N $(g\,N\,m^{-2}\,y^{-1})$	10.5	10.5	4.4	13.9	19.2	9.8	13.5	10.8	14.5	18.9	13.2
Worm turnover (% of uptake from org. N)	9.7	10.0	27.3	7.0	4.7	9.2	7.9	12.3	9.7	6.2	8.9
Residues recirculated $(g\,N\,m^{-2}\,y^{-1})$	5.1	5.5	6.4	3.9	5.4	2.7	5.2	5.9	4.1	5.3	3.7
Org. N from manures $(g\,N\,m^{-2}\,y^{-1},\,2y^{-1},\,4y^{-1})$	0	$9.4\,y^{-1}$	$18.7\,y^{-1}$	$35.4\,2y^{-1}$	$69.6\,4y^{-1}$	$169.6\,4y^{-1}$	$20.7\,y^{-1}$	$41.2\,y^{-1}$	$86.0\,2y^{-1}$	$178.0\,4y^{-1}$	$178.0\,4y^{-1}$
Org. matter from manures $(g\,(dry\,wt.)\,m^{-2}\,y^{-1},\,2y^{-1},\,4y^{-1})$	0	325	650	1300	2600	2600	750	1500	3000	6000	6000
Worm consumption $(g\,org.\,matter\,(dry\,wt.)\,m^{-2}\,y^{-1})$	295	325	370	295	264	264	322	404	440	235	235
Worm consumption (% of org. matter from manures $y^{-1})$	30.0	100	57.0	45.4	40.1	40.1	43.0	26.9	29.3	15.6	15.6

11.5 DISCUSSION

It can be seen from Table 11.4 that, by fertilizer treatment, $10.5\,g\,N\,m^{-2}$ is needed from mineralization of organic bound N to meet the demands of the crop. This is more than the $5.1\,g\,N\,m^{-2}$ which is recirculated in crop residues. It thus appears that there is a net loss of humus N by this treatment, even if the amount of fertilizer N which is not taken up by the crop, $3.6\,g\,N\,m^{-2}$, is subtracted. Some of this is lost by denitrification and leaching.

By application of both 50 and 100 t manure annually, there will be a net gain of organic N content of the soil. The organic N input from FYM is approximately twice the input from SLU. The effect on crop growth is clearly seen (Table 11.4) where manures were applied each 2nd and 4th year. The 2nd year after application of 200 and 400 t, crop production by 400 t SLU is 28 % greater than that of 200 t SLU but in the case of FYM only 23 % greater. The fourth year after application of 400 t manure, crop production by SLU treatment was only 51 % of that of the 2nd year after application and, for FYM, 70 %. This demonstrates the long-term effect of the higher organic matter content of FYM.

The contribution by earthworms to the turnover of organic N needed to satisfy the crop demand is estimated as % of the difference between total plant uptake and NH_3–N uptake. Because of the very high NH_3 content of SLU this proportion is very high by annual SLU application; 10.0 % and 27.3 % for 50 and 100 t SLU respectively. For 50 and 100 t FYM the amounts are only 7.9 % and 12.3 % respectively.

The N turnover by earthworms is the same from both 50 and 100 t SLU and FYM, $1.07\,g\,N\,m^{-1}\,y^{-1}$. By application of 100 t FYM annually turnover is $1.33\,g\,N\,m^{-2}\,y^{-1}$, 10.8 % greater than by application of 100 t SLU. The greatest turnover by the worms, $1.40\,g\,N\,m^{-2}\,y^{-1}$, was on the plots receiving 200 t FYM each 2nd year. In the 2nd year after application this amounted to 9.7 % of the plant uptake.

A secondary effect of FYM application is an increase in the proportion of the deep burrowing species *A. longa* and *L. terrestris*.

The earthworms probably prefer the manures which contain 2.8 % N to crop debris which contains only 0.52 % (Table 11.4).

The micro-organism content of the earthworm gut is consistently reported as exceeding that of the surrounding soil, e.g. Parle (1963). An estimate of microbial N turnover in earthworm faeces after deposition may be obtained from the reduction in carbon content by 20 % (Table 11.2) in earthworm faeces up to 14 days old. This suggests that microbial activity will have been responsible for a turnover of 15 kg or more $N\,ha^{-1}\,y^{-1}$ in *A. caliginosa* faeces. The annual N turnover attributable to earthworm activity, directly by earthworm metabolism and indirectly by stimulation of microbial activity may therefore have been for example about $28\,kg\,N\,ha^{-1}\,y^{-1}$ by application of $100\,t\,FYM\,ha^{-1}\,y^{-1}$.

The C:N ratio in the casts fell from 13.0 to 10.4 (Table 11.2), a ratio at which mineralized N can be taken up by the plants. Measurements of liberation of N_2O and N_2 have shown that anaerobic denitrification in animal manured plots may be very high, and in the long term, one of the most important effects of earthworms on N turnover may be to reduce denitrification by improving soil aeration and drainage and by stimulating microbial decomposition.

11.6 REFERENCES

Andersen, C. (1980) The influence of farmyard manure and slurry on the earthworm population (Lumbricidae) in arable soil. In *Soil Biology as Related to Land use Practices* (ed. D. L. Dindal), *Proc. VIIth Int. Soil Zool. Coll., EPA Washington DC*, pp. 325–334.

Bolton, P. J. and Phillipson, J. (1976) Burrowing, feeding, egestion and energy budgets of *Allolobophora rosea* (Savigny) (Lumbricidae). *Oecologia (Berl.)*, **23**, 225–245.

Byzova, J. B. (1965) Comparative rate of respiration in some earthworms (Lumbricidae, Oligochaeta). *Rev. Ecol. Biol. Sol.*, **2**, 207–216.

Crossley, D. A., Reichle, D. E. and Edwards, C. A. (1971) Intake and turnover of radioactive cesium by earthworms (Lumbricidae). *Pedobiologia*, **11**, 71–76.

Heilbrunn, L. V. (1947) *An Outline on General Physiology*. 2nd edn., Saunders, Philadelphia and London. 184 pp.

Krogh, A. (1914) The quantitative relation between temperature and standard metabolism in animals. *Int. Z. Phys.-Chem. Biol.*, **1**, 491–508.

Lakhani, K. H. and Satchell, J. E. (1970) Production by *Lumbricus terrestris*. *J. Anim. Ecol.*, **39**, 473–492.

Lofs-Holmin, A. (1980) Measuring growth of earthworms as a method of testing sublethal toxicity of pesticides. Experiments with benomyl and trichloroacetic acid (TCA). *Swed. J. Agric. Res.*, **10**, 25–33.

McNeill, S. and Lawton, J. H. (1970) Annual production and respiration in animal populations. *Nature (London)*, **225**, 472–474.

Needham, A. E. (1957) Components of nitrogenous excreta in the earthworms *Lumbricus terrestris* L. and *Eisenia foetida* (Savigny). *J. Exp. Biol.*, **34**, 425–445.

Parle, J. N. (1963) Micro-organisms in the intestines of earthworms. *J. Gen. Microbiol*, **31**, 1–11.

Petrusewicz, K. and MacFadyen, A. (1970) Productivity of terrestrial animals. Principles and methods. *IBP Handbook No. 13*. Blackwell, Oxford. 190 pp.

Phillipson, J. and Bolton, P. J. (1976) The respiratory metabolism of selected lumbricids. *Oecologia (Berl.)*, **27**, 141–155.

Phillipson, J., Abel, R., Steel, J. and Woodell, S. R. J. (1978) Earthworm numbers, biomass and respiratory metabolism in a beech woodland – Wytham Woods, Oxford. *Oecologia (Berl.)*, **33**, 291–309.

Satchell, J. E. (1967) Lumbricidae. In *Soil Biology* (eds. A. Burges and F. Raw), Academic Press, London, pp. 259–322.

Satchell, J. E. (1969) I. Studies on methodical and taxonomical questions. Methods of sampling earthworm populations. *Pedobiologia*, **9**, 20–25.

Syers, J. K., Sharpley, A. N. and Keeney, D. R. (1979) Cycling of nitrogen by surface casting earthworms in pasture ecosystems. *Soil Biol. Biochem.*, **11**, 181–185.

Chapter 12

Earthworm population dynamics in different agricultural rotations

A. LOFS-HOLMIN

12.1 INTRODUCTION

Few investigations of earthworms have been made in agricultural rotations and most have been occasional samplings. The disadvantage of occasional sampling in a rotation is that insufficient consideration is given to the previous history of the field. Population sizes may be correlated with the present crop when they should have been correlated with a previous crop or treatment. If earthworms are sampled in a rotation during different years it is difficult to separate effects of differences in weather from those of agricultural treatments.

The present study was made in four long-term husbandry trials in Sweden. These are designed so that all crops in each 6–8 year rotation can be sampled simultaneously, eliminating the effects of weather variations between years.

12.2 FIELD SITES AND ROTATIONS

Three 8-year rotations with different treatments of crop residues were investigated at Lönhult, Höganäs in south Sweden, and at Ultuna, Uppsala. Tables 12.1 and 12.2 list the crops, treatments and soil types.

The three rotations are intended to represent different types of farming. One is typical of cattle farming where the straw is removed and used for the animals. The straw is to some extent returned to the field with the farmyard manure (FYM) which is spread once or twice per rotation. The second rotation is typical of arable farming where crop residues are burnt on the field. In the third rotation the residues are ploughed in.

In the north of Sweden at Ås, Östersund, and at Röbäcksdalen, Umeå,

Table 12.1 Rotations and treatment of residues.

Lönhult

Rotation A	Winter wheat	Sugarbeet[a]	Barley	Ley I[b]	Ley II	Winter rape	Winter wheat	Peas
Residues	R	R	R	R	R	P	R	R
Rotation B	Winter wheat	Sugarbeet	Barley	Winter rape	Ley II	Winter wheat	Spring wheat	Barley
Residues	B	B	B	B	B	B	B	B
Rotation C	Winter wheat	Sugarbeet	Barley	Ley	Winter rape	Winter wheat	Spring wheat	Peas
Residues	P	P	L	P	P	P	P	P

Ultuna

Rotation D	Fallow	Winter turnip rape[c]	Winter wheat	Peas	Barley[d]	Ley I[e]	Ley II	Oats
Residues	—	P	R	R	R	R	R	R
Rotation E	Fallow	Winter turnip rape	Winter wheat	Barley	Spring wheat	Peas	Winter wheat	Oats
Residues	—	P	B	B	B	P	B	B
Rotation F	Fallow	Winter turnip rape	Winter wheat	Barley	Ley	Winter wheat	Peas	Oats
Residues	—	P	P	L	P	P	P	P

Ås and Röbäcksdalen

Rotation G	Barley[c]	Ley I[e]	Ley II	Ley III	Ley IV[c]	Ley V
Residues	R	—	—	—	—	P
Rotation H	Barley	Ley I[e]	Ley II	Ley III[c]	Oats + peas	Forage of rape[c]
Residues	R	—	—	P	R	R

[a] FYM, 25 t ha^{-1}.
[b] Meadow fescue for seed production.
[c] FYM, 30 t ha^{-1}.
[d] FYM, 20 t ha^{-1}.
[e] Red clover and timothy.
B, burnt; L, left in field; P, ploughed in; R, removed.

Table 12.2 Field and sampling data.

Trial	Soil type	Variables	Rotation length (y)	Started	Replicates	Plot size (m)	Pots for earthworm sampling		
							Put down	Taken up	No. per treatment
Ultuna	Heavy clay	Crops, FYM, treatment of residues	8	1959	2	10 × 20	22–23/5	19/7	6
Lönhult	Moraine on clay	Crops, FYM, treatment of residues	8	1957	2	23 × 9.25	7–8/6	27/7	6
Ås	Loamy moraine	Leys, FYM	6	1955	2	21 × 9	28/6	17/8	10
Röbäcksdalen	Loamy fine sand	Leys, FYM	6	1958	2	22.5 × 7.5	27/6	16/8	10

3–5 year leys were included in two different 6-year rotations. FYM was supplied twice in each rotation.

12.3 METHODS

Earthworms were sampled with 1-litre plastic pot-traps filled with an FYM–clay mixture (Lofs-Holmin, 1979). This was used to minimize disturbance of the small (Table 12.2) plots. It also facilitated uniform sampling of large trials within a very short period of time at low cost. Its drawback is that it fails to sample adult *Lumbricus terrestris* L. effectively; only juveniles of this species are recorded in Fig. 12.2. The data were subjected to analysis of variance.

12.4 RESULTS AND DISCUSSION

As the trapping method samples only a proportion of the population, the results are presented as numbers of worms trapped per plot (Figs 12.1–12.5). The main population differences (Table 12.3) are clearly related to the agricultural treatments.

Table 12.3 Significant differences between population densities under successive crops.

Rotation	Crops	Species	Population change	Significance (P) of difference
A	Barley after sugarbeet	*A. caliginosa*	Increase	a < 0.05, j < 0.05
A	Ley 1 after barley	*A. caliginosa*	Decrease	a < 0.01, j < 0.05
A	Peas after wheat	*A. caliginosa*	Increase	j < 0.02
A	Peas after wheat	*A. caliginosa*	Decrease	j < 0.02
B	Rape after sugarbeet	*A. caliginosa*	Decrease	a $+$ j < 0.001
C	Ley after barley	*A. caliginosa*	Decrease	a $+$ j < 0.001
C	Peas after wheat	*A. caliginosa*	Increase	a $+$ j < 0.02
D	Wheat after turnip rape	*L. terrestris*	Increase	j < 0.05
F	Wheat after turnip rape	*L. terrestris*	Increase	j < 0.05
G	Ley 1 after barley	*A. rosea*	Decrease	j < 0.05
G	Barley after ley V	*D. subrubicunda*	Increase	a $+$ j < 0.05
H	Barley after rape	*A. rosea*	Decrease	j < 0.002
H	Barley after rape	*D. subrubicunda*	Increase	a $+$ j < 0.05

a, adults; j, juveniles.

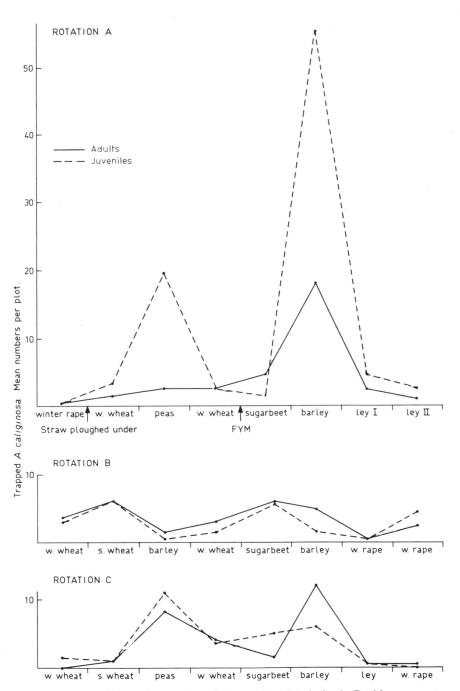

Fig. 12.1 *A. caliginosa* in rotations A, B and C at Lönhult. A, Residues except rape straw removed. FYM ploughed in after winter wheat. B, Residues burnt. C, Residues ploughed in.

12.4.1 Effects of crop residues

In rotation A the larger peak (Fig. 12.1) is caused by the FYM and the smaller by ploughed-in rape straw. Fluctuations in the earthworm population (Table 12.3), at Lönhult, dominated by *Allolobophora caliginosa* Sav., in the complete rotation can be explained as follows: FYM is supplied between winter wheat and sugarbeet. The adults present produce numerous cocoons which hatch mainly in the autumn or in the following spring. These new individuals remain undetected if the plots are sampled in summer, but in the following year under barley a large increase, especially of juveniles, will be noticed. The sugarbeet, however, returns a minimum of plant residues to the soil, as both tops and roots are harvested. The barley is undersown with fescue ley so only the roots are returned. The large population initiated by the FYM cannot be maintained and in the first year of the ley the population again falls. Numbers are even lower in the second year of ley when only dead grass roots are returned to the soil. In the winter rape that follows, the population reaches a minimum, but receives a new supply of food when the ley is ploughed under. The population starts to build up in the winter wheat and increases further in the peas after the rape straw is ploughed in. A decline then follows until FYM is supplied again.

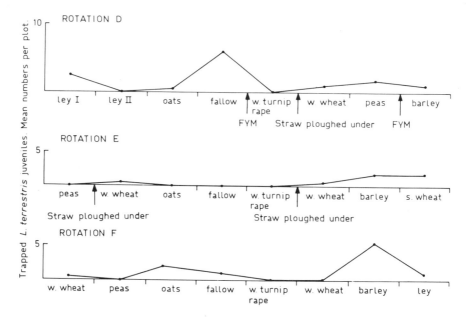

Fig. 12.2 Juvenile *L. terrestris* in rotations D, E and F at Ultuna. D, Residues except turnip rape straw removed. FYM ploughed in after fallow and after peas. E, Residues except turnip rape and peas straw burnt. F, Residues ploughed in.

In rotation B, the residues are burnt in the field but there is a small and uniform supply of roots. This reduces the amplitude of the population fluctuations and, compared with rotation A, the production of young is very small in relation to the number of adults. As in A, the population is lowest $1\frac{1}{2}$ years after sugarbeet.

In rotation C, fluctuations are larger than in B and may be explained by differences in amount and quality of the residues ploughed under. Again the population is very small $1\frac{1}{2}$ years after the sugarbeet and as the barley straw is not ploughed in, the population reaches a minimum in the winter rape following the ley.

At Ultuna only *L. terrestris* juveniles were numerous. Rotation E, with burning of residues, had the lowest numbers ($P < 0.05$) (Fig. 12.2). In the year after fallow all three rotations had their lowest juvenile populations but ploughing in turnip rape straw seemed to initiate a population increase.

12.4.2 Effects of leys

Figure 12.3 shows rotation G with five and rotation H with three years of ley at Röbäcksdalen. *Lumbricus rubellus* Hoffm. is clearly favoured by the

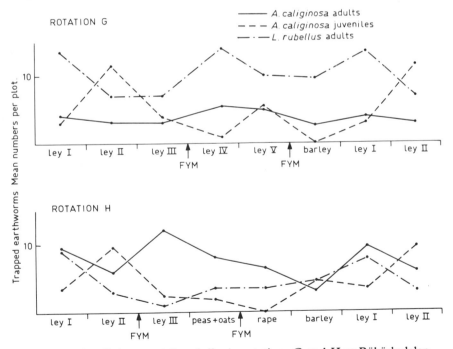

Fig. 12.3 *A. caliginosa* and *L. rubellus* in rotations G and H at Röbäcksdalen. G, Five years of ley. H, Three years of ley.

longer lasting ley, $P < 0.01$, while *A. caliginosa* dominates when ploughing is more frequent ($P < 0.01$). FYM is spread on the ley once and
ploughed in once per rotation. Production of *A. caliginosa* juveniles is
most pronounced when FYM is worked into the soil while *L. rubellus*
responds faster to FYM spread on the ley.

Figures 12.4 and 12.5 show rotations G and H at Ås. *A. caliginosa* and
A. longa Ude have more fluctuating populations in rotation H. *Octalasion
lacteum* Örley is favoured by the longer lasting ley ($P < 0.002$). Adult *A.
longa* and *O. lacteum* were most abundant in rotations G and H, juveniles
comprising averages of 15–40 % of the total. *A. rosea* Sav. seems to build
up a large population in the ley, especially if FYM is spread on the surface.

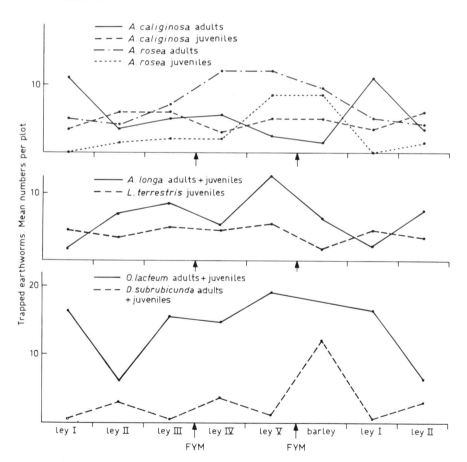

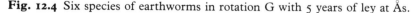

Fig. 12.4 Six species of earthworms in rotation G with 5 years of ley at Ås.

Ploughing itself does not appear to reduce numbers, but ploughing in combination with FYM is followed by a decrease in both rotations. *Dendrobaena subrubicunda* Eisen, which is brought into the field with the FYM, seems to maintain a small population in the soil and increases when FYM is ploughed in. As this species grows quickly and matures fast, adults and juveniles follow the same fluctuation pattern.

The return of organic matter to the soil is one of the most important factors that affect earthworm populations. This is most apparent in the curves representing the juveniles. Those for adult worms reflect not only the rate of recruitment from the juvenile population but, *inter alia*, winter survival and the longevity of each species. Farmyard manure has the

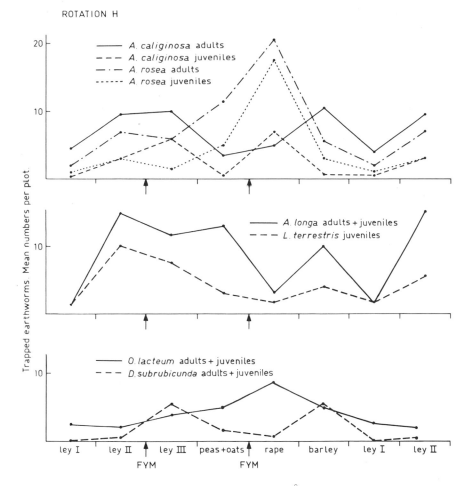

ROTATION H

A. caliginosa adults
A. caliginosa juveniles
A. rosea adults
A. rosea juveniles

A. longa adults + juveniles
L. terrestris juveniles

O. lacteum adults + juveniles
D. subrubicunda adults + juveniles

Trapped earthworms. Mean numbers per plot.

ley I   ley II   ley III   peas+oats   rape   barley   ley I   ley II
         FYM   FYM

Fig. 12.5 Six earthworm species in rotation H at Ås.

largest influence but favours different species depending on whether it is spread on the surface or ploughed in. Litter-dwelling species like *L. rubellus* are favoured by spreading and *A. caliginosa* by ploughing in. Straw residues ploughed into the soil also stimulate earthworm populations and rape and turnip rape residues seem to be especially effective.

Practices which do not bring any organic matter into the soil have the opposite effect: the most adverse being root crops, fallow, undersown ley and burning of residues.

The negative effect of fallow on earthworms has been reported by Barley (1958) and Atlavinytė (1964). Deficiency of organic matter, frequent soil cultivation and the lack of a crop protecting the surface from summer drought probably all contribute. Removal of straw and especially burning, caused a decline in the numbers of most species as reported by Edwards and Lofty (1979).

Several years of ley stabilize the population numbers of most species, while more frequent ploughing causes fluctuations. Species like *L. rubellus*, being dependent on surface litter for food, are favoured by longer lasting ley. *A. caliginosa* needs organic matter mixed into the mineral soil and is often the dominating species in arable land ploughed annually. Edwards and Lofty (1975) also found that cultivation favoured this species.

This investigation demonstrates that numbers and composition of earthworm populations in arable land are highly dependent on agricultural practices and that it is necessary to know the history of a field to be able to explain the density and composition of its earthworm population.

12.5 REFERENCES

Atlavinytė, O. (1964) Distribution of earthworms (Lumbricidae) and larvae of insects in the eroded soil under cultivated crops. *Pedobiologia*, **4**, 245–250.

Barley, K. P. (1958) The influence of earthworms on soil fertility. I. Earthworm populations found in agricultural land near Adelaide. *Austr. J. Agric. Res.*, **10**, 171–178.

Edwards, C. A. and Lofty, J. R. (1975) The influence of cultivations on soil animal populations. In *Progress in Soil Zoology* (ed. J. Vanek), Academia, Prague, pp. 399–407.

Edwards, C. A. and Lofty, J. R. (1979) The effect of straw residues and their disposal on the soil fauna. In *Straw Decay and its Effect on Utilization and Disposal* (ed. E. Grossbard), J. Wiley & Sons, Chichester, pp. 37–44.

Lofs-Holmin, A. (1979) A pot method for field experiments with earthworms. Swedish University of Agricultural Sciences. *Department of Ecology and Environmental Research, Report 6*, Uppsala. 41 pp.

Chapter 13

Earthworm ecology in forest soils

J. E. SATCHELL

13.1 SPECIES COMPOSITION OF FOREST EARTHWORM FAUNAS

As early as 1908, Arldt concluded from considerations of palaeographic changes in land and sea distribution of present day earthworm species that terricole oligochaetes first appeared in the Upper Triassic. Michaelsen (1921), interpreting earthworm distribution in terms of Wegner's theory of continental drift, put their origin in the Carboniferous period and Wilcke (1955), from considerations of earthworm ecology, concluded that mull-forming, soil-dwelling worms could not have evolved before deciduous forests existed in the late Cretaceous.

Despite this surge of lumbricine evolution in the forest environment, relatively few modern species are restricted to forest habitats. Tree- and log-inhabiting species occur in superhumid rain-forests and some large litter-feeding forms, discussed in the following chapter, are characteristic of forested areas of central and eastern Europe. Nevertheless, the distribution of the peregrine species of the glaciated parts of the northern hemisphere reflects the availability of food resources and the properties of the local soil environment rather than the prevailing plant association, actual or potential. Thus *Eisenia foetida*, thought to have originated as a corticolous species of mountain forests south of the Caspian Sea (Omrani, 1973) but now a typically anthropophyllic dunghill and garden compost species, occurs in forest habitats, e.g. under *Alnus glutinosa* in accumulations of wet decomposing leaf litter. *Eiseniella tetraedra*, a limnic or amphibious species, occurs in boggy places both in forests and moorlands and the corticolous *Dendrobaena rubida* complex inhabits similar habitats in forest and heathland ecosystems. Some earthworms are generally absent from woodland when present in adjacent soils of similar character, e.g. the unpigmented form of *Allolobophora chlorotica* is common in woodland, gardens and arable land. The green form, though perhaps abundant in adjacent pastures, is rarely found in woodland (Satchell,

1967), although the nuances of the ecology of the two forms are as yet too fine for this to be convincingly explained.

In 1884, the Danish forester P. E. Müller drew a basic edaphic distinction between mull soils with an abundance of earthworms and mor soils, as he thought, without earthworms. From studies of a range of soil and vegetation types in Danish forests and heaths, Bornebusch (1930) developed this concept and defined two characteristic species associations, in mull *Lumbricus terrestris, Allolobophora longa, A. turgida, A. trapezoides, A. chlorotica, Eisenia rosea* and *Octolasium cyaneum* and, in beech raw humus, *Dendrobaena octaedra* and, more rarely, *Lumbricus rubellus, L. castaneus* and *Dendrobaena arborea*. With some revisions and nomenclatural changes this division has stood the test of time. Satchell (1967) reported eight characteristic species including *L. terrestris, A. caliginosa* and *A. longa* from woodlands in the English Lake District with mull soils. In moder sites there were usually only four species of small size of which the largest was *L. rubellus*. In the most extreme acid mor sites only *D. octaedra* and *Bimastos eiseni* were found. Nordström and Rundgren (1974) divided the earthworm faunas of Swedish forests into two major associations and two minor intermediate types. The main types were (a) an *A. rosea–A. caliginosa–L. terrestris* association found in deciduous woods dominated by ash (*Fraxinus*) or elm (*Ulmus*) with a well-developed field layer and brown forest soil with almost neutral pH and 'adequate' soil moisture, and (b) a *D. octaedra* association found in conifer woods with a needle litter layer on podsols or podsolic brown earths with low pH and periodic soil moisture deficits. Analysis of the earthworm faunas in prealpine and alpine woods by Martinucci and Sala (1979) showed good agreement with the associations described by Nordström and Rundgren but these authors place more emphasis on the close correlation between available water and earthworm abundance. More recently, Satchell (1980) interpreted the contrasting behaviour, morphology and physiology of these two groups, predominantly soil dwelling non-pigmented forms and surface-living red-pigmented forms, as representing evolutionary poles arising from r and K selection with subsidiary adaptive radiation.

13.2 POPULATION DENSITY AND BIOMASS

The population density of earthworms in forests, like its species composition, falls within the same range as that of other ecosystems presenting similar soil conditions and food resources. This is illustrated by Table 13.1 which presents estimates of population density and biomass of earthworms in a global range of forest types (Zajonc, 1971). Zisci (Chapter 14) gives additional data for Hungarian forests and Lee

Table 13.1 Population density and biomass of earthworms in various forest types (based on Zajonc, 1971).

	Population density (No. m^{-2})			Biomass (g m^{-2})		
	Range	$\overline{X}$	(n)	Range	$\overline{X}$	(n)
Taiga	< 1–111	22.8	10	< 1–45	7.4	8
Mountain spruce and pine forests	4–28	12.33	3	2–17	8.4	3
Pine forest	2–12	48.3	5	< 1–9	21.8	5
Mixed mountain forests	1–250	53.2	20	< 1–300	84.0	20
Mixed lowland forests	5–439	140.0	13	< 1–221	68.3	13
Beech forests	0–356	62.0	16	0–142	30.1	16
Oak forests	0–192	92.4	20	0–153	59.6	20
Steppe forests	3–184	55.7	6	< 1–112	30.0	6
Deciduous sub-mediterranean forests	4–37	16.3	3	1–15	5.0	5
Xerophilous mediterranean forests	—	—	—	1–2	1.3	2
Subtropical forests	60–172	112.5	2	7–60	33.6	2
Tropical forests	0–39	12.9	6	0–17	3.7	6

(Chapter 15) for tropical sites. Highest values are found in mixed forests where density and biomass estimates are comparable to those for grassland sites other than certain pastures, e.g. those cited by Stout (Chapter 4) and Syers and Springett (Chapter 7) which were pure white clover swards or contained a high proportion of clover. The dependence of species diversity, population density and biomass on soil type rather than litter of the dominant tree is illustrated (Table 13.2) by data compiled by Phillipson *et al.* (1978) for beechwoods of which the earthworm faunas fall into distinct mull and mor facies.

13.3 LITTER SELECTION AND BREAKDOWN

13.3.1 Litter selection

The main mechanism of defence against herbivores by grassland plants is storage of the larger part of their nutrient reserve in underground organs and rapid replacement of grazed photosynthetic tissue by tillering. Trees follow in general an alternative strategy of raising most of their photo-synthetic tissue above the level of grazing mammals on woody stems and concentrating distasteful tannins and related polyphenolics in the young foliage when its vulnerability to attack by phytophagous insects is greatest. This legacy of herbivore chemical repellants is received by surface-feeding earthworms when the leaves eventually fall as litter. The

Table 13.2 Comparison of earthworm densities, biomasses and species numbers in beechwoods on mor/'acid' and mull/'basic' soils (from Phillipson et al., 1978).

Authority	Mor soils or pH < 5.5			Authority	Mull soils or pH > 5.5		
	No. m^{-2}	g m^{-2} (fresh wt.)	No. spp.		No. m^{-2}	g m^{-2} (fresh wt.)	No. spp.
Baltzer (1956)	6	1.40	2	Baltzer (1956)	54	14.59	6
	24	9.22	3		51	14.78	5
	4	4.39	1		52	14.23	4
	22	13.62	2		41	11.65	5
	4	1.83	1		33	18.78	4
	25	10.78	1		54	42.57	4
	38	4.42	3		38	22.14	3
	37	13.77	3		28	11.16	3
	22	10.61	2		99	37.40	4
	12	4.83	2		86	30.73	7
	18	6.54	3	Bornebusch (1930)	177	53.10	3
	30	4.38	2		93	27.90	—
Bornebusch (1930)	29	1.45	1		73	5.90	4
	81	5.40	4		220	—	—
	23	1.15	1		155	—	1
					110	—	1
					90	—	2

Reference			
van der Drift (1951)	30	—	2
Füller (1953)	36	16.00	4
Nordström and Rundgren (1973)	67.4	12.39	6
	89.8	25.60	8
Rabeler (1960)	7.0	} 7.20*	2
	10.5		6
	3.5		2
	14.0		5
	10.5		4
	19.5		5
Sergienko (1969)	8.0	6.10	5
Volz (1962)	—	3.05	—
	—	2.86	—
Zajonc (1967)	18.5	12.00	4
Mean	25.54	7.78	3.11
Standard deviation	22.19	5.93	1.83

Reference			
Lindquist (1938)	90	16.89	5
	116	75.58	5
	94	28.71	7
	150	—	6
	127	—	—
	141	—	5
	184	48.04	5
Phillipson et al. (1978)	164.6	41.02	10
	117.5	38.63	10
Mean	101.47	29.15	4.91
Standard deviation	51.86	17.78	2.20

* Estimated by Zajonc (1971).

concentrations of these substances and of other constituents vary widely between leaves of different ages and different species such that litter-feeding worms in mixed woodlands are presented with a choice of food materials between which they exhibit marked preferences. These preferences, already well known in the nineteenth century to Charles and Francis Darwin and their contemporaries, were studied by Bornebusch (1953) and Satchell and Lowe (1967). The latter authors demonstrated relationships between the palatability to *L. terrestris* of leaf discs and their N, soluble carbohydrate and phenolic contents but concluded that microbial action on the leaves was also important. Subsequent attempts to separate microbial phagostimulants from fractionated litter extracts were unsuccessful and in view of the difficulties of defining, by chemical means, unknown organic compounds of which only a few molecules may suffice to elicit a feeding response, this approach was abandoned. In 1972, Wright showed a positive attraction to earthworms of *Pseudomonas aeruginosa* and Cooke and Luxton (1980) demonstrated the palatability to *L. terrestris* of filter paper discs contaminated with *Mucor hiemalis* and a species of *Penicillium*. An extension of this work is described in Chapter 32.

13.3.2 Litter breakdown

On the basis of earthworm feeding experiments by Franz and Leitenberger (1948) on *Corylus* leaf litter, by van Rhee (1963) on alder and orchard leaf litter, and by Needham (1957) on elm leaves, Satchell (1967) calculated that litter-feeding earthworms, notably *L. terrestris* and *L. rubellus*, consumed in excess of $27 \, mg \, g^{-1}$ fresh wt. day^{-1}. The weight of leaves falling annually in forests outside the tropics is generally about $2-5 \, t \, ha^{-1}$ and in temperate deciduous forests about $3 \, t \, ha^{-1}$ (Galoux, 1953). A population of $10 \, g \, m^{-2}$ biomass, consuming $30 \, mg \, g^{-1} \, day^{-1}$ for 200 days in the year would consume annually 600 kg or about 20 % of the annual leaf fall. The populations of taiga, montane spruce and pine forests, mediterranean and submediterranean forests (Table 13.1) and of beech forests on mor soils (Table 13.2) have average biomass values less than this and thus lack the potential to prevent the accumulation of a permanent litter layer. In oak, beech and mixed forests in favourable environments, with earthworm biomass values up to $300 \, g \, m^{-2}$ (Table 13.1), the annual litter fall could be entirely consumed within a few months at this rate of ingestion.

Under favourable conditions, leaves may be removed from the ground surface, if not actually ingested at rates similar to this. Raw (1962) placed apple leaves under netting in amounts equivalent to $2 \, t \, ha^{-1}$ in an orchard containing approximately $168 \, g \, m^{-2}$ of *L. terrestris*. After two months all but 14 out of 1000 leaves had been buried by earthworms, only 0.5 % by weight remaining on the surface. In a study of leaf litter disappearance in a deciduous woodland with abundant *L. terrestris*, Bocock *et al.* (1960) found that, after 6 months, only a few midribs remained from *Fraxinus* litter. *Quercus* litter disappeared more slowly, about half the original weight remaining after one year. A similar experiment by Edwards and Heath (1963) is discussed in detail in Satchell (1974). These observations are consistent with the pattern of disappearance of leaf litter from deciduous woodlands on mull soils. In temperate climates, the more palatable leaves disappear during the winter and spring following leaf fall and recognizable fragments of the least palatable types of leaves remain, though patchily distributed, until the second year when they too disappear in the spring peak of earthworm activity.

In a review of the literature on litter selection experiments and of their own observations on *L. terrestris*, Perel and Sokolov (1964) note the relative unpalatability of *Quercus* litter compared with that of *Corylus* or *Tilia* but conclude that 'as in an oak forest mixed with lime, maple and hazel, where oak-leaf litter constitutes the bulk of the forest floor, *L. terrestris* worms do not sustain any dietary deficiency and play an active role in decomposing forest litter'. Conifer needles are generally even less palatable to earthworms than oak litter but similarly are disintegrated by earthworm activity. In commercial *Azalea* production, the plants are cultured in pine litter which provides optimal root growth conditions of aeration and available water when the needles are fresh. Feeding activity by *Dendrobaena* spp. and *L. rubellus* can reduce this material in a few months to a fragmented condition in which the particles settle and the air and water characteristics are unsuitable for its continued use (Huengens, 1969). A comparison of bast fibres of basswood with faeces of *Dendrobaena rubida* cultured in this material by Striganova (1968) showed that the excrement had a substantially higher content of fulvic and humic acids than the uneaten fibres and that the worms had not merely fragmented the material but had contributed directly to its humification. Neuhauser and Hartenstein (1978) demonstrated that five species of earthworms were unable to degrade lignin but were able to convert the ring carbon atoms of phenolics, such as vanillin, into respiratory CO_2. In view of the mechanism of the reaction, they suggest that earthworm peroxidases increase humification by polymerization of aromatic compounds.

13.4 CONTRIBUTION TO SOIL METABOLISM

Satchell (1967) calculated the oxygen consumption of a population of *L. terrestris* with a biomass of 140 g m^{-2} in *Fraxinus–Quercus* woodland to be about 22.9 litres m^{-2}, equivalent to about 460 kJ m^{-2} y^{-1}. Assuming an annual litter fall of 3–5 t ha^{-1} for this type of forest, this would have been equivalent to about 5–8 % of the total soil respiration. A similar estimate of 4–5 % was obtained by Phillipson *et al.* (1978) from a *Fagus* woodland on a mull soil with a somewhat smaller earthworm biomass.

A carbon flow model for a *Liriodendron* forest in Tennessee developed by Reichle *et al.* (1973), postulates an above-ground litter input of 464 g m^{-2} y^{-1} dry wt. and a root litter input of 100 g m^{-2} y^{-1}. The carbon respired by a population of *Octolasion lacteum* of 14 g m^{-2} dry wt. (approx. 77 g fresh wt.) biomass is estimated as 34.5 g m^{-2} y^{-1}, about 6 % of the input. All these estimates are consistent with Barley's (1964) estimate of 4 % for an Australian pasture and lead to the conclusion that the direct contribution of earthworms to soil metabolism is too small to have a significant effect on the C:N ratio of the organic horizons. However, ingestion of organic matter by earthworms in the *Liriodendron* forest model amounted to 208 g m^{-2} y^{-1} which was equivalent to a turnover of 75 % of the mean annual standing crop of the O$_2$ horizon. The importance of litter feeding by earthworms is thus the processing of the organic matter and the enhancement of microbial activity when it is brought into the more humid and nutrient-rich environment of the surface soil.

Krivolutzky and Pokarzhevsky (1977) have investigated the role of earthworms in nutrient turnover in oak forest on chernozem in the European part of the USSR. A population comprising *L. terrestris*, *D. octaeda* and *Eisenia nordenskioldii* was analysed for titrometric Ca, Mg and photometric K content and on the basis of its biomass and estimated ingestion rate, flow rates for these elements for the active period May–September were calculated. The amounts passing through the earthworm population were for Ca, Mg and K respectively 157, 122 and 57 % of the amount in the leaf fall and 2.1, 1.2 and 3.6 % of the amount in the 0–5 cm soil layer. These turnover rates are impressive but the physical prevention of the accumulation of a permanent litter in forest soils with high earthworm activity probably has a more fundamental influence on the physicochemistry of the soil and hence the conditions in the arena in which germinating seeds and microbial propagules compete. In forest soils in which they flourish, earthworms are fundamental to the dynamics of the ecosystem.

13.5 REFERENCES

Arldt, T. (1908) Die Ausbreitung der terricolen Oligochäten in Laufe der erdgeschichtlichen Entwicklung des Erdreliefs. *Zool. Jb. Syst.*, 26.

Baltzer, R. (1956) The earthworms of Westphalia. *Zool. Jb. Syst.*, **84**, 355–414.

Barley, K. P. (1964) Earthworms and the decay of plant litter and dung – a review. *Proc. Austr. Soc. Anim. Prod.*, **5**, 236–240.

Bocock, K. L., Gilbert, O., Capstick, C. K., Twinn, D. C., Waid, J. S. and Woodman, M. J. (1960) Changes in leaf litter when placed on the surface of soils with contrasting humus types. I. Losses in dry weight of oak and ash leaf litter. *J. Soil Sci.*, **11**, 1–9.

Bornebusch, C. H. (1930) The fauna of forest soil. *Forst. ForsVaes. Danm.*, **11**, 158 pp.

Bornebusch, C. H. (1953) Laboratory experiments on the biology of worms. *Dansk. Skorforen. Tidssk.*, **38**, 557–559.

Cooke, A. and Luxton, M. (1980) Effect of microbes on food selection by *Lumbricus terrestris*. *Rev. Ecol. Biol. Sol.*, **17**, 365–370.

Edwards, C. A. and Heath, G. W. (1963) The role of soil animals in breakdown of leaf material. In *Soil Organisms* (eds. J. Doeksen and J. van der Drift), North Holland Publ. Co., Amsterdam, pp. 76–83.

Franz, H. and Leitenberger, L. (1948) Biological-chemical investigations into the formation of humus through soil fauna. *Öst. Zool. Z.*, **1**, 498–518.

Füller, H. (1953) Tiergeographisch-ökologisch Untersuchung über die Lumbriziden des mittleren Saalatales. *Wiss. Z. 1952–1953, Jena.* 50–51.

Galoux, A. (1953) *Trav. Stn. Rech. Gorenendael, Ser. A., No. 8.* 235 pp.

Heungens, A. (1969) The physical decomposition of pine litter by earthworms. *Pl. Soil*, **31**, 22–30.

Krivolutzky, D. A. and Pokarzhevsky, A. D. (1977) The role of soil animals in nutrient cycling in forest and steppe. In *Soil Organisms as Components of Ecosystems.* (eds. U. Lohm and T. Persson), *Ecol. Bull. (Stockholm)*, **25**, 253–260.

Lindquist, B. (1938) Dalby Soderskog: en *skansk lövskog i forntid och nutid.*, Stockholm, 273 pp.

Martinucci, G. B. and Sala, G. (1979) Lumbricids and soil types in prealpine and alpine woods. *Bull Zool.*, **46**, 279–297.

Michaelsen, W. (1921) Die Verbreitung der Oligochäten im Lichte der Wgener'schen Theorie der Kontinentalverschiebung und anderer Fragen zur Stammeschichte und Verbreitung dieser Tiergruppe. *Verh. Ver. Naturw. Unterh. Hamburg*, **29**.

Needham, A. E. (1957) Components of nitrogenous excreta in the earthworms *Lumbricus terrestris* L. and *Eisenia foetida* (Savigny). *J. Exp. Biol.*, **34**, 425–446.

Neuhauser, E. F. and Hartenstein, R. (1978) Reactivity of soil macro-invertebrate peroxidases with lignin and lignin model compounds. *Soil Biol. Biochem.*, **10**, 341–342.

Nordström, S. and Rundgren, S. (1973) Associations of lumbricids in southern Sweden. *Pedobiologia*, **13**, 301–326.

Nordström, S. and Rundgren, S. (1974) Environmental factors and lumbricid associations in southern Sweden. *Pedobiologia*, **14**, 1–27.

Omrani, G. A. (1973) *Bodenzoologische Untersuchungen uber Regenwurmer im Zentral-und Nordiran.* Inaugural Dissertation, Justus-Liebig University.

Perel, T. S. and Sokolov, D. F. (1964) A quantitative assessment of the role of earthworms *Lumbricus terrestris* L. (Lumbricidae, Oligochaeta) in processing forest litter. *Zool Zh.*, **43**, 1618–1624.

Phillipson, J., Abel, R., Steel, J. and Woodell, S. R. J. (1978) Earthworm numbers, biomass and respiratory metabolism in a beech woodland – Wytham Woods, Oxford. *Oecologia (Berl.)*, **33**, 291–309.

Rabeler, W. (1960) Die Artenbestande der Regenwurmer in Laubwald-Biozonosen (Querce-Fageta) des oberen und mittleren Wesgebietes. *Mitt. Flor.-soz. ArbGemein.*, N.F. 8, 333–337.

Raw, R. (1962) Studies of earthworm populations in orchards. 1. Leaf burial in apple orchards. *Ann. Appl. Biol.*, **50**, 389–404.

Reichle, D. E., Crossley, D. A., Jr., Edwards, C. A., McBrayer, J. F. and Solling, P. (1973) Organic matter and 137 Cs turnover in forest soil by earthworm populations: application of bioenergetic models to radionuclide transport. In *Radionuclides in Ecosystems*, USAEC CONF-710501 (ed. D. J. Nelson), pp. 240–246.

Satchell, J. E. (1967) Lumbricidae. In *Soil Biology* (eds. A. Burges and F. Raw), Academic Press, London and New York, pp. 259–322.

Satchell, J. E. (1974) Litter – interface of animate/inanimate matter. In *Biology of Plant Litter Decomposition* (eds. C. H. Dickinson and G. J. F. Pugh), Academic Press, London and New York, pp. xiii–xliv.

Satchell, J. E. (1980) r worms and K worms: a basis for classifying lumbricid earthworm strategies. In *Soil Biology as Related to Land Use Practices* (ed. D. L. Dindal), *Proc. VIIth Int. Soil Zool. Coll., EPA, Washington DC*, pp. 848–864.

Satchell, J. E. and Lowe, D. G. (1967) Selection of leaf litter by *Lumbricus terrestris*. In *Progress in Soil Biology* (eds. O. Graff and J. E. Satchell), North Holland Publ. Co., Amsterdam, pp. 102–119.

Sergienko, M. I. (1969) Spreading of earthworms in biocoenosis of Tcernogoria along ecological profile of Borochta-Goverla. *Pedobiologia*, **9**, 112–113. [In Russian.]

Striganova, B. R. (1968) Study of the role of woodlice and earthworms in the humification of decomposing wood. *Pochvovedenie*, **8**, 85–90.

van der Drift, J. (1951) Analysis of the animal community of a beech forest floor. *Tijdschr. Ent.*, **94**, 1–168.

van Rhee, J. A. (1963) Earthworm activities and the breakdown of organic matter in agricultural soils. In *Soil Organisms* (eds. J. Doeksen and J. van der Drift), North Holland Publ. Co., Amsterdam, pp. 55–59.

Volz, P. (1962) Pedozoologische Untersuchungen in der Umgebung von Banyuls sur Mer. *Vie et Milieu*, **13**, 545–563.

Wilcke, D. E. (1955) Bemerkungen zum Problem des erdzeitlichen Alters der Regenwurmer (Oligochaeta Opisthopora). *Zool. Anz.*, **154**, 149–156.

Wright, M. A. (1972) Factors governing ingestion by the earthworm *Lumbricus terrestris* with special reference to apple leaves. *Ann. Appl. Biol.*, **70**, 175–188.

Zajonc, I. (1967) Uber die Saisondynamik der Humusbildung durch Regenwürmer in einem Buchenwald der Karpathen. In *Progress in Soil Biology* (eds. O. Graff and J. E. Satchell), North Holland Publ. Co., Amsterdam, pp. 397–498.

Zajonc, I. (1971) Synusia analysis of earthworms (Lumbricidae Oligochaeta) in the oak-hornbeam forest in south-west Slovakia. In *Productivity of Forest Ecosystems* (ed. P. Duvigneaud), U.N.E.S.C.O., Paris, pp. 443–452.

Chapter 14

Earthworm ecology in deciduous forests in central and southeast Europe

A. ZICSI

14.1 INTRODUCTION

Darwin's studies (1881) on earthworm ecology concentrated on those species endemic to north and northwest Europe. These species, now described as peregrine, are those which repopulated the area after the withdrawal of the glacial ice sheets. In Wilcke's classification (1953), they are divided into two ecological groups: humus inhabiting forms (litter inhabiting, Zicsi, 1968) and mineral soil inhabiting forms. This classification indicates that the significance of lumbricids in a soil complex is determined by their species composition. In areas once affected by glaciation there are, with the exceptions of *Lumbricus terrestris* and *Allolobophora longa*, only small-bodied species of both ecological groups with more-or-less shallow burrows, and the significance of earthworm activity in all soils has been judged on the basis of these particular population compositions. Recently, the use of vermifuges in research into the indigenous ranges of the Lumbricidae in Europe and Asia Minor, has revealed the occurrence of large-bodied, deep-burrowing species. These were either new to science or previously reported only sporadically but now show areas of continuous distribution.

Until recently, no data were available on the activities of these deep-burrowing species in the soil and the observations reported here were made to determine the significance of these species in the turnover of organic material. This is of particular importance in Hungary where afforestation of 850 000 ha of land useless for agriculture is planned, of which 600 000 ha will not be used for economically productive forestry but for environmental purposes and nature conservation. Our aim was to follow the zootic sequences of decomposition of forest litter in order to give the forestry service advice for selecting tree species to be used in afforestation programmes. Information was obtained from a research project launched 10 years ago to establish the activity of soil animals and their long-term role in decomposition processes under the climatic conditions prevailing in two associations of *Quercus petraea–Carpinetum*. Our work on these large-bodied, deep burrowing species concerns their occurrence and abundance in natural forest associations and their feeding ecology under laboratory and field conditions.

14.2 OCCURRENCE AND ABUNDANCE OF DEEP-BURROWING SPECIES IN NATURAL FOREST ASSOCIATIONS

Results of investigations of more than 1500 localities in Hungary reveal the presence of six large-bodied species: *Lumbricus polyphemus* (Fitzinger, 1833); *L. terrestris* L., 1758; *Fitzingeria platyura depressa* (Rosa, 1893); *F. p. montana* (Černosvitov, 1932); *Octodrilus transpadanus* (Rosa, 1884); *O. gradinescui* (Pop, 1938) (Table 14.1).

Table 14.1 Occurrence, abundance and biomass of large-bodied species in experimental areas.

	Relative abundance	Mean population density (m^{-2})	Mean weight (g) per animal
L. polyphemus	5.06 (44)*	3.18	7.64
L. terrestris	4.66	—	—
F.p. depressa	12.93 (55)	7.24	2.26
F.p. montana	4.60 (29)	8.10	5.68
O. transpadanus	5.54 (41)	4.54	3.16
O. gradinescui	1.46	—	—

* Number of samples. *L. terrestris* and *O. gradinescui* do not occur in indigenous Hungarian forests but were compared in feeding experiments.

L. terrestris has been introduced in Hungary and occurs mainly in the vicinity of human settlements. Occasionally it has spread into neighbouring plots of indigenous forest where other earthworm species have been less successful.

Apart from *L. terrestris*, the distributions of large-bodied, deep-burrowing species in Hungary coincide with the occurrence of brown forest loam. According to the climazonal vegetation map, these species might be found in the zone containing the hornbeam–oak forests (*Quercus petraea–Carpinetum*) but should be completely lacking in stands of closed beech (*Melitti-Fagetum*) and oak (*Quercus petraea-cerris*). Their absence from these associations under the prevailing climatic condition may be explained by the results of the feeding experiments.

Little is known about the distributions of the large-bodied, deep-burrowing species in central and southeast Europe but *L. polyphemus* and *Fitzingeria* species are located in the southern regions of the German Federal Republic, in Austria, Yugoslavia, Slovakia, Romania and the Carpathian Ukraine. Large-bodied forms of *O. transpadanus* have been found only in scattered localities in Slovakia, Yugoslavia and Romania although the species has a wider distribution. *O. gradinescui* has been reported only from Hungarian and Romanian localities although other large-bodied members of this genus include *O. complanatus*, widely distributed in southern Europe, and the *O. mima* species group confined to Italy and Yugoslavia.

These large-bodied species can differ considerably in size and weight depending on biotope conditions at different altitudes. The mean weight of *L. polyphemus* was 4.09 g in meadows in the Grossglockner region of Austria (Zicsi, 1981) and 8.12 g in Hungary in a *Quercus petraea–Carpinetum*. Size and weight are closely connected with the continuity of food supply. Under starvation conditions the worms may lose 40–50 % of their weight without dying.

14.3 FEEDING ECOLOGY

The experimental work was carried out in the biological cave laboratory in the cave Baradla at Aggtelek (Hungary) where the relative humidity is 97 ± 2–3% and the temperature $10 \pm 1\,°C$ throughout the year. Earthworms were kept singly in columns of soil 25 cm × 25 cm × 0.5 m, 1 m or 1.5 m high, or in plastic pots. The following species were introduced: *Lumbricus polyphemus, L. terrestris, Fitzingeria p. platyura, F. p. depressa, F. p. montana, Octodrilus transpadanus, O. gradinescui, O. complanatus* and *O. rucneri*. Food was provided in the form of litter of linden (*Tilia platyphyllos* Scop.), ash (*Fraxinus excelsior* L.), maple (*Acer platanoides* L.), hornbeam (*Carpinus betulus* L.), beech (*Fagus sylvatica* L.) and three kinds of oak (*Quercus robur* L., *Q. petraea* Matt. and *Q. cerris* L.). Litter samples were given to the animals at monthly intervals from November onwards. In Table 14.2 the amounts of litter consumed monthly by four specimens of each of four species investigated in separate experiments in 1975/6 and 1976/7 are presented. The monthly means for litters on which the worms had commenced feeding in the same month each year are combined as the mean for the 2 years, those for litters on which feeding commenced in different months in the 2 years are presented separately.

Table 14.2 Mean litter consumption of four earthworm species (mg day^{-1} g^{-1} live wt.)

	Nov. 75/77	Dec. 75/77	Jan. 75/77	Feb. 75/77	Mar. 75/77	Apr. 75/77	May 75/77	June 75/77	July 75/77	Aug. 75/77	Sept. 75/77
L. polyphemus											
T. platyphyllos	12.7	22.6	17.7	22.7	31.7	—	—	—	—	—	—
F. excelsior	11.8	20.5	19.6	24.6	32.4	—	—	—	—	—	—
A. platanoides	11.6	18.5	12.9	23.9	27.3	—	—	—	—	—	—
C. betulus	6.4	8.7	10.7	11.9	15.9	15.1	21.9	21.4	22.5	27.2	34.9
Q. petraea	0.0	0.0	0.0	0.0	2.1	6.4	10.1	13.5	17.3	23.3	25.1
Q. cerris	0.0	0.0	0.0	1.5	2.1	4.2	12.5	15.8	22.2	25.2	29.0
Q. robus	0.0	0.0	0.0	0.0	3.5	4.5	8.0	13.1	14.7	19.7	23.9
F. sylvatica	0.0	0.0	0.0	1.8	4.4	6.6	6.5	6.6	26.2	16.2	27.4
L. terrestris											
T. platyphyllos	15.4	27.3	26.8	39.8	33.0	—	—	—	—	—	—
F. excelsior	14.2	20.3	24.9	29.3	29.8	—	—	—	—	—	—
A. platanoides	12.9	22.2	26.6	27.9	32.6	—	—	—	—	—	—
C. betulus	8.0	10.8	14.7	19.0	18.0	16.9	23.4	24.2	22.6	30.7	34.1
Q. petraea	0.0	0.0	0.0	3.2	2.7	19.7	9.1	13.3	14.4	21.5	25.6
Q. cerris	0.0	0.0	0.0	3.4	2.3	5.4	8.2	14.3	17.9	20.3	27.0
Q. robus	0.0	0.0	2.5	2.0	3.3	7.9	9.8	12.9	29.6	24.0	28.0
F. sylvatica	0.0	0.0	0.0	2.4	6.2	3.7	6.9	10.5	16.1	19.6	24.3

Table 14.2 *continued*

	Nov. 75/77	Dec. 75/77	Jan. 75/77	Feb. 75/77	Mar. 75/77	Apr. 75/77	May 75/77	June 75/77	July 75/77	Aug. 75/77	Sept. 75/77
F. p. depressa											
T. platyphyllos	16.0	24.9	28.8	31.5	34.2	—	—	—	—	—	—
F. excelsior	13.0	19.6	25.4	31.3	34.3	—	—	—	—	—	—
A. platanoides	8.5	16.9	20.7	25.5	27.7	—	—	—	—	—	—
C. betulus	6.9	8.9	14.1	15.0	16.2	16.8	18.6	22.5	26.0	26.9	33.0
Q. petraea	0.0	0.0	0.0	0.0	1.6	4.5	4.6	11.6	20.5	22.1	26.1
Q. cerris	0.0	0.0	0.0	0.0	1.8	1.7	3.7	13.8	21.0	22.6	28.9
Q. robus	0.0	0.0	0.0	3.3	3.4	3.3	4.4	15.9	18.0	21.7	25.6
F. sylvatica	0.0	0.0	0.0	0.0	0.0	2.9	2.1	13.8	10.1	13.1	18.0
F. p. montana											
T. platyphyllos	9.2	14.1	18.0	15.6	22.6	—	—	—	—	—	—
F. excelsior	10.3	16.4	14.4	18.6	23.9	—	—	—	—	—	—
A. platanoides	3.8	4.1	4.0	11.3	10.1	—	—	—	—	—	—
C. betulus	4.2	5.4	6.6	7.8	7.8	7.8	11.3	16.7	19.5	21.5	21.8
Q. petraea	0.0	0.0	0.0	0.0	2.3	2.1	2.1	11.0	13.0	12.8	17.3
Q. cerris	0.0	0.0	0.0	0.0	0.0	1.3	1.7	12.2	11.3	14.9	14.5
Q. robus	0.0	0.0	0.0	0.0	1.0	1.8	3.6	11.2	14.8	12.9	18.6
F. sylvatica	0.0	0.0	0.0	0.0	0.0	3.4	5.6	4.4	10.3	16.5	19.3

Experiments in which only a single litter was given showed that preference was determined by the degree of decomposition of the litter. As much of the readily decomposing litter of linden, ash and maple was consumed during the first months after leaf fall, November to February, as of the slower decomposing litters of hornbeam from March to June, and of beech and the species of oak from June to September.

14.4 DISCUSSION

In forests where these earthworms occur, the easily decomposing litter of linden, ash and maple, subject to climatic variations, completely disappears at the latest by April whereas the litter of hornbeam does not disappear completely until the end of June. Continuity of food supply is of cardinal importance in the survival of these large-bodied earthworm populations, and provides an explanation for their absence from closed stands of oak or beech forest under the same climatic conditions. In recently afforested stands where the composition of the tree species has been carefully controlled, large-bodied earthworms if introduced could be expected to survive and to have a profound effect on the structure of the soil through their cast production.

The soil column experiments (Table 14.3) established that the cast production of *Lumbricus polyphemus* reached 100.8 mg day^{-1} g^{-1} live weight, *L. terrestris* 54.7 mg, *F.p. depressa* 35.3 mg, *F.p. montana* 78.3 mg, *Octodrilus transpadanus* 107.8 mg, and *O. gradinescui* 124.0 mg. *L. polyphemus* alone, at a density of 4 m^{-2} would transport 8125 kg dry weight of soil ha^{-1} annually.

Table 14.3 Cast production in one year of large earthworm species in experimental monoliths.

| | Cast production (g dry wt.) | | | | | | |
	Soil surface	0–50 cm	50–100 cm	100–150 cm	Total	In soil (%)	mg day^{-1} g^{-1} live wt.
L. polyphemus	69.0	121.5	70.3	20.3	281.1	75.4	100.8
L. terrestris	78.3	14.4	—	—	92.7	15.5	54.7
F.p. montana	4.4	76.5	60.6	21.7	163.2	97.4	78.5
F.p. depressa	3.2	15.7	10.1	—	29.0	88.9	35.1
O. gradinescui	14.1	178.4	35.2	—	227.7	93.8	124.0
O. transpadanus	31.2	68.9	61.3	—	161.4	80.6	107.8

Field experiments in two hornbeam–oak forests with similar climatic conditions confirmed the calculated consumption rates. The two forests differed only in the presence of the two large-bodied species

L. polyphemus ($5 \, \mathrm{m}^{-2}$, $35.7 \, \mathrm{g} \, \mathrm{m}^{-2}$) and *F.p. depressa* ($5 \, \mathrm{m}^{-2}$, $11.8 \, \mathrm{g} \, \mathrm{m}^{-2}$) in one stand (Szendehely, Cserhát hills) and their absence from the other (Vinyabükk, Vértes hills). Litter was determined monthly in both stands and the food consumption values were obtained from the feeding experiments. These data established that where the two large-bodied species occurred, 76% of the litter was consumed by earthworms while in the other stand only 31% of the litter was consumed by the species – *Lumbricus rubellus, Dendrobaena octaedra, D. rubida* and *D. auriculata*. (Further details are described in Pobozsny, 1975, 1977; Zicsi, 1972, 1975, 1977; Zicsi *et al.*, 1971; Zicsi and Pobozsny, 1977.)

The results of this work indicate that these species, which are widely distributed in central and southeast Europe, should receive greater consideration by forest management authorities.

14.4 REFERENCES

Darwin, C. (1881) *The Formation of Vegetable Mould through the Action of Worms, with Observations on their Habits.* John Murray, London. 326 pp.

Pobozsny, M. (1975) Die Bedeutung zweier Regenwurm-Arten für Humifizierungsprozesse. *Pedobiologia*, **15**, 439–445.

Pobozsny, M. (1977) Veranderungen einiger chemischer Eigenschaften in den Exkrementen von *Lumbricus polyphemus* Fitz. (Oligochaeta: Lumbricidae). *Opusc. Zool., (Budapest)*, **14**, 99–103.

Wilcke, D. E. (1953) Über die vertikale Verteilung der Lumbriciden im Boden. *Z. Morphol. Ökol. Tiere*, **41**, 372–385.

Zicsi, A. (1968) Ein zusammenfassendes Verbreitungsbild der Regenwürmer auf Grund der Boden- und Vegetationsverhaltnisse Ungarns. *Opusc. Zool., (Budapest)*, **8**, 99–164.

Zicsi, A. (1972) The work of the Aggteleki Cave Biological Laboratory, *Allatt, Közl.*, **59**, 155–160. [In Hungarian.]

Zicsi, A. (1975) Zootische Einflüsse auf die Streuzersetzung in Hainbuchen-Eichenwäldern Ungarns. *Pedobiologia*, **15**, 432–438.

Zicsi, A. (1977) Die Bedeutung der Regenwürmer bei der Streuzersetzung in mesophilen Laubwäldern Ungarns. *P. Cent. Pir. Biol. Exp.*, **9**, 75–84.

Zicsi, A. (1981) Regenwürmer des Grossglocknergebietes. Veröffentlichung des Österreich MAB-Hochgebirgsprogramms Hohe Tauern, **B4**, 91–94.

Zicsi, A. and Pobozsny, M. (1977) Einfluss des Zersetzungsverlaufes der Laubstreu auf die Konsumintensität einiger Lumbriciden-Arten. In *Soil Organisms as Components of Ecosystems* (eds. U. Lohm and T. Persson), *Proc. VIth Int. Soil Zool. Coll., Ecol. Bull. (Stockholm)*, **25**, 229–239.

Zicsi, A., Hargitai, L. and Pobozsny, M. (1971) Über die Auswirkung der Tätigkeit des Regenwurmes *Lumbricus polyphemus* Fitz. auf die Veränderungen der Humusqualität im Boden. *Ann. Zool. Ecol. Anim. (Hors-ser)*, **71**, 397–408.

Chapter 15

Earthworms of tropical regions – some aspects of their ecology and relationships with soils

K. E. LEE

15.1 INTRODUCTION

In tropical and temperate regions alike earthworms are among the most widespread of invertebrate animals and are found mainly in the soils of forests, woodlands, shrublands and grasslands, which together cover *ca.* 80 million km², or *ca.* 54% of the land surface of the earth (Whittaker and Likens, 1973). Their principal effects on soils, as recognized by Darwin (1881), result from their ingestion of organic detritus and soil, partial digestion of the organic detritus, deposition of casts at or beneath the surface, and the construction of burrows in the soil.

Most of our knowledge of earthworms is based on the Eurasian family Lumbricidae, which includes *ca.* 10% of the total of *ca.* 3000 species of earthworms. Some 15 lumbricid species have been spread by European man and are now well established throughout the temperate regions of the world. They are rare in tropical regions, mainly confined to small areas at high altitudes, and the dominant earthworms of the tropics are representatives of the families Almidae, Kynotidae, Glossoscolecidae, Megascolecidae, Eudrilidae and Ocnerodrilidae (as defined by Jamieson, 1978).

15.2 LITTER FALL, ORGANIC MATTER AND EARTHWORMS

The production of some tropical ecosystems is compared with temperate and boreal ecosystems in Table 15.1. Production and organic detritus are

179

Table 15.1 Areas of tropical forests and savannas, temperate and boreal forests, woodlands, shrublands and grasslands, with estimates of annual primary production, energy fixed and organic detritus input to soil, and standing crop of organic detritus. (Based on data of Whittaker and Likens (1973), Reiners (1973) and Schlesinger (1977)).

	Area (10^6 km^2)	Net primary production		Net energy fixed		Organic detritus input to soil		Standing crop of organic detritus	
		Total (10^9 t C y^{-1})	Mean (g C m^{-2} y^{-1})	Total (10^{15} kJ y^{-1})	Mean (kJ m^{-2} y^{-1})	Total (10^9 t C y^{-1})	Mean (g C m^{-2} y^{-1})	Total in soil and litter (g C m^{-2})	Litter layer (g C m^{-2})
Tropical ecosystems									
Forests	24.5	20.4	850	779	31 800	12.5	510	10 400	150
Savannas	15.0	4.7	315	176	11 700	3.5	230	3 700	100
Total	39.5	25.1	—	955	—	16.0	—	—	—
Mean	—	—	650	—	24 200	—	405	7 900	130
Temperate and boreal ecosystems									
Temperate forests	12.0	6.7	560	293	24 400	2.5	210	11 800	1200
Boreal forests	12.0	4.3	360	192	16 000	2.0	170	14 900	2000
Wood- and shrub-lands	8.0	2.2	270	96	11 300	2.8	350	6 900	280
Temperate grasslands	9.0	2.0	225	75	8 300	3.4	380	19 200	200
Total	41.0	15.2	—	656	—	10.7	—	—	—
Mean	—	—	370	—	15 800	—	265	13 400	1040

given in terms of their organic carbon contents. The total areas of the tropical ecosystems are about equal to those of the temperate ecosystems; mean net primary production, energy fixed and organic detritus input to the soil are 50–60 % higher per unit area in tropical than in temperate ecosystems, while the standing crop of surface litter is only *ca.* 12.5 %, and the total organic detritus in soil and litter of tropical ecosystems is only *ca.* 60 % of corresponding estimates for temperate ecosystems. High rates of primary production in tropical environments are apparently more than compensated by very high rates of decomposition in tropical compared with temperate ecosystems. Tropical soils generally have low contents of organic matter; the mean organic carbon content of tropical forest soils is 10.4 kg m^{-2} compared with 11.8 kg m^{-2} in temperate forests and 14.9 kg m^{-2} in boreal forests, while the corresponding figure for tropical savannas is 3.7 kg m^{-2} compared with 19.2 kg m^{-2} for temperate grasslands (Schlesinger, 1977).

Many tropical soils are strongly leached and weathered and are very low in plant nutrients. In the African, American, Asian and Australian tropics there is predominantly a pattern of ancient land surfaces mantled with deep layers of weathered and nutrient-deficient polygenetic soils (cf. e.g. Beckmann (in press)), while in much of the temperate zones the retreat of the ice caps, within the last 15 000 years, left large areas of unweathered rock, on which the first soils are still forming. Maldague (1970) distinguished between 'inherent mineralogical fertility', which depends upon the weathering of nutrients from mineral materials in the soil, and 'biological fertility', which depends upon the rate of cycling of nutrients from the primary producers, through the decomposers, back to the primary producers. Nutrient cycling is almost entirely due to biological processes, and the efficiency of the decomposers is the principal regulator of the rate of return of nutrients from litter to maintain plant growth. The activities and efficiency of the decomposer organisms are themselves regulated mainly by temperature, moisture and rate of litter input.

In tropical forests, where organic C in detritus inputs is higher than in any other ecosystem type, and averages 510 g C m^{-2} y^{-1}, temperatures vary little throughout the year and average 25–30 °C. Soil moisture deficit is rare and potential decomposition rates are usually more than sufficient to dispose of the litter that falls so that nutrient cycling is rapid and efficient. This permits high levels of primary production on soils that have inherently low fertility; primary production averages 850 g C m^{-2} y^{-1} in tropical forests on soils with 10.4 kg C m^{-2}, compared with 560 g C m^{-2} y^{-1} on soils with 11.8 kg C m^{-2} in temperate forests and 360 g C m^{-2} y^{-1} on soils with 14.9 kg C m^{-2} in boreal forests. This represents a specific annual production rate of 82 g of fixed C kg^{-1} of soil

C for tropical forests, 47 g of fixed C kg^{-1} of soil C in temperate forests, and 24 g of fixed C kg^{-1} of soil C in boreal forests. Corresponding figures for tropical savanna are 85 g of fixed C kg^{-1} of soil C and for temperate grasslands 12 g of fixed C kg^{-1} of soil C.

Earthworms account for a substantial proportion of the biomass of large decomposer organisms in many tropical and temperate soils. Table 15.2, which is based on data from Kitazawa (1971), provides some examples of their populations, biomass and effects on organic matter decomposition in a range of boreal to tropical forest ecosystems. Numbers and biomass are very low in the example from lowland tropical rain forest. Few population studies have been made in such forests, but it seems likely that they usually have low earthworm numbers. Madge (1965) recorded 34.2 m^{-2}, with a biomass of 10.2 g wet weight m^{-2} in a Nigerian rain forest, and Edwards (1977) found 68, 107 and 0 earthworms per m^2 at three sites in lower montane forests in New Guinea; earthworms were virtually absent from lowland rain forests in the Solomon Islands (Lee, 1969) and were very rare in similar forests in Vanuatu (Lee, 1975, in press). The soils of lowland tropical forests are often acidic (pH 3–5), and are frequently saturated with water and poorly oxygenated; earthworms find refuge in logs, under bark, in the leaf bases of some subcanopy trees, and in the bases of epiphytes, and are common in the soil only in relatively well-drained sites. The rate of litter fall in tropical rain forests is very high; Lee (1974) calculated the mean dry weight of litter fall in 18 tropical rain forests from Africa, S.E. Asia, Australia and central America as 1.024 kg m^{-2} y^{-1}, with a mean energy content of 18 700 kJ m^{-2}, which is not much different from the figure in Table 15.2.

Comparison of specific rates of respiration (kJ g^{-1} of earthworms y^{-1}) illustrates the disproportionate contribution of a small earthworm biomass to litter decomposition in lowland tropical forest (Table 15.2). Constant high temperatures and adequate moisture apparently ensure that high rates of decomposition are maintained. In temperate forests and deciduous tropical forests, decomposition processes are seasonal, due to seasonal variations in litter fall, ambient temperatures and rainfall.

Results of a study of the energy budget and soil consumption of a population of earthworms (Megascolecidae and Eudrilidae) in a sparsely wooded savanna ecosystem at Lamto, Ivory Coast (Lavelle, 1974) are summarized in Table 15.3. About 50 kg of soil m^{-2} y^{-1} is ingested by an earthworm population totalling 20 m^{-2}, with a mean biomass of 20.5 g wet weight m^{-2}. Their respiratory output is 975 kJ m^{-2}y^{-1}, i.e. specific rate of respiration is 47.6 kJ g^{-1} y^{-1}, which is much greater than that of the earthworms in the forest ecosystems listed in Table 15.2. Total energy utilized by the earthworms in respiration and production is 1050 kJ

Table 15.2 Contribution of oligochaetes to litter decomposition in some boreal, temperate and tropical forest ecosystems. (Data derived from Kitazawa (1971).)

	Subalpine coniferous (boreal) forest	Temperate deciduous forest	Subtropical rain forest	Lowland tropical rain forest	Equatorial highland forest	Tropical deciduous forest
Location	Japan	Japan	Japan	Sabah	Kenya	Thailand
Altitude (m)	1700–2400	1000–1500	30–600	10	1936	400
Mean annual temperature (°C)	4	7	15	26	17	25
Soil	Podsols and Brown forest soils	Brown forest soils	Yellowish-brown forest soils	Weathered tropical yellow loams	Lateritic soils	Lateritic soils
Litter fall (kJ m^{-2} y^{-1})	9000	7300	23 000	96 000	—	—
OLIGOCHAETES						
Mean (no. m^{-2})	72.0	68.0	52.8	2.8	156.8	19.2
Mean biomass (g m^{-2} (wet wt.))	2.16	17.76	6.83	2.58	17.39	0.27
Biomass as % total macrodecomposers	53	93	74	60	78	24
Respiration (kJ m^{-2} y^{-1})	6.53	41.48	78.91	60.40	209.63	—
Respiration as % total macrodecomposers	26	48	50	18	53	—
Respiration as % total soil fauna respiration	1.1	6.0	3.9	6.9	13.5	—
Specific rate of respiration (kJ g^{-1} y^{-1})	3.0	2.3	11.6	23.4	12.1	—
Ratio respiration: litter fall (kJ m^{-2} y^{-1})	0.0007	0.006	0.003	0.0006	—	—

$m^{-2}y^{-1}$, which is about 9% of the energy available from the organic matter of the ingested soil (Table 15.3).

Table 15.3 Soil consumption and energy budget of earthworms in a sparsely wooded savanna ecosystem at Lamto, Ivory Coast. (Based on data from Lavelle (1974).)

Earthworm population data	
Numbers	$20 \, m^{-2}$
Biomass	$20.5 \, g\,m^{-2}$ (fresh wt.)
Production	$43 \, g\,m^{-2}y^{-1}$ (fresh wt.)
Total soil ingested	$50 \, kg\,m^{-2}y^{-1}$
Energy budget	$(kJ\,m^{-2}y^{-1})$
Total available in ingested soil	11 750
Earthworms	
Biomass	35
Production	73
Ratio Production/Biomass	2.1
Earthworm food utilization	1 050
Earthworm faeces	10 700
Loss via earthworm respiration	975
	= 8.3% of available food energy
Assimilation efficiency	
$\dfrac{\text{Total energy utilized}}{\text{Total energy in food}}$ or Respiration + Production	= 9%

Low rates of energy utilization are characteristic of all earthworms and of most saprophagous soil animals; the few figures that are available indicate that assimilation efficiency is generally $< 10\%$. The most important contribution of earthworms to decomposition processes appears to be in the comminution and mixing of plant litter and soil in the gut, and the effects of this process on further decomposition by the soil microflora in earthworm casts. Some $50 \, kg\,m^{-2}\,y^{-1}$ of casts are produced by the Lamto earthworms, and these contain organic matter with an energy content of $10\,700 \, kJ\,m^{-2}\,y^{-1}$. A population of seven lumbricid species (biomass $96.5 \, g\,m^{-2}$) in a pasture in Germany were shown by Graff (1971) to produce casts at the rate of $25.75 \, kg\,m^{-2}\,y^{-1}$, with an energy content of *ca.* $96\,000 \, kJ\,m^{-2}\,y^{-1}$; the organic C content of the casts from this pasture was *ca.* 8.3%, compared with *ca.* 0.5% in the Lamto casts, reflecting the very much higher organic C content of temperate than of tropical grassland soils (Table 15.1).

15.3 CASTS

Darwin (1881) was aware of the wide geographical distribution of earthworms that cast at the ground surface and he described many forms of casts that had been observed in temperate regions, and also in India, Ceylon and Burma. Surface-casting species are known among all the families of earthworms, and they are important agents of pedogenesis and soil fertility.

The various forms of surface casts are made up from (1) ovoidal, subspherical or spherical pellets, ranging in size from < 1 mm to > 1 cm ϕ, varying with the size of the species that excretes them, and (2) paste-like slurries that form generally rounded but less regular shapes. Composite casts, consisting of aggregated masses of the two basic types, are common, especially among tropical species, and may be of very large size. Madge (1969) described casts of the eudrilids *Eudrilus eugeniae* (fine granular pellets, piled in heaps $3-5$ cm ϕ and > 3 cm high) and of *Hyperiodrilus africanus* (pipe-like composite casts, *ca.* $2.5-8$ cm high and $1-2$ cm ϕ, with a vertical hole extending up from the burrow opening through them, but closed at the top) from a site at Ibadan in Nigeria. Nye (1955) found pipe-like casts up to 6 cm high and 1.5 cm ϕ, similar to those of *H. africanus*, but produced by the glossoscolecid *Hippopera nigeriae*, also at Ibadan. Darwin (1881) was sent massive composite casts from Nice (France) up to > 8 cm high and 2.5 cm ϕ, produced by earthworms identified by E. Perrier as introduced megascolecids from India or S. E. Asia, and similar casts from India, one of them 15 cm high and 3.75 cm ϕ, and large irregularly shaped casts that weighed up to 90 g. Casts produced by the megascolecid *Metapheretima jocchana* in the savanna lands of northern New Guinea were up to 5 cm high and 5 cm ϕ, and weighed $25-45$ g (Lee, 1967), while in the rain forests of northern Queensland I have found composite casts up to 10 cm high and 10 cm ϕ, weighing > 400 g, produced by an unidentified megascolecid. Gates (1961) reported similar composite casts in Burma that weighed up to 1600 g.

Casts usually contain more organic matter than the surrounding soil, reflecting the selection of plant detritus in preference to inorganic soil particles by earthworms. They also usually contain more of the smaller particle size fractions and less of the larger size fractions of the inorganic soil materials than the surrounding soil; the maximum size of particles ingested is related to the size of the worms. Nye (1955) found that the casts of *Hippopera nigeriae*, which is *ca.* 250 mm long and up to *ca.* 8 mm ϕ, contained no mineral particles > 1 mm ϕ, 1% of $0.5-1$ mm ϕ, and 15% of $0.2-0.5$ mm ϕ, while the soil at $2.5-60$ cm depth contained $32-50\%$ particles > 1 mm, $8-19\%$ $0.5-1$ mm and $12-29\%$ $0.2-0.5$ mm. For a larger earthworm, *Metapheretima jocchana* from New Guinea,

Table 15.4 Annual rate of surface cast production of some tropical earthworm populations.

Location	Vegetation	Earthworms			Weight of casts ($kg\,m^{-2}\,y^{-1}$)	Period of cast production	Depth of casts produced (mm)	Reference
		Taxa	Biomass ($g\,m^{-2}$)	Population (no. m^{-2})				
Nigeria	Grassland	*Hyperiodrilus africanus* *Eudrilus eugeniae*	—	—	17.58	Wet season (6 months)	15–20	Madge (1969)
Nigeria	Secondary forest	*Hippopera nigeriae*	—	—	5.12	Wet season (6 months)	2–4	Nye (1955)
Ivory Coast	(a) Grass savanna	*Eudrilidae* and *Megascolecidae*	39.7	202	21.7	Inactive during 'big dry' (3 months)	*ca.* 25	Lavelle (1978)
	(b) Shrub savanna		54.4	350	27.8			
	(c) Shrub savanna (protected from fire)		35.9	400	26.6			
Cameroon	Mountain savanna	n.d.*	—	—	21.0	—	*ca.* 20	Kollmannsperger (1956)
Sudan	Gezira Grassland	n.d.*	—	—	26.8	—	*ca.* 25	Beaugé (1912)
India	Pasture	*Eutyphoeus waltoni* *Metaphire californica*	—	—	1.3	Wet season (7 months)	*ca.* 1	Roy (1957)

* n.d. = not determined.

500–600 mm long and 9–10 mm ϕ, Lee (1967) showed from chemical, physical and clay mineralogical analyses of casts and soil, that the casts were derived from the 0–25 cm layer of the soil, but contained only 3 % of particles > 2 mm ϕ compared with 15 % in the 0–25 cm soil layer. Casts of *Millsonia anomala* and of Eudrilidae in the savannas of Ivory Coast contained respectively 46.5 % and 32 % of coarse sand, compared with 50.5–56.5 % coarse sand in the 0–40 cm soil layer (Lavelle, 1978). Similarly, Bolton and Phillipson (1976) found that the maximum sizes of mineral particles in the intestines of *Allolobophora rosea, A. caliginosa* and *Octolasion cyaneum* in an English woodland soil were respectively *ca.* 0.1, 0.2 and 0.5 mm; the three species are respectively *ca.* 3, 4 and 7–8 mm ϕ; their intestinal ϕ must be roughly proportional to the maximum sizes of soil particles ingested.

Surface casts, containing higher levels of organic matter and a selected range of particle sizes, accumulate to form the surface horizon, the *vegetable mould* of Darwin, at similar rates in tropical and temperate soils. Rates of accumulation in some tropical environments are summarized in Table 15.4, and range from *ca.* 1 to *ca.* 28 kg m^{-2} y^{-1}. Casting rates of lumbricids in temperate climates range from *ca.* 1 to *ca.* 25 kg m^{-2} y^{-1}. These amounts represent a depth of accumulation of *ca.* 1–25 mm y^{-1} on the soil surface. In moist tropical regions with little seasonality in rainfall, casting may be almost continuous through the year. In tropical regions with wet and dry seasons casting is generally restricted to the wet season; similarly in southern Australia and in New Zealand the surface-casting lumbricids are inactive during the hot, dry summer months, while in Europe and Japan some lumbricids are inactive during the summer and also during the coldest winter months. In most tropical and temperate regions surface-casting species, of all major taxonomic groups, are active for no more than *ca.* 6–7 months of the year.

15.4 EARTHWORM-MODIFIED SOIL HORIZONS

'The share which worms have taken in the formation of the layer of vegetable mould, which covers the whole surface of the land in every moderately humid country' was the stated subject of Darwin's volume on earthworms.

Surface soil horizons made by earthworms are widely distributed in the tropics, though the activities of termites as a dominant soil-forming element of the soil fauna in many tropical regions complicates the pattern of earthworm pedogenesis. Nye (1955) described a catenary sequence of soils in Nigeria where a surface layer, from *ca.* 10–60 cm deep, consisted of material transported to the surface by termites; the uppermost 2.5 cm of the termite-transported layer was further modified by earthworms

(*Hippopera nigeriae*), which had physically stirred and incorporated organic matter into the superficial soil. Nye distinguished the shallow earthworm-modified layer as the CrW horizon and the deeper unmodified termite-transported material as the CrT horizon. The CrW horizon had *ca.* 1% and the CrT horizon *ca.* 0.5% organic C. Such soil profiles illustrate an important difference between the effects of earthworms and those of termites on soil profile development. The earthworms that deposit surface casts are in most cases active only in the uppermost soil layers and their effect is to stir the uppermost horizon and to mix plant detritus from the surface with a shallow superficial soil layer. Termites seek clay-rich soil materials with a high moisture content, and use it to construct surface mounds and feeding galleries, or to pack into galleries in logs and trees that form their food and often their nests. They often bring this material from deep (50 cm or more) in the soil profile, so that in the long term their effect is to overturn soil profiles, especially bringing the finer particle size materials from deep soil layers to the surface (see Lee and Wood, 1971). The range of particle sizes moved by termites relates to their ability to grasp them in their mandibles and carry them, in contrast to that moved by earthworms, which relates to their ability to ingest them. Some African termites, the humivores (especially *Cubitermes* spp.), ingest soil in the manner of earthworms and apparently digest its included organic matter, but most termites feed on plant tissue from above the ground surface and require soil only for construction. Drummond (1886) and many authors since have considered termites to be the tropical analogues of earthworms, but the effects of most termites on pedogenesis are rather different from those of earthworms. Soil profiles with a superficial layer that is probably due to termites cover much of the ancient landscapes of tropical Africa, Australia and South and central America. In the more humid portions of these regions earthworms impose a further modifying influence, as described by Nye (1955) and by Lavelle (1978), who found surface horizons 15–20 cm thick, formed from earthworm casts, in the savanna lands of the Ivory Coast. Termites have the advantage, owing to their social organization, that they live in nests and forage through galleries of their own construction, within which they have some control over the physical environment (Lee and Wood, 1971). They are thus less vulnerable than earthworms to the seasonally wet and dry climates of much of the tropical regions of the world. Nye (1955) found the distribution of earthworms to be 'very erratic' in Ghana and Nigeria; many parts of the forested regions of the two countries show little sign of earthworm activity.

15.5 MICRORELIEF ATTRIBUTED TO EARTHWORM ACTIVITIES

Pickford (1926) described crater-like mounds *ca.* 1 m ϕ, with a central bowl-like depression 30–100 cm deep, in a South African grassland, and attributed their formation to earthworms. Very large tower-like casts, up to *ca.* 75 mm high, produced by *Microchaetus* sp. (Glossoscolecidae), cover the sides of mounds and Pickford proposed that the worms built the mound walls and were thus provided with unsaturated soil in which they could live during the wet season, when the surrounding land is inundated. Similar microrelief features in tropical regions have since been described and attributed to surface-casting earthworms associated with them.

In northern Uganda there are large areas of seasonally inundated swamps with steep-sided flat-topped mounds in which live three species of glossoscolecid earthworms, *Glyphidrilus* sp., *Alma stuhlmanni* and *A. emini*. In the wet season the flat tops of the mounds are just beneath the water. Harker (1955) suggested that the mounds were formed from the earthworms' casts, and this was also the conclusion of Wasawo and Visser (1959) who analysed soil from the tops of mounds and showed that it closely resembled the soil in the casts. Pitted and trenched areas of grassland in New Guinea, where a large earthworm, *Metapheretima jocchana*, was found to cast on the edges of the pits or on the higher ground between the trenches, were similarly attributed by Haantjens (1965) to the earthworms.

Darwin (1881) noted the susceptibility of the fine soil material in worm casts to down-slope movement during heavy rain and to removal by the wind during dry seasons, and there is much evidence that these are important factors in the loss of soil by erosion where surface-casting earthworms are numerous. In the Ugandan swamps and the New Guinea grasslands described above, earthworms are concentrated in the higher mounds and ridges where their burrows and surface casts are found, and there is little or no evidence (burrows or casts) that they are active in the soil beneath the intervening lower areas. Casts are deposited on the mounds and ridges only during the wet season, and the worms are inactive during the dry season. Soil must be lost from the higher to the lower ground by erosion of the casts, so unless the earthworms burrow beneath the lower ground and move soil laterally and upward to cast it on the higher ground, and there is no evidence for this, the net effect of the earthworms' casting must be to accelerate the levelling by erosion of the raised surfaces. Further, the proposition that earthworms combine their efforts to build a refuge for the wet season implies a degree of social organization in earthworm communities for which there is no evidence, and would require them to build the refuges during the dry season, when they are known to be inactive. The processes that result in the microrelief features are not known, but it must be concluded that it is unlikely that they are due to earthworms.

15.6 EARTHWORM BURROWS AND WATER INFILTRATION

The porosity of soils is increased by earthworm burrows, but because of their relatively large diameter (mostly in the range 1–10 mm) burrows can conduct water only when it is in a tension-free state. They behave as an additional drainage system that operates only when the amount of surface water exceeds the capacity of capillary intake by the soil, i.e. mainly during periods of heavy rain, a common feature of tropical climates, or when soils are irrigated. Lavelle (1978) estimated that earthworms in the soils of a tropical savanna removed to the surface *ca.* 30–40 m^3 of soil ha^{-1} y^{-1}; this must represent a minimum figure for the volume of pore space represented by their excavations, and it corresponds to a space of 3 litres m^{-2} y^{-1}. Satchell (1967) estimated that up to two-thirds of the air-filled space in soils may be earthworm burrows, while Edwards and Lofty (1977) estimated that earthworm burrows constitute *ca.* 5 % of total soil volume.

Traditional models of water infiltration into soils do not allow for the effect of surface-opening holes of the dimensions of earthworm burrows, but recently Edwards *et al.* (1979) and Germann and Beven (1981a, b, c) have provided models that demonstrate the great significance of such holes in increasing infiltration rates, while Ehlers (1975) and Tisdall (1978) have measured their effects in the field in temperate regions.

Rates of water infiltration in the tropics were measured by Wilkinson (1975) in Nigeria where three years' cropping (cotton, sorghum, ground nuts) followed two, three or six years' fallow under Gamba grass. Highest infiltration rates (up to 13–14 cm of water h^{-1}) were found under fallow, where earthworms were very numerous and the surface soil consisted almost entirely of earthworm casts, while the lowest rates were found in cultivated soil, where earthworms and their casts were few or absent. One year of cultivation reduced infiltration to *ca.* 18 % of that after six years' fallow. Lal (1974) compared ploughed with zero-tillage plots under maize in Nigeria; after two years' cropping there were *ca.* 100 earthworm casts m^{-2} on the surface of ploughed plots, and the water infiltration rate was 21 cm h^{-1}, compared with 2400 casts m^{-2} and an infiltration rate of 36 cm h^{-1} in zero-tilled plots.

15.7 CONCLUSIONS

There is great taxonomic diversity among tropical earthworms, but in their relationships with soils they show many resemblances to the lumbricids, on which most knowledge of earthworms is based. The predominance of ancient land surfaces that were not glaciated in recent geological times and have had a long history of warm climates and diverse vegetation has allowed the development and survival of species that are adapted to narrowly limited niches. This has led, for instance, to the dominance of tree- and log-inhabiting species in superhumid rain forests, the widespread presence of very large species in the deep layers of many forest and savanna soils, behavioural adaptations to alternate wet and dry seasons and diurnal temperature changes, physiological adaptations for water conservation, including resorption in the gut of urinary water and thickening of the cuticle in some species exposed to drought; convergence of morphological, physiological and behavioural adaptations is found in widely divergent taxonomic groups and geographically isolated regions.

Within the limitations of the basic pattern of morphological and physiological 'design' of earthworms, tropical earthworms have diverged and converged to take advantage of a very wide range of subterranean, arboreal and aquatic habitats, in all but the most arid tropical ecosystems.

15.8 REFERENCES

Beaugé, C. (1912) Les vers de terre et la fertilité du sol. *J. Agric. Prat., Paris*, **23**, 506–507.

Beckmann, G. G. (in press) The development of old landscapes and soils. In *Soils: an Australian Viewpoint.* Division of Soils, CSIRO. CSIRO, Melbourne/Academic Press, London.

Bolton, P. J. and Phillipson, J. (1976) Burrowing, feeding, egestion and energy budgets of *Allolobophora rosea* (Savigny) (Lumbricidae). *Oecologia (Berl.)*, **23**, 225–245.

Darwin, C. R. (1881) *The Formation of Vegetable Mould through the Action of Worms with Observations on their Habits.* John Murray, London. 298 pp.

Drummond, H. (1886) On the termite as the tropical analogue of the earthworm. *Proc. R. Soc. Edinburgh*, **13**, 137–146.

Edwards, C. A. and Lofty, J. R. (1977) *Biology of Earthworms.* 2nd edn., Chapman and Hall, London. 333 pp.

Edwards, P. J. (1977) Studies of mineral cycling in a montane rain forest in New Guinea. II. The production and disappearance of litter. *J. Ecol.*, **65**, 971–992.

Edwards, W. M., van der Ploeg, R. R. and Ehlers, W. (1979) A numerical study of the effects of noncapillary sized pores upon infiltration. *Soil Sci. Soc. Am. J.*, **43**, 851–856.

Ehlers, W. (1975) Observations on earthworm channels and infiltration on tilled and untilled loess soil. *Soil Sci.*, **119**, 242–249.

Gates, G. E. (1961) Ecology of some earthworms with special reference to seasonal activity. *Am. Midl. Nat.*, **66**, 61–86.

Germann, P. and Beven, K. (1981a) Water flow in soil macropores. I. An experimental approach. *J. Soil Sci.*, **32**, 1–13.

Germann, P. and Beven, K. (1981b) Water flow in soil macropores. II. A combined flow model. *J. Soil Sci.*, **32**, 15–29.

Germann, P. and Beven, K. (1981c) Water flow in soil macropores. III. A statistical approach. *J. Soil Sci.*, **32**, 31–39.

Graff, O. (1971) Stickstoff, Phosphor und Kalium in der Regenswurmlosung auf der Wiesenversuchsfläche des Sollingsprojektes. In *C. R. IV Coll. Pedobiol. Dijon, 1970,* I.N.R.A. Publ. 71–7, 503–511.

Haantjens, H. A. (1965) Morphology and origin of patterned ground in a humid tropical lowland area, New Guinea. *Austr. J. Soil Res.*, **3**, 111–129.

Harker, W. (1955) Pasture agronomist report no. 34. Appendix B. In *Water Resources Survey of Uganda.* A. Gibb and Partners, Uganda Govt. Publ.

Jamieson, B. G. M. (1978) Phylogenetic and phenetic systematics of the opisthoporous Oligochaeta (Annelida: Clitellata). *Evol. Theor.*, **3**, 195–233.

Kitazawa, Y. (1971) Biological regionality of the soil fauna and its function in forest ecosystem types. In *Productivity of Forest Ecosystems* (ed. P. Duvigneaud), *Proc. Brussels Symp. 1969. Ecol. Conserv.*, **4**, 485–498.

Kollmannsperger, F. (1956) Über die Bedeutung der Regenwürmer fur die Fruchtbarkeit der Bergsavannen Kameruns. *Zool. Anz.*, **157**, 216–219.

Lal, R. (1974) No-tillage effects on soil properties and maize (*Zea mays* L.) production in western Nigeria. *Pl. Soil*, **40**, 321–331.

Lavelle, P. (1974) Les vers de terre de la savane de Lamto. In *Analyse d'un écosystème tropical humide: la savane de Lamto. Bull. de Liaison des Chercheurs de Lamto.* No. spéc. **5**, 133–166.

Lavelle, P. (1978) Les vers de terre de la savane de Lamto (Côte d'Ivoire). Peuplements, populations et fonctions de l'écosystème. *Publ. Lab. Zool. E.N.S.*, **12**, 301 pp.

Lee, K. E. (1967) Microrelief features in a humid tropical lowland area, New Guinea, and their relation to earthworm activity. *Austr. J. Soil Res.*, **5**, 263–274.

Lee, K. E. (1969) Earthworms of the British Solomon Islands Protectorate. *Philos. Trans. R. Soc. London, Ser. B*, **255**, 245–354.

Lee, K. E. (1974) Soil fauna energetics in tropical forest. In *Soil Fauna and Decomposition Processes* (ed. D. Parkinson), Report IBP/PT Theme Meeting, Louvain, 1972, pp. 54–68.

Lee, K. E. (1975) Introduction. In *A discussion of the 1971 Royal Society-Percy Sladen expedition to the New Hebrides* (eds. E. J. H. Corner and K. E. Lee), *Philos. Trans. R. Soc. London, Ser. B*, **272**, 269–276.

Lee, K. E. (in press) Soil animals and pedological processes. In *Soils: an Australian Viewpoint.* Division of Soils, CSIRO. CSIRO, Melbourne/Academic Press, London.

Lee, K. E. and Wood, T. G. (1971) *Termites and Soils.* Academic Press, London and New York, 251 pp.

Madge, D. S. (1965) Leaf fall and litter disappearance in a tropical forest. *Pedobiologia*, **5**, 273–288.

Madge, D. S. (1969) Field and laboratory studies on the activities of two species of tropical earthworms. *Pedobiologia*, **9**, 188–214.

Maldague, M. E. (1970) Rôle des animaux édaphiques dans la fertilité des sols forestiers. *Publs. Inst. Natn. Étude Agron. Congo, Sér. Sci.*, **112**. 245 pp.

Nye, P. H. (1955) Some soil-forming processes in the humid tropics. III. Laboratory studies on the development of a typical catena over granitic gneiss. *J. Soil Sci.*, **6**, 63–72.

Pickford, G. E. (1926) The Kommetjie flats. *Blytheswood Review*, Cape Province, S. Africa. **3**, 57–58.

Reiners, W. A. (1973) A summary of the world carbon cycle and recommendations for critical research. In *Carbon and the Biosphere. Brookhaven Symposium on Biology*, **24**, 368–382.

Roy, S. K. (1957) Studies on the activities of earthworms. *Proc. Zool. Soc. Calcutta*, **10**, 81–98.

Satchell, J. E. (1967) Lumbricidae. In *Soil Biology* (eds. A. Burges and F. Raw), Academic Press, London. 259–322.

Schlesinger, W. H. (1977) Carbon balance in terrestrial detritus. *Ann. Rev. Ecol. Syst.*, **8**, 51–81.

Tisdall, J. M. (1978) Ecology of earthworms in irrigated orchards. In *Modification of Soil Structure* (eds. W. W. Emerson, R. D. Bond and A. R. Dexter), Wiley, New York, pp. 297–303.

Wasawo, D. P. S. and Visser, S. A. (1959) Swampworms and tussock mounds in the swamps of Teso, Uganda. *East Afr. Agric. J.*, **25**, 86–90.

Whittaker, R. H. and Likens, G. E. (1973) Carbon in the biota. In *Carbon and the Biosphere. Brookhaven Symposium on Biology*, **24**, 281–302.

Wilkinson, G. E. (1975) Effect of grass fallow rotations on the infiltration of water into a savanna zone soil of northern Nigeria. *Tropic. Agric. Trinidad*, **52**, 97–103.

Chapter 16

The ecology of earthworms in·southern Africa

A. J. REINECKE

16.1 INTRODUCTION

The changes in the environment caused by man's industrial and agricultural activities have influenced earthworm populations in many parts of southern Africa. These changes have been considerable and many former specimen localities of fifty or even less years ago cannot be found today (Ljungström, 1972; Du Plessis, 1978). This chapter relates the present distribution of endemic and exotic species to climate, changes in vegetation following development for agriculture, and the physiology and ecology of individual species.

The taxonomic status of many species is unclear; that of the Microchaetinae is discussed in recent contributions by Ljungström and Reinecke (1969) and Pickford (1975), with a major contribution by Du Plessis and Reinecke (in preparation) on the genus *Microchaetus* in southern Africa (Du Plessis, 1978). Pickford (1937) contributed extensively to our knowledge of the acanthodrilines.

The endemic megadrile fauna consists of the following subfamilies:

1. Alluroidinae (based on the monotypic genus *Standeria*)
2. Microchaetinae (Pickford, 1975) consisting of the genera *Microchaetus* and *Tritogenia*. The genus *Geogenia* is included under *Microchaetus* since it was established by Kinberg on an erroneous conclusion regarding the diagnostic value of the calciferous glands (Du Plessis, 1978)
3. Ocnerodrilinae with two genera
4. Acanthodrilinae with four genera (Pickford, 1937).

There are more than twenty species in the exotic megadrile fauna belonging to at least four families (Ljungström, 1972). There can be little doubt that man was the main agent in introducing these species into southern Africa. The Lumbricidae came from Europe and North America with European settlers at various times after 1500. Numerous trading contacts were established between Asia and southern Africa after 1600, but the oriental megascolecid fauna could have been introduced even earlier by Arab traders visiting the east coast. The time and manner of introduction of the exotic acanthodriline species and the originally South American species, *Eukerria saltensis*, is very speculative. European trading activities could have resulted in the wide distribution of this species in the southern hemisphere.

Ecological research on earthworms of southern Africa started as late as 1968 (Reinecke and Ljungström, 1969) on introduced lumbricids. This was followed by ecophysiological studies on the upper lethal temperatures of lumbricids (Reinecke, 1974) and the temperature preferences of lumbricids and microchaetines. Reinecke and Ryke (1970a) studied the water relations of microchaetines and their casting activity and also completed a seasonal study of the population densities of microchaetines in the Transvaal area (Reinecke and Ryke, 1972). This was followed by a quantitative and qualitative survey of irrigation areas along the Mooi River in Transvaal (Visser and Reinecke, 1977).

16.2 DISTRIBUTION

16.2.1 Endemic species

The indigenous vegetation of southern Africa is correlated to a great extent with rainfall, with forested areas to the east, primary grasslands in the interior plateau, and a western semi-desert region with rainfall < 100 mm y^{-1}. The endemic acanthodrilines, monographed by Pickford (1937), are forest dwellers and were probably the most affected by man's destruction of forested areas. Most of these species feed on decaying organic material present on the floors of the temperate mountain forests in the Eastern Cape. They are absent from exotic *Eucalyptus* and *Pinus* plantations. The acanthodrilines are typically found in areas with precipitation > 2000 mm y^{-1} and with small seasonal and diurnal changes in soil temperature. Population densities do not vary much seasonally as is the case with other groups on the interior temperate plateau. Only a few records of rather small populations of *Udeina* and *Microscolex* species exist for the temperate interior, mostly in heavy clays and under thornbush savanna.

Macro-invertebrates are largely absent from most of the Transvaal and Free State and also along the plantations of exotic trees in the Natal coastal area (Sims, 1978) although there has been very little intensive collecting in these areas. Areas of indigenous vegetation like the Kruger National Park produced no endemic species in a preliminary survey (Reynolds and Reinecke, 1976). Subsequent collecting over a wider area after the rainy season yielded new species of octochaetines in the northernmost part of the park (Reinecke and Ackerman, 1977). Endemics of this subfamily are known to be present in neighbouring Zimbabwe, Malawi and Mozambique. The low and erratic rainfall may be the major factor preventing earthworms (even the exotic lumbricids) from colonizing these areas. The lumbricids are frequently found in agricultural lands under irrigation.

The primary grasslands of the interior plateau are the natural habitat of the microchaetines and they are mostly found in sandy soils although some records of *Tritogenia* are from heavy clays. The microchaetines are now recorded from more than 200 localities of which the vast majority are south of the Vaal and Orange Rivers. The genus *Microchaetus* occurs most abundantly east of the 25° meridian, especially in the more coastal areas. Species occurring in the eastern areas such as the Transkei and Eastern Cape had higher population densities. The more advanced proandric forms were only found to the east of the 25° meridian and the more primitive holandric forms are distributed as relict groups in natural mountainous habitats to the west of this line. From this survey of the genus *Microchaetus*, it seems clear that agricultural practices have reduced the area it formerly inhabited. It is seldom, if ever, found in large numbers in intensively cultivated areas but can still be found in moderate numbers in natural habitats in areas such as Ciskei and Transkei. Few were found in the Cape and Natal but large numbers can still be found in the Eastern Cape.

16.2.2 Exotic species

Exotic earthworms in southern Africa are generally found in agricultural soils where the water regime allows survival. Lumbricidae are numerically dominant in irrigated soils and abound in garden soils throughout southern Africa. The oriental megascolecids are abundant in sugar cane plantations in Natal and Eastern Transvaal while *Pontoscolex corethrurus* is prominent in brackish soils in the eastern coastal plains.

Visser and Reinecke (1977) surveyed the earthworm fauna of irrigated areas and demonstrated the dominance of introduced species in population density, biomass and distributional range. Subsequent work (Reinecke and Visser, 1980) showed that the destruction of soil structure by cultivation was a major factor in decreasing earthworm populations, although crop rotation and the application of NPK fertilizer may also have played a minor role.

16.3 ECOLOGY

16.3.1 Population densities of endemic and exotic species

The mean population densities of various endemic and exotic species, and the frequency of occurrence of the various species in 32 localities in an

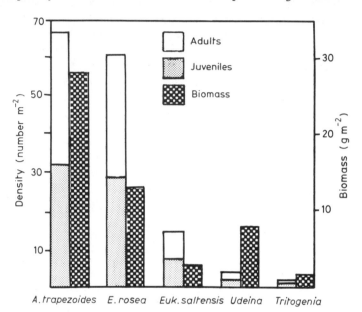

Fig. 16.1 Population densities and biomass in irrigated agricultural soils.

Table 16.1 Mean frequency of occurrence and population density of endemic and exotic species in southern Africa.

	32 Irrigated agricultural sites		28 'natural' sites	
	Occurrence (%)	Numbers m⁻²	Occurrence (%)	Numbers m⁻²
Endemic spp.				
Tritogenia	6.3	2.3	36.4	21.5
Udeina	3.1	3.4	14.6	6.1
Microchaetus modestus	—	—	41.3	42.3
Exotic spp.				
Allolobophora trapezoides	40.6	67.1	4.2	54.1
Eisenia rosea	65.6	61.3	6.1	38.7
Eiseniella tetraedra	—	—	2.1	11.3
Eukerria saltensis	15.6	14.5	—	—

irrigated area around Potchefstroom, are presented in Fig. 16.1 and Table 16.1. The population densities of endemic species were very low compared with exotic species. *Allolobophora trapezoides* had the highest population density for the area as a whole with a mean of 67.1 worms m^{-2} (calculated for 32 localities of which it occurred in only 13).

Conversely, in 'natural' vegetation adjacent to the agricultural fields (Table 16.1) *Microchaetus modestus* had the highest density and the widest distribution in 28 localities. The exotic species occurred sparsely but their population densities where they did occur were not much lower than those of the endemic species. These results illustrate the ability of the exotic lumbricids to flourish in irrigated agricultural soils. The dominance of endemic species in 'natural' localities adjacent to agricultural land

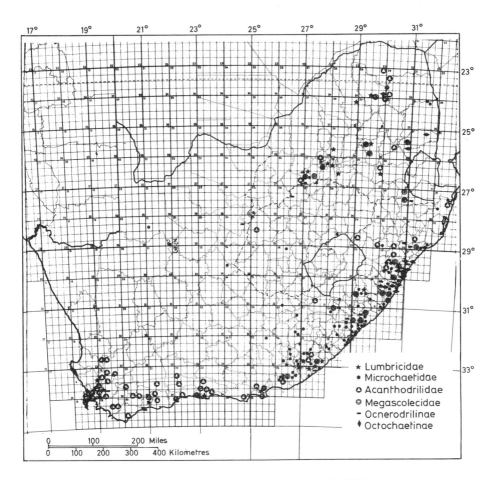

Fig. 16.2 Distribution of the earthworm fauna of South Africa.

indicates that the exotic species were not able to extend their distribution much beyond the borders of agricultural and garden soils.

The population densities of both endemic and exotic species reach a maximum during autumn and spring and are closely related to rainfall pattern. The spring population peak is heavily dependent upon the first summer rains in the northern parts of the country. Rainfall is, however, erratic, and the start of the rains varies much from year to year. During the dry, relatively cold winters, there is a marked decrease in the number of worms in areas with summer precipitation. In the southernmost tip of Africa which experiences an evenly distributed winter rainfall, the Acanthodrilinae occur in fairly high numbers during most of the year.

Estimates of biomass, based on collection by hand-sorting, suggest that earthworm biomass in natural grasslands is higher for endemic species and for exotic species in irrigated soils. Weights exceeding 100 g m^{-2} are found in some natural localities, partly attributable to large-sized species. The endemics of forest, mainly acanthodrilines, lack large deep-burrowing species and have a much lower biomass. A very patchy distribution is shown by both endemic and introduced species, although the latter may be fairly evenly distributed in irrigated areas (Fig. 16.2).

16.3.2 Casting

Amongst the endemic species in southern Africa, *Microchaetus modestus* is an active surface caster (Reinecke and Ryke, 1970b). Members of the genus *Tritogenia* deposit their casts below the surface. One of the most interesting features of grassland soils in the south-eastern parts of Transvaal, Eastern Cape and Natal is the casts produced by the endemic Microchaetinae. Figure 16.3 illustrates the casting activity (dry wt.) of *M. modestus* in a 'natural' area along the banks of the Vaal River over a period of one year and shows its close relation with rainfall. The temperature varied between 9°C and 30°C but cast production continued unless the soil moisture fell below tolerance limits. In a nearby irrigated potato field (Fig. 16.3) the casting activity was much lower despite the favourable moisture conditions, probably because the population density was lower. Observations in many parts of southern Africa (Reinecke and Ryke, 1970a) indicate that lumbricid and other exotic species, although found commonly in isolated places, contribute little to surface cast production.

Casts of the giant earthworm *Microchaetus microchaetus* reach a height of 20 cm or 'several inches' (Pickford, 1926) (Fig. 16.4). Ljungström and Reinecke (1969) estimated an annual cast production of approximately 50 t ha^{-1} for this species. This extremely long earthworm, reaching lengths of 1.5 m, is supposedly responsible for the highly irregular topography of the Kommetjie Flats where little mounds are formed by

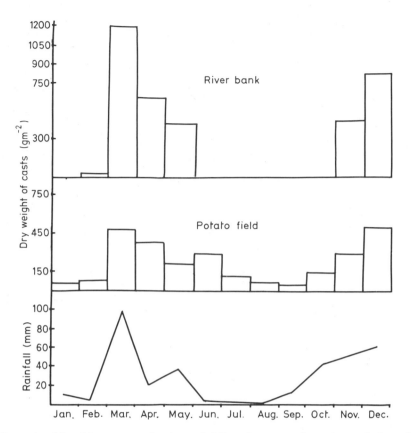

Fig. 16.3 Monthly cast production of *Microchaetus modestus* in an irrigated potato field and an adjacent site on the banks of the Vaal River.

Fig. 16.4 A single cast of *Microchaetus microchaetus* from the Kommetjie Flats.

Fig. 16.5 Casts produced by endemic microchaetines in the Eastern Cape.

earthworm casts (Fig. 16.5) causing bowls (kommetjies) to develop between them.

16.3.3 Survival mechanisms

The earthworm fauna exhibits a variety of specializations adapted to changing environmental conditions. Microchaetines possess enteronephric excretory organs and lack dorsal pores, facilitating moisture conservation. Under experimental conditions they show a higher resistance to desiccation than do lumbricid species (Reinecke, 1969). *Pontoscolex corethrurus* has a wide tolerance for saline conditions.

Many of the endemic species go into diapause under adverse conditions and juveniles are commonly found rolled up in drier soils. The role of temperature in inducing this condition is uncertain. *Microchaetus modestus* were found in diapause at soil temperatures of 8°C while soil moisture conditions were still favourable (Reinecke and Ryke, 1972). Microchaetines are usually found in diapause during the dry winter months of June and July and in August and September, depending on the onset of the summer rains. Survival mechanisms of this nature are essential for earthworms in most parts of southern Africa where the rainfall is erratic.

The endemic species can tolerate marginally higher temperatures than introduced species. Reinecke and Ryke (1974) have shown that *Microchaetus modestus* has a slightly wider range of temperature preferences than some introduced lumbricids although the success of introduced species in agricultural soils might suggest the opposite.

Soil temperatures in many areas in southern Africa may reach 40°C and more for short periods. The endemic microchaetines burrow into the subsoil where soil moisture and temperature are more stable. The endemic acanthodrilines occupy topsoil but occur in forested areas where moisture and temperature conditions remain favourable for most of the year. The ability to migrate deep into the soil is also found in the introduced Megascolecidae in southern Africa. Winter temperatures are not low enough to affect earthworm populations adversely.

In various species of the genus *Microchaetus*, the hindmost part of the slender body is easily autotomized under stress situations. This defence mechanism against predators has not been observed in introduced species.

16.3.4 Relation to soil type

Figure 16.6 illustrates the distribution of five species in an irrigated area based on 32 samples. The relation between the distributions and soil texture supports observations in other areas. Introduced species are

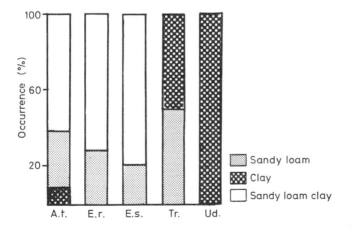

Fig. 16.6 Distribution in relation to soil type in 32 sites. A.t., *Allolobophora trapezoides*; E.r., *Eisenia rosea*; E.s., *Eukerria saltensis*; Tr. *Tritogenia*; Ud., *Udeina*.

found mostly in sandy clay loam, endemics in a variety of soils from heavy clays to sandy loams, and the majority of microchaetines have been collected from sandy loams in Eastern Cape and Transkei. The distribution of earthworms in relation to soil type could reflect food availability and soil moisture conditions.

16.3.5 Vertical migration

Response to temperature and moisture changes is probably the main cause of seasonal movement in the soil profile. Lethal soil temperatures are frequently recorded in soils where earthworms occur. The bigger microchaetine species are capable of avoiding these by deep burrowing and have a clear advantage over the small introduced species which occur in the upper 20 cm. The distribution of the worms is correlated with the depth of the water table in areas such as the Transkei and Eastern Cape. Wherever the water table rises to approximately 50 cm depth, microchaetines can be found regardless of the moisture content of the surface soil (Ljungström and Reinecke, 1969). This explains the occurrence of large endemic species in such semi-desert areas as the Karoo.

The giant *Microchaetus microchaetus*, occurring in the Eastern Cape, migrates to depths of 70 cm and more (Fig. 16.7). The megascolecid species *Metaphire* and *Amynthas* are frequently found at 30 to 40 cm in moist soils along river beds or in irrigated lands. The microchaetines are

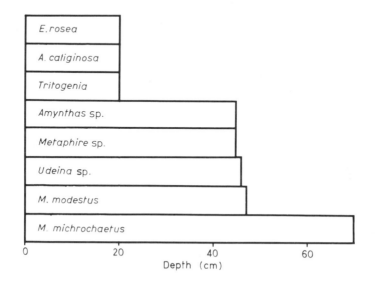

Fig. 16.7 Vertical distribution of southern African earthworms.

affected adversely by ploughing despite the depths they can penetrate. The endemic *Udeina* have also been found deep in the soil but differ from other acanthodrilines which are smaller and occupy the upper horizons of forest soils.

16.3.6 Reproduction and life cycles

Cross-fertilization occurs in many endemic species although partheno-genesis is expected in some *Tritogenia* species. Laboratory studies suggest that endemic species produce fewer cocoons per individual than intro-duced species. The number of hatchlings per cocoon is usually only one or two. The introduced species, notably the lumbricids, seem to be more prolific than the endemic microchaetines but as they occur in gardens and irrigated soils their fecundity may reflect richer food supplies. The introduced lumbricids have shorter maturation times than the endemic microchaetines.

Reproduction in microchaetines occurs throughout the year when conditions are favourable. Microchaetines exhibit a remarkable ability to recolonize areas after prolonged drought, although fewer cocoons are produced than in lumbricids. The observation that the deeper-burrowing endemics produce fewer cocoons than the subsurface lumbricids supports Satchell's (1967) interpretation of Evans and Guild's (1948) data on cocoon production by lumbricids in terms of the hazards to which the cocoons are exposed.

The cocoons of endemics have a flexible incubation period ensuring that hatchlings are not produced during dry periods. The growth period to sexual maturity depends on environmental conditions. For *Microchaetus modestus*, a period of 16 weeks at 20°C has been recorded under laboratory conditions but a much longer period would be expected in natural conditions. Shorter life cycles and high fecundity clearly favour introduced species which in agricultural soils are frequently exposed to physical damage, predation by birds and desiccation of the soil surface.

16.4 CONCLUSIONS

1. There are areas of low-rainfall grassland which have never had earthworms, and in general the distribution of earthworms is closely correlated with precipitation.
2. Where native vegetation is undisturbed the endemic fauna is not being replaced by exotic species.
3. Areas of native vegetation, when cleared lose their endemic earthworms and, if irrigated, may support a population of exotic species.

4. Although the endemic species are better adapted to climatic conditions in South Africa they are not as well able to survive the pressures of cultivation.

16.5 REFERENCES

Evans, A. C. and Guild, W. J. Mc (1948) On the life cycles of some British Lumbricidae. *Ann. Appl. Biol.* **35**, 471–484.

Du Plessis, S. F. (1978) 'n Ondersoek na die verspreiding en taksonomie van die Microchaetinae (Oligochaeta) in Suid-Afrika. Unpublished D.Sc. Thesis. Potchefstroom University for CHE.

Ljungström, P. O. (1972) Introduced earthworms of South Africa. On their taxonomy, distribution, history of introduction and on the extermination of endemic earthworms. *Zool. Jb. Syst.*, **99**, 1–81.

Ljungström, P. O. and Reinecke, A. J. (1969) Variations in the morphology of *Microchaetus modestus* (Microchaetidae, Oligochaeta) from South Africa with notes on its biology. *Zool. Anz.*, **182**, 216–224.

Pickford, G. E. (1926) The Kommetjie Flats. *Blythswood Rev.*, **3**, 57–58.

Pickford, G. E. (1937) *A Monograph of the Acanthodriline Earthworms of South Africa.* Heffer, Cambridge, 612 pp.

Pickford, G. E. (1975) Contributions to a study of South African Microchaetinae (Annelida: Oligochaeta). *Trans. Conn. Acad. Arts Sci.*, **46**, 13–76.

Reinecke, A. J. (1969) *Die Ekologie van Microchaetus modestus (Oligochaeta).* Unpublished M.Sc. Thesis. Potchefstroom University for CHE. 77 pp.

Reinecke, A. J. (1974) The upper lethal temperature of *Eisenia rosea* (Oligochaeta). *Wetensk. Bydraes P.U. Vir C.H.O., B*, **62**, 1–15.

Reinecke, A. J. and Ackerman, D. (1977) New earthworm species (Octochaetinae) from Northeastern Transvaal, South Africa. *Wetensk, Bydraes P.U. Vir C.H.O., B*, **90**, 1–12.

Reinecke, A. J. and Ljungström, P. O. (1969) An ecological study of the earthworms from the bank of the Mooi River in Potchefstroom, South Africa. *Pedobiologia*, **9**, 106–111.

Reinecke, A. J. and Ryke, P. A. J. (1970a) On the water relations of *Microchaetus modestus* Mich. (Microchaetidae: Oligochaeta). *Wetensk. Bydraes P.U. Vir C.H.O., B*, **25**, 1–21.

Reinecke, A. J. and Ryke, P. A. J. (1970b) Casting activity of a South African endemic earthworm, *Microchaetus modestus*. *Wetensk. Bydraes P.U. Vir C.H.O., B*, **16**, 1–14.

Reinecke, A. J. and Ryke, P. A. J. (1972) Research into the ecology of *Microchaetus modestus* (Oligochaeta), a South African endemic species. *Wetensk. Bydraes P.U. Vir C.H.O., B*, **42**, 1–42.

Reinecke, A. J. and Ryke, P. A. J. (1974) The upper lethal temperature of *Microchaetus modestus* (Oligochaeta). *Rev. Ecol. Biol. Sol.*, **11**, 333–351.

Reinecke, A. J. and Visser, F. A. (1980) The influence of agricultural land use practices on the population densities of *Allolobophora trapezoides* and *Eisenia rosea* in southern Africa. In *Soil Biology as Related to Land Use Practices* (ed. D. L. Dindal), *Proc. 7th Int. Soil Zool. Coll., EPA, Washington D.C.*, pp. 310–324.

Reynolds, J. W. and Reinecke, A. J. (1976) A preliminary survey of the earthworms of the Kruger National Park, South Africa. *Wetensk. Bydraes P.U. Vir C.H.O., B*, **62**, 1–19.

Satchell, J. E. (1967) Lumbricidae. In *Soil Biology* (eds. A. Burges and F. Raw), Academic Press, London and New York, 259–322.

Sims, R. W. (1978) Megadrilacea (Oligochaeta). In *Biogeography and Ecology of Southern Africa* (ed. M. J. A. Werger) Dr. W. Junk, The Hague, pp. 663–676.

Visser, F. A. and Reinecke, A. J. (1977) The earthworms of the Mooi River irrigation area in Potchefstroom, South Africa (Oligochaeta:Lumbricidae, Acanthodrilidae, Microchaetidae and Ocnerodrilidae). *P. Cent. Pir. Biol. Exp.,* **9**, 95–108.

Chapter 17

Effects of fire on the nutrient content and microflora of casts of *Pheretima alexandri*

M. V. REDDY

17.1 INTRODUCTION

Pheretima alexandri, one of the largest and dominant species of earthworms at Medziphema, Nagaland, North Eastern India, deposits a thick layer of castings on the soil surface. This report describes (a) the status of certain macronutrients in these casts and in the underlying soils both with and without litter cover, (b) the microflora of the casts, and (c) the effect of natural wild fire on the casts.

17.2 METHODS

Casts of *P. alexandri* and the underlying surface soil (0–5 cm layer) and subsoil (5–10 cm layer) were collected in replicate for analysis from a forested site of 50 m² in which the trees (*Calicarpa arborea* Roxb., *Schima wallichii* Choist and *Dillenia indica* Winu) were the main source of litter. The undergrowth was cleared from an area of 1 m² and the soil was removed to 10 cm depth to eliminate the litter and nutrients originating from it. Five worms of *P. alexandri* were inoculated into the area. The casts ejected by these worms, and samples of soil from the 10–15 cm layer and underlying soil of the 15–20 cm layer were collected for analysis. Casts ejected during the first day were discarded to avoid contamination from the worms' original habitat. Burnt casts from another adjacent area affected by natural wild fire were also collected.

The casts and the underlying soils were air-dried, passed through a 2.0 mm mesh sieve, and the pH was measured. Organic carbon, available phosphorus (P_2O_5) and potassium (K_2O) were estimated by the methods of Wilde *et al.* (1979). The microflora of the casts collected from the area covered with litter and the area affected by fire were examined by the dilution plate technique.

17.3 RESULTS

The pH, percentage of organic carbon and available P_2O_5 and K_2O of the unburnt and burnt casts and the underlying soils with and without litter are summarized in Table 17.1. The casts on the soils with litter were less acid than were the underlying soils, but the burnt casts were more acid than the unburnt ones. In the soil without litter, the underlying soils were acid, but the organic carbon and available K_2O of the casts and their underlying soils with litter were almost double those of the casts and their underlying soils without litter. The concentrations of available P_2O_5 were lower in the soils without litter than in those with litter but those of the casts from the two areas were similar. The organic carbon content was lower in the burnt casts than the unburnt casts whereas available K_2O was higher. Available P_2O_5 in the burnt casts was double the amount in the unburnt casts.

Table 17.1 Some properties of burnt and unburnt casts and their underlying soils with and without litter cover.

	Soil without litter			Soil with litter			
	Sub-soil	Surface soil	Worm-casts	Sub-soil	Surface soil	Worm-casts	Burnt wormcasts
pH	5.40	5.65	7.70	5.75	6.25	6.30	6.15
Organic carbon (%)	1.42	1.52	1.70	2.44	2.66	3.36	2.95
Available P_2O_5 (mg 100 g^{-1})	0.07	0.15	0.24	0.12	0.19	0.22	0.43
Available K_2O (mg 100 g^{-1})	2.58	3.31	4.78	4.42	5.98	7.36	10.18

Table 17.2 Microflora of unburnt and burnt wormcasts.

	Numbers g^{-1} of cast (dry wt. $\pm$ SE)	
	Unburnt casts	Burnt casts
Bacteria ($\times 10^5$)	120.00 ± 10.97	86.00 ± 16.46
Actinomycetes ($\times 10^5$)	46.67 ± 8.11	40.67 ± 5.78
Fungi ($\times 10^3$)	30.33 ± 2.03	9.00 ± 3.51

Table 17.2 summarizes the microfloral properties of the burnt and unburnt casts. Fewer bacteria, actinomycetes and fungi were recorded in the burnt than in the unburnt casts. The unburnt casts were rich in fungi amongst which the following eight genera were identified: *Penicillium, Cladosporium, Fusarium, Cephalosporium, Acremonium, Trichoderma, Aspergillus* and *Humicola*. A *Penicillium* species formed 27% and a *Cladosporium* species 22% of the total fungi. *Humicola* was the least abundant genus forming approximately 2% of the total. A red yeast and sterile mycelia were also recorded in unburnt casts. In the burnt casts only the *Penicillium* and an *Aspergillus* were found. Application of *t* tests to the data in Table 17.2 showed that the numbers of bacteria and actinomycetes in the burnt and unburnt casts were not significantly different but that the numbers of fungi were significantly lower in the burnt casts at the 95% probability level.

17.4 DISCUSSION

The pH of the casts of *P. alexandri* was significantly higher than that of the underlying soils either with or without litter cover. This is consistent with findings of Darwin (1881) and many subsequent workers (Lunt and Jacobson, 1944; Nye; 1955, Sharpley and Syers, 1976; Reddy (in press)) that the wormcasts deposited on acid soils have a higher pH than the underlying soil.

The percentages of organic carbon and available P_2O_5 and K_2O in the soil covered with litter and without litter were significantly higher in casts than in the underlying soils. This is in agreement with the findings of Madge (1969), who recorded three times more organic carbon in casts compared with the neighbouring soils, and with Satchell (1958), who stated that wormcasts are not only richer in organic carbon than the underlying soils but also have higher available phosphorus and exchangeable potassium. Gupta and Sakal (1967) and Reddy (in press) reached similar conclusions. They found little difference between the nutrient contents of casts and the underlying soils but as the soils they studied were cultivated, no litter layer was present. More available phosphorus (Graff, 1970; Vimmerstedt and Finney, 1973) and readily exchangeable phosphorus (Sharpley and Syers, 1976, 1977) have also been recorded in casts than in the underlying soils.

As might be expected, the percentages of organic carbon, available P_2O_5 and K_2O were higher in the casts and the soils covered with litter than in the casts and soils without litter. This is consistent with the findings of Nye (1955) that the proportions of exchangeable cations in casts agree well with the relative amounts in the leaf litter overlying them and confirms the interpretation of Satchell (1958, 1967) that the higher concentrations of mineral nutrients found in wormcasts in the underlying soil are attributable to the worms feeding selectively on organic matter.

Fire decreased the pH and percentage of organic carbon, and increased the available P_2O_5 and K_2O content in the burnt casts. There is no literature on the effects of fire on wormcasts but investigations on the effect of burning on soils of North Eastern India indicate decreases in organic carbon content and increases in available P_2O_5 and K_2O (Anon., 1977). The decrease in organic carbon in burnt casts is associated with a reduction in numbers of micro-organisms but the relationship is not fully understood. The spores of the *Penicillium* and *Aspergillus* found in the burnt casts may be more fire-resistant than the thin-walled spores of some of the other fungi, such as *Acremonium* found in unburnt casts. A reduction in the species diversity of micro-organisms is also reported to occur in soil following fire (Swift, 1976). Widden and Parkinson (1975) also found that *Penicillium* spp. were reduced in burnt soil, partly as a direct result of the heat of the fire and thereafter by inhibition of spore germination by chemical products of burning. *Penicillium* and *Trichoderma* spp. were nevertheless the first colonizers of burnt ground. The microbial succession in burnt earthworm casts doubtless reflects similarly complex environmental changes.

17.5 ACKNOWLEDGEMENTS

I am grateful to my wife who assisted me and to Mr N. K. Verma who analysed the microflora. The casts and soil samples were analysed in the soil testing laboratory, Government of Meghalaya, Shillong.

17.6 REFERENCES

Anonymous (1977) *Annual report*, ICAR Research Complex for North Eastern Hill Region, Shillong. 240 pp.

Darwin, C. (1881) *The Formation of Vegetable Mould through the Action of Worms, with Observations on their Habits.* John Murray, London, 326 pp.

Graff, O. (1970) Phosphorus contents of earthworm casts. *LandbForsch-Völkenrode*, **20**, 33–36.

Gupta, M. L. and Sakal, R. (1967) Role of earthworms in the availability of nutrients in garden and cultivated soils. *J. Indian Soc. Soil Sci.*, **15**, 149–151.

Lunt, H. A. and Jacobson, M. G. M. (1944) The chemical composition of earthworm casts. *Soil Sci.*, **58**, 367–375.

Madge, D. S. (1969) Field and laboratory studies on the activities of two species of tropical earthworms. *Pedobiologia*, **9**, 188–214.

Nye, P. H. (1955) Some soil forming processes in the humid tropics. IV. The action of soil fauna. *J. Soil Sci.*, **6**, 73–83.

Reddy, M. V. (in press) Some chemical properties of the casts of the earthworm, *Drawida willsi* Michaelsen (Moniligastridae). *Indian Biologist*.

Satchell, J. E. (1958) Earthworm biology and soil fertility. *Soils Fertil.*, **21**, 209–219.

Satchell, J. E. (1967) Lumbricidae. In *Soil Biology* (eds. A. Burges and F. Raw), Academic Press, London and New York, pp. 259–322.

Sharpley, A. N. and Syers, J. K. (1976) Potential role of earthworm casts for the phosphorus enrichment of run-off water. *Soil Biol. Biochem.*, **8**, 341–346.

Sharpley, A. N. and Syers, J. K. (1977) Seasonal variation in casting activity and in the amounts and release to solution of phosphorus forms in earthworm casts. *Soil Biol. Biochem.*, **9**, 227–231.

Swift, M. J. (1976) Species diversity and the structure of microbial communities in terrestrial habitats. In *The Role of Terrestrial and Aquatic Organisms in Decomposition Processes* (eds. J. M. Anderson and A. Macfadyen), Blackwell Scientific Publications, Oxford, pp. 185–222.

Vimmerstedt, J. P. and Finney, J. H. (1973) Impact of earthworm introduction on litter burial and nutrient distribution in Ohio strip mine spoil banks. *Proc. Soil Sci. Soc. Am.*, **37**, 388–391.

Widden, P. and Parkinson, D. (1975) The effects of a forest fire on soil microfungi. *Soil Biol. Biochem.*, **7**, 125–138.

Wilde, S. A., Corey, R. B., Iyer, J. G. and Voigt, G. K. (1979) *Soil and Plant Analysis for Tree Culture*, Oxford and IBH Publishing Co., New Delhi. 224 pp.

Chapter 18

Earthworms and land reclamation

J. P. CURRY and D. C. F. COTTON

18.1 INTRODUCTION

The objectives of land reclamation range from restoration of derelict industrial wasteland for amenity purposes to improvement of infertile soils for agricultural production. The success of conventional reclamation techniques may be limited by poor soil structure and low inherent soil fertility. Evidence linking earthworm activity with good soil structure and fertility has stimulated research into earthworm ecology in reclaimed soils. This chapter reviews published work on this topic and presents preliminary data on the effects of introducing earthworms to cut-over peat sites in Ireland.

18.2 EARTHWORMS IN IMPROVED MINERAL SOILS

The impact of agricultural development on the indigenous flora and fauna is illustrated dramatically in New Zealand where 50% of the total land area has been brought into agricultural production in the past 100 years (Brougham, 1979). Most of this is sown pasture in which the commonest and most widespread earthworms are European lumbricids. Of the fifteen species known from New Zealand, the topsoil-mixing species *Aporrectodea caliginosa* (Sav.) and the organic matter-feeding species *Lumbricus rubellus* Hoff. are widespread and abundant (Martin, 1977). There are still considerable areas from which lumbricid earthworms are absent (Stockdill, 1966), including extensive areas of improved upland. Improved production has been linked with the appearance of earthworms in such pasture. Hamblyn and Dingwall (1945) and Nielson (1951) estimated the rate of spread of *A. caliginosa* introduced by transplanting turves as 10–13 m y^{-1}. Earthworm establishment was accompanied by an increase in the proportion of higher fertility plants such as ryegrass and clover and increased stock-carrying capacity.

Stockdill (1959, 1966) likewise introduced *A. caliginosa* into hill pastures in New Zealand. It failed to become established in unimproved tussock grassland of low natural fertility and pH 4.8–5.2, but readily became established in improved pasture which had been cultivated, limed, fertilized and resown six years previously. Following earthworm establishment, the organic mat disappeared and compacted soil of poor structure was transformed into deep, friable top soil (see Chapter 7).

Barley and Kleinig (1964) described the successful introduction of *A. caliginosa* and the megascolecid species *Microscolex dubius* (Fletcher) into sown, irrigated pasture on sandy loam soil in N.S.W., Australia. Within eight years populations of $300 \, \text{m}^{-2}$ were present. There was a noticeable improvement in soil structure and the surface mat of litter and dung became incorporated into the soil. Earthworms failed to become established in similarly treated clay soils. Subsequently, Noble *et al.* (1970) demonstrated an inverse relationship between mat development and earthworm populations in these sites. Soil bulk density was reduced and N content and C/N ratios were higher in inoculated pastures. Another example of successful earthworm introductions into newly reclaimed irrigated land in Uzbekistan is provided by Ghilarov and Mamajev (1966), who considered *Aporrectodea trapezoides* (Duges) and *A. prashadi* to be the most suitable species for introduction to advance soil formation.

Atlavinyte (1976) described the changes that occurred in the earthworm community of a peaty gley sandy loam soil in Lithuania following reclamation and cultivation. Shrub elimination and cultivation were accompanied by a decrease in worm population. Drainage eliminated the wet soil species *Eiseniella tetraedra* (Sav.), *Dendrobaena octaedra* (Sav.) and *Octolasion lacteum* (Oerley). *A. caliginosa* became the dominant species in the arable reclaimed lands, accompanied by *Aporrectodea rosea* (Sav.), *L. rubellus*, *Allolobophora chlorotica* (Sav.) and *Lumbricus terrestris* L. which came in 3–4 years after reclamation.

Some experiments on improving impoverished soils by stimulating earthworm populations have been reported. Bosse (1967) reported large increases in indigenous populations of impoverished vineyard soils after the addition of organic matter. Soil structure improved within five years. Huhta (1979) tested the influence of added deciduous litter (birch and alder) and/or lime on earthworm populations in a dry heath coniferous forest soil in Finland. The indigenous earthworm fauna consisted of small populations of the surface-living species *D. octaedra*, *Dendrodrilus rubidus* (Sav.) and *Lumbricus rubellus*. In lime-treated plots, introduced *A. caliginosa* became established and indigenous species, particularly *L. rubellus* and *D. rubidus*, increased. However, the long-term benefits of these changes have yet to be demonstrated. Langmaid (1964) reported marked improvement in afforested virgin podzols in New Brunswick after earthworm invasion and subsequent incorporation of surface organic matter and soil mixing. *Allolobophora tuberculata* (Eisen), *Lumbricus festivus* (Sav.) and *L. terrestris* had been deliberately introduced into some sites, but in others they had probably become established accidentally during logging operations. On the other hand, Satchell (1980a, b) found that periodically water-logged heather moor in Yorkshire which had been afforested with birch supported only an insignificant population of a surface-living species and was unlikely ever to support deep-burrowing species without drastic reclamation measures.

18.3 EARTHWORMS IN POLDER SOILS

Van Rhee (1969a) pointed out that reclaimed polder soils in the Netherlands take 50–100 years to mature and that the process could be considerably advanced by earthworm activity. Natural invasion of polder soils may be rapid (Meijer, 1972) and significant populations can be expected to develop without intervention. In orchards on the site of a former lake, van Rhee (1969a) reported densities exceeding $200\,m^{-2}$ 26 years after drainage and reclamation, with *L. rubellus* accounting for 55 %, *A. caliginosa* 18 %, *A. rosea* 21 % and *A. chlorotica* 6 %. However, the rate of natural spread is only about $4–6\,m\,y^{-1}$ (van Rhee, 1969b) and should be accelerated by artificial introduction.

While populations may recover from periodic flooding by sea water within a few years (Piearce, 1982), prolonged flooding completely eliminates earthworms. Van Rhee (1963) described several attempts to re-establish earthworm populations by inoculation in a polder where earthworm populations had been killed by inundation by the sea. *Aporrectodea longa* (Ude) became established and, within a year, surface accumulation of litter decreased. Populations of $280\,m^{-2}$ (72.5 % *A. caliginosa*, 18.5 % *A. longa* and 8.5 % *L. rubellus*) became established following introductions in a seven year-old pasture on newly reclaimed polder. An orchard soil inoculated 13 years previously had 140–500 worms m^{-2}, entirely *A. caliginosa*. In both cases mull soils had developed compared with mor in adjacent areas without worms.

Van Rhee (1969a, b, 1977) inoculated grassland with *A. caliginosa* and *A. chlorotica*, and orchard sites with *A. caliginosa* and *L. terrestris*. Population densities reached $750\,m^{-2}$ in the grass areas, with *A. caliginosa* multiplying twice as fast and spreading more rapidly than *A. chlorotica*. In orchard sites multiplication was much slower, possibly because of pesticides, and after eight years the mean density was $140–250\,m^{-2}$ (80 % *A. caliginosa* and 20 % *L. terrestris*). Increased litter incorporation and soil aggregation and improved aeration and water relationships were observed. Van Rhee concluded that *A. caliginosa* is a particularly successful early colonizer and may play an important role in the early stages of soil formation in polder soils. The beneficial effects on soil structure were confirmed by the soil morphological studies of Rogaar and Boswinkel (1978). Van Rhee did not detect any influence of earthworm activity on grass or fruit yield, but Hoogerkamp *et al.* (Chapter 8) report increased grass yields from polder soils where worms are present.

18.4 EARTHWORMS IN RECLAIMED MINERAL WASTES

Total mineral wastes generated in the European Economic Community exceed 1800 million t y^{-1} and this is expanding at 2–3% per year (Klein, 1980). Some 70–80% of all wastes are currently disposed of by tipping and, despite increasing interest in waste recycling and waste incineration for energy production, landfilling is likely to remain a major method of disposal in many countries. The rehabilitation of the resulting dumps is a problem of considerable biological interest. Acidity tends to be the principal factor limiting biological activity in colliery wastes, but this is not always so and pH values may range from 1.8 to 8.5 (Gemmell, 1977). Intense acidity is associated with weathering of iron pyrites. Metal mine spoils are usually not highly pyritic and are often less acid than colliery wastes; the principal limiting factor in this case may be metal toxicity. Factors likely to inhibit earthworm establishment in the early stages of reclamation include unfavourable moisture conditions, excessive fluctuation of surface temperature and lack of suitable food.

Invertebrate colonization of land reclaimed from open-cast coal mining has been studied by Dunger (1969a, b) and Hutson (1980a, b). Microarthropods become established rapidly once the vegetation cover is developed and in the initial stages these are the dominant invertebrate decomposers. Dunger (1969a) reported earthworm colonization after 5 years in reclaimed brown coal–lignite opencast workings in East Germany planted with *Alnus glutinosa*, but it took 10 years for earthworms to become dominant and on more acid sites they were still at the edges after 15 years. Litter incorporation was strongly influenced by faunal development (Dunger, 1969b); in the initial stages when microarthropods were dominant less than 5% of the available litter was actually decomposed but, after 7 years when lumbricids had become established, total litter breakdown had risen to 27%, and the soil had changed from moder to mull. After 10 years earthworms removed 70% of the litter fall and accounted for 94% of invertebrate respiration.

Standen *et al.* (1982) found that earthworm numbers were still low

($<33\,\mathrm{m}^{-2}$) 11 years after restoration of opencast coal mining sites for agriculture in Co. Durham. They report populations of the surface-living *L. rubellus* up to $162\,\mathrm{m}^{-2}$ from 14-year-old colliery spoil heaps naturally colonized by scrub woodland. Significant numbers of *L. terrestris* ($11–22\,\mathrm{m}^{-2}$) were present in a few sites which were more than 50 years old. Site age, shade and moisture content were found to be significantly correlated with earthworm numbers or biomass.

While a period of maturation may be required before reclaimed waste becomes habitable by earthworms, early colonizers may be found soon after the establishment of the vegetation. We found no earthworms in 100-year-old acidic and barren copper mine spoil at Avoca, Co. Wicklow, but in a small area which had been limed and sown with grass and clover four years previously, a low density ($12\,\mathrm{m}^{-2}$) population of *Lumbricus friendi* Cog. had become established. Topsoiling greatly facilitates earthworm establishment: in a base metal spoil heap at Tynagh, Co. Galway which had been topsoiled and sown with grass and clover six years previously, populations of the soil-dwelling *A. caliginosa* and *A. rosea* were present. Surface application of organic matter to reclaimed sites favours earthworms by increasing food supply and by stabilizing temperature and moisture conditions. Dunger (1969a) successfully introduced *A. caliginosa* into reclaimed coal dumps after mulching the soil surface with leaf compost, and Hutson (1972) reported earthworm colonization of the edges of newly reclaimed coal mining waste where grass clippings were left on the surface. Notable success in introducing *L. terrestris* into reclaimed coal mine spoil has been reported by Vimmerstedt and Finney (1973) and by Hamilton and Vimmerstedt (1980). Rates of population increase and spread, and effects on soil properties are outlined by Vimmerstedt (Chapter 19).

Satchell and Stone (1977) reported that *L. rubellus*, *L. castaneus* and *Dendrobaena* spp. were early colonizers of reclaimed pulverized fuel ash, a waste product of coal-fired electrical power stations. Small populations of *A. caliginosa*, *A. rosea* and *L. terrestris* were found around the perimeters of older ash fields but their viability was doubtful. Where topsoil was added, normal agricultural earthworm populations developed.

Brockmann *et al.* (1980) studied invertebrate colonization and community development in municipal refuse tips covered by soil. They reported an immigration rate for *L. rubellus* and *D. octaedra* of $4\,\mathrm{m}\,\mathrm{ha}^{-1}\mathrm{y}^{-1}$.

18.5 EARTHWORMS IN RECLAIMED PEAT

Earthworms in virgin peat are scarce, with populations ranging from $<0.1\,\mathrm{m}^{-2}$ (Svendsen, 1957a) to $12\,\mathrm{m}^{-2}$ (Guild, 1948). The burrowing species of mull soils are absent and surface-dwelling species such as *L. rubellus*, *D. rubidus*, *D. octaedra* and *Lumbricus eiseni* Levinsen predominate (Guild, 1948; Boyd, 1956, 1957; Svendsen, 1957a, b). Factors inimical to earthworm activity in peat include low pH, high moisture content, and poor-quality litter of high C/nutrient ratio, low N content and low palatability attributable to phenolic compounds (Swift *et al.*, 1979).

Reclamation of peat enhances biological activity in the surface horizons. Kipenvarlitz (1953) (cited by Skoropanov, 1961) reported that the numbers of earthworms in cultivated peat soils approach those in highly fertile soils, with maximum numbers occurring under grass and legumes. Guild (1948) found 50–100 earthworms m^{-2} in improved hill pasture in Scotland compared with only 12 m^{-2} in unimproved, acid peaty soil. The numbers of species increased from 5–6 in unimproved to 9–10 in improved pastures, with the establishment of typical pasture species such as *A. caliginosa*, *A. chlorotica* and, in smaller numbers, *L. terrestris*, *A. longa* and *A. rosea*. Persson and Lohm (1977) reported five lumbricid species, *A. caliginosa*, *A. rosea*, *D. octaedra*, *D. rubidus* and *L. rubellus*, from a fen peat in Sweden which had been reclaimed for more than 60

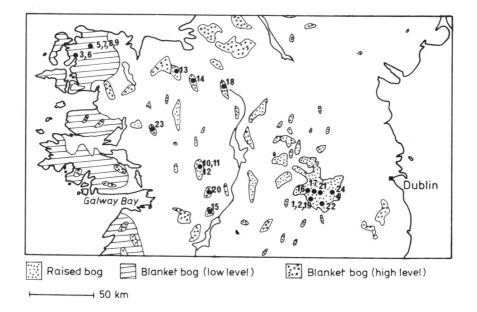

Fig. 18.1 Location of sampling sites. Sites 1, 2 and 4 were on shallow cutaway, the others were on deeper peat. All sites were under grass.

years. The mean density was $133 \, \mathrm{m}^{-2}$, with *A. rosea* contributing 94.5 %
of the number.

Currently, large-scale industrial extraction of peat is in progress in
Ireland. The residual 50 cm peat layer, known as 'cutaway', is reclaimed
mainly for grass production, but future crops may include fast growing
trees for biomass production. In October 1978, 24 reclaimed sites of
various types (Fig. 18.1) were surveyed to assess the rate and extent of
natural colonization by earthworms.

The sites can be divided into young sites reclaimed for less than 25 years
with low population densities ($< 25 \, \mathrm{m}^{-2}$) and mature sites reclaimed for
25 years or longer with populations of $100–200 \, \mathrm{m}^{-2}$ (Table 18.2). Sixteen
species were recorded which can be divided into three groups (Table
18.1). The first, characteristic of young sites, is of relict, pre-reclamation
species typical of acid bog peat and raw humus horizons in other habitats.
D. octaedra, *D. rubidus* and *L. eiseni* are typical. A wider survey might have
revealed more *E. tetraedra* in wet sites. *L. castaneus* and *L. rubellus* are also
widespread in moorland habitats as highly mobile coprophages. At the
opposite extreme in the mature sites where the organic horizons and the
soil pH have been radically changed, typical soil dwelling forms are found
such as *A. rosea*, *A. longa* and *O. cyaneum* which are intolerant of acid
conditions and *L. terrestris* which, though more pH tolerant, requires a
considerable depth of aerobic soil and is slow to colonize new habitats. A
third vicariant group comprises species in both young and old sites and
this includes *A. caliginosa* and *A. chlorotica*, both of which have been
reported elsewhere as early colonists. Our results confirm in particular the
colonizing capacity of *A. caliginosa* reported as a pioneer in such diverse
habitats as irrigated desert (El-Duweini and Ghabbour, 1965) restored
coal mining wastes (Dunger, 1969a, b) and reclaimed polders (van Rhee,
1977). Its evident capacity for rapid establishment no doubt contributes
to its status as the world's most abundant lumbricid.

The data indicate that all the earthworm species normally found in
fertile mineral soils can become established in reclaimed peat, but sizeable
populations take at least 25 years to develop, possibly less under optimum
conditions of drainage and management. Abundance in mature sites is low
compared with that in grassland on productive mineral soils (Curry, 1976;
Cotton and Curry, 1980a, b). This may be due to inadequate drainage
(Baker, 1983) and low fertility.

Earthworms were introduced into a six year-old pasture on cutover peat
at Clonsast, Co. Offaly, where about 40 cm of peat remained; this was
predominantly of mixed forest origin and overlay a well-developed relict
soil derived from limestone till. Reclamation had raised the pH to 7.
Preliminary sampling revealed *A. chlorotica* at low density ($6 \, \mathrm{m}^{-2}$) and a
few specimens of *A. tuberculata*, *D. octaedra*, *L. castaneus* and *L. rubellus*

Table 18.1 Occurrence of earthworm species in grassland on reclaimed peat soils.

	Young sites												Mature sites											
Site	1	2	3	4	5	6	7	8	9	10	11	12	13	14	15	16	17	18	19	20	21	22	23	24
Years since restoration	1–2	5–6	7–8	10	12	14–16	17–20	17–20	17–20	24	24	24	← <25									>25 →		
D. octaedra (Sav.)	X																							
D. rubidus (Sav.)	X		X					X		X	X	X											X	
E. tetraedra (Sav.)								X	X	X	X													
L. castaneus (Sav.)						X			X															
L. eiseni (Lev.)																								
L. rubellus (Hoff.)	X						X	X	X	X	X	X	X											
L. festivus (Sav.)			X	X		X	X	X										X		X	X		X	X
O. tyrtaeum (Sav.)			X	X					X		X	X							X	X	X		X	X
A. chlorotica (Sav.)	X	X	X	X		X		X	X					X	X	X	X	X	X	X	X	X	X	X
A. caliginosa (Sav.)			X	X		X	X		X	X		X	X	X	X	X	X	X	X	X	X	X	X	X
A. tuberculata (Eisen)									X								X					X		
A. rosea (Sav.)												X	X	X	X	X	X	X	X	X	X	X		X
A. longa (Ude)															X								X	
L. terrestris L.													X	X		X		X	X	X	X	X		X
O. cyaneum (Sav.)																		X				X		
S. mammalis (Sav.)																								X

Table 18.2 Abundance of earthworms in eight mature reclaimed peat soils.

	Population density (no. m^{-2})		Proportion of adult population (%)	
	Mean	Max.	Mean	Max.
A. caliginosa	41	105	43.6	68
A. chlorotica	16	39	20.0	60
A. rosea	15	30	17.7	34
A. tuberculata	4	26	3.7	28
L. festivus	3	10	2.6	8
L. terrestris	2	6	2.6	9
O. cyaneum	2	5	2.7	13
L. rubellus	2	5	2.8	13
O. tyrtaeum	2	14	2.2	15
D. rubidus	1	5	1.0	5
Other spp.	1	5	1.0	5
Total no. m^{-2}	145	220		

were found in mineral spoil on the bank of an adjacent ditch. Some 500 earthworms were introduced at each of two points 100 m apart in April 1978. They consisted mainly of *A. chlorotica*, *A. caliginosa* and *L. terrestris* with some *A. rosea*, *A. longa* and *O. cyaneum*. *A. chlorotica* became established (Table 18.3). Other species were apparently less successful but may have survived as cocoons (Table 18.4). Subsequently

Table 18.3 Population densities (no. m^{-2} ± SE) in sites with and without introduced earthworms.

	Earthworms introduced		Control area‡
	After 1 year*	After 2 years†	
A. chlorotica	113 ± 14	72 ± 15	21 ± 12
A. longa	12 ± 7	1	
A. sp. juveniles		1	
L. festivus		16 ± 8	
Lumbricus juveniles	11 ± 9	27	
Total	136 ± 6	117 ± 17	21 ± 12
Cocoons	161 ± 71	223 ± 52	26 ± 10

Samples were hand-sorted:
* $n =$ 3(50 × 50 × 20 cm)
† $n =$ 16(22 × 22 × 10 cm) at 50 cm intervals along zones intersecting at inoculation point.
‡ $n =$ 4(22 × 22 × 10 cm) 50 m from inoculation point.

Table 18.4 Distribution of earthworms 2 years after introduction.

Distance from point of introduction (cm)	50	100	150	200
Number of worms per sample	5	4	5	8
Number of cocoons per sample	5	5	16	17

Means not underscored by the same line differ significantly ($P < 0.05$) based on ANOVA of $\log_{10}(X + 1)$ data.

Table 18.5 Earthworms recovered from two cages 6 months after introduction, and from an equal area ($0.25 \, m^2$) of adjacent soil.

	Cages		Adjacent soil
A. chlorotica	12	16	3
A. caliginosa	2	1	
A. juveniles	5	2	
D. rubidus	9	4	
D. juveniles	5	4	
L. castaneus	3	4	
L. rubellus	10	8	
L. terrestris	0	1	
L. juveniles	14	15	
Totals	60	55	3
A. chlorotica/ A. caliginosa cocoons	260	199	11
L. rubellus cocoons	80	70	2

introduced populations were confined in open-topped cubic cages with 0.5 m sides of nylon mesh (0.5 mm aperture). Twenty-six worms collected from adjacent peaty grassland were added to two cages buried to a depth of 25 cm. Six months later the number and species diversity of the earthworms in the cages was similar to that of mature sites (Table 18.5).

18.6 CONCLUSIONS

A. caliginosa, A. chlorotica and *L. rubellus* are the most successful early colonizers in many situations. *L. rubellus* has a high reproductive rate (Evans and Guild, 1948), disperses rapidly, is an important litter and dung consumer (Keogh, 1979; Martin and Charles, 1979), and is the main surface-casting species in some situations (Syers *et al.*, 1979). The deeper-working species such as *L. terrestris* are less successful colonizers of new sites. This may be due to high mortality caused by adverse soil conditions

(Luff and Hutson, 1977) coupled with low reproductive potential (Evans and Guild, 1948). If soil physical factors are limiting, introduction of the deep-working species should be deferred until the site has matured. If food is limiting, addition of organic material such as municipal sludge or animal wastes may allow earlier establishment.

Large-scale introduction programmes will require large numbers of earthworms. Stockdill's (1982) sod machine has proved useful in New Zealand hill pasture and may be suitable for other situations. Its value is restricted where deep-burrowing species are required and where sods on the surface may impede grass harvesting. Currently, most worms for introduction are hand-picked. This is a laborious and expensive method (cf. Tomlin, Chapter 29), and is only suitable for small-scale applications. Mass rearing of soil-dwelling species has so far not proved successful and the scope of introduction projects seems likely to depend on developments in vermiculture technology.

18.7 ACKNOWLEDGEMENTS

We thank Bord na Mona and An Foras Taluntais for allowing us to use their experimental sites at Clonsast. This work was financed in part by a research grant from the National Board for Science and Technology.

18.8 REFERENCES

Atlavinyte, O. (1976) The effect of land reclamation and field management on the change in specific composition and densities of Lumbricidae. *Pol. Ecol. Stud.*, **2**, 147–152.

Baker, G. H. (1983) The distribution, abundance and species associations of earthworms (Lumbricidae) in a reclaimed peat soil in Ireland. *Holarct. Ecol.*, (in press).

Barley, K. P. and Kleinig, C. R. (1964) The occupation of newly irrigated lands by earthworms. *Austr. J. Sci.*, **26**, 290–291.

Bosse, J. (1967) Restoration of biologically impoverished Weinberg soils, an example of earthworm colonization. (German, English summary). In *Progress in Soil Biology* (eds. O. Graff and J. E. Satchell), North Holland Publishing Co., Amsterdam, pp. 299–309.

Boyd, J. M. (1956) The Lumbricidae in the Hebrides. II Geographical distribution. *Scott. Nat.*, **68**, 165–172.

Boyd, J. M. (1957) Ecological distribution of the Lumbricidae in the Hebrides. *Proc. R. Soc. Edinburgh*, **66**, 311–338.

Brockmann, W., Koehler, H. and Schriefer, T. (1980) Recultivation of refuse tips: soil ecological studies. In *Soil Biology as Related to Land Use Practices* (ed. D. L. Dindal), *Proc. VIIth Int. Soil Zool. Coll. EPA, Washington D.C.*, pp. 161–168.

Brougham, R. W. (1979) The grasslands of New Zealand: an overview. In *Proc. 2nd Austr. Conf. Grassld. Invertebrate Ecol.* (eds. T. K. Crosbie and R. P. Pottinger), Government Printer, Wellington, pp. 16–23.

Cotton, D. C. F. and Curry, J. P. (1980a) The effects of cattle and pig slurry fertilizers on earthworms (Oligochaeta, Lumbricidae) in grassland managed for silage production. *Pedobiologia*, **20**, 181–188.

Cotton, D. C. F. and Curry, J. P. (1980b) The response of earthworm populations (Oligochaeta, Lumbricidae) to high applications of pig slurry. *Pedobiologia*, **20**, 189–196.

Curry, J. P. (1976) Some effects of animal manures on earthworms in grassland. *Pedobiologia*, **16**, 425–438.

Dunger, W. (1969a) Fragen der naturlichen und experimentellen Besiedlung kulturfeindlicher Boden durch Lumbriciden. *Pedobiologia*, **9**, 146–151.

Dunger, W. (1969b) Über den Anteil der Arthropoden an der Umsetzung des Bestandesabfalles in Anfangs Bodenbildungen. *Pedobiologia*, **9**, 366–371.

El-Duweini, A. K. and Ghabbour, S. I. (1965) Population density and biomass of earthworms in different types of Egyptian soils. *J. Appl. Ecol.*, **2**, 271–287.

Evans, A. and Guild, W. J. McL. (1948) Studies on the relationships between earthworms and soil fertility. IV. On the life cycles of some British Lumbricidae. *Ann. Appl. Biol.*, **35**, 471–484.

Gemmell, R. P. (1977) *Colonization of Industrial Wasteland. Inst. Biol., Stud. Biol.* No. 80. Edward Arnold, London. 72 pp.

Ghilarov, M. S. and Mamajev, B. M. (1966) Über die Ansiedlung von Regenwürmern in den artesisch bewässerten Oasen der Wüste Kysyl-Kum. *Pedobiologia*, **6**, 197–218.

Guild, W. J. McL. (1948) Studies on the relationship between earthworms and soil fertility. III. The effect of soil type on the structure of earthworm populations. *Ann. Appl. Biol.*, **35**, 181–192.

Hamblyn, C. J. and Dingwall, A. R. (1945) Earthworms. *N. Z. J. Agric.*, **71**, 55–58.

Hamilton, W. E. and Vimmerstedt, J. P. (1980) Earthworms on forested spoil banks. In *Soil Biology as Related to Land Use Practices* (ed. D. L. Dindal), *Proc. VIIth Int. Soil Zool. Coll., EPA, Washington D.C.*, pp. 409–417.

Huhta, V. (1979) Effects of liming and deciduous litter on earthworm (Lumbricidae) populations of a spruce forest, with an inoculation experiment on *Allolobophora caliginosa*. *Pedobiologia*, **19**, 340–345.

Hutson, B. R. (1972) The invertebrate fauna of a reclaimed pit heap. *Landscape Reclamation*, Vol. 2, IPC Press, Guildford, pp. 64–69.

Hutson, B. R. (1980a) Colonization of industrial reclamation sites by Acari, Collembola and other invertebrates. *J. Appl. Ecol.*, **17**, 255–275.

Hutson, B. R. (1980b) The influence on soil development of the invertebrate fauna colonizing industrial reclamation sites. *J. Appl. Ecol.*, **17**, 277–286.

Keogh, R. G. (1979) Lumbricid earthworm activities and nutrient cycling in pasture ecosystems. In *Proc. 2nd Austr. Conf. Grassld. Invertebrate Ecol.* (eds. T. K. Crosby and R. P. Pottinger), Government Printer, Wellington, pp. 49–51.

Klein, L. (1980) Waste management policy in the European Communities. In *Proceedings of Seminar: Today's and Tomorrow's Wastes* (ed. J. Ryan), National Board for Science and Technology, Dublin, pp. 14–20.

Langmaid, K. K. (1964) Some effects of earthworm invasion in virgin podzols. *Can. J. Soil Sci.*, **44**, 34–37.

Luff, M. L. and Hutson, B. R. (1977) Soil faunal populations. In: *Landscape Reclamation Practice* (ed. B. Hackett), IPC Science and Technology Press, Guildford, pp. 125–147.

Martin, N. A. (1977) Guide to the lumbricid earthworms of New Zealand pastures. *N. Z. J. Exp. Agric.*, **5**, 301–309.

Martin, N. A. and Charles, J. C. (1979) Lumbricid earthworms and cattle dung in New Zealand pastures. In *Proc. 2nd Austr. Conf. Grassld. Invertebrate Ecol.* (eds. T. K. Crosby and R. P. Pottinger), Government Printers, Wellington, pp. 52–54.

Meijer, J. (1972) An isolated earthworm population in the recently reclaimed Lauwerszeepolder. *Pedobiologia*, **12**, 409–411.

Nielson, R. L. (1951) Earthworms and soil fertility. *Proc. N. Z. Grassld. Assoc.*, **13**, 158–167.

Noble, J. C., Gordon, W. T. and Kleinig, C. R. (1970) The influence of earthworms on the development of mats of organic matter under irrigated pastures in Southern Australia. In *Plant Nutrition and Soil Fertility, Proc. XIth Int. Grassld. Congr.*, Queensland, Australia. Queensland Univ. Press, pp. 465–468.

Persson, T. and Lohm, U. (1977) Energetic significance of the annelids and arthropods in a Swedish grassland soil. *Ecol. Bull. (Stockholm)*, **23**, 211 pp.

Piearce, T. G. (1982) Recovery of earthworm populations following saltwater flooding. *Pedobiologia*, **24**, 91–100.

Rogaar, H. and Boswinkel, J. A. (1978) Some soil morphological effects of earthworm activity; field data and X-ray radiography. *Neth. J. Agric. Sci.*, **26**, 145–160.

Satchell, J. E. (1980a) Potential of the Silpho Moor experimental birch plots as a habitat for *Lumbricus terrestris*. *Soil Biol. Biochem.*, **12**, 317–323.

Satchell, J. E. (1980b) Earthworm populations of experimental birch plots on a *Calluna* podzol. *Soil Biol. Biochem.*, **12**, 311–316.

Satchell, J. E. and Stone, D. A. (1977) Colonization of pulverized fuel ash sites by earthworms. *P. Cent. Pir. Biol. Exp.*, **9**, 59–74.

Skoropanov, S. G. (1961) *Reclamation and Cultivation of Peat Bog Soils*. Israel Programs for Scientific Translations, Jerusalem, 1968.

Standen, V., Stead, G. B. and Dunning, A. (1982) Lumbricid populations in open cast reclamation sites and colliery spoil heaps in County Durham, U.K. *Pedobiologia*, **24**, 57–64.

Stockdill, S. M. J. (1959) Earthworms improve pasture growth. *N. Z. J. Agric.*, **98**, 227–233.

Stockdill, S. M. J. (1966) The effect of earthworms on pastures. *Proc. N. Z. Ecol. Soc.*, **13**, 68–75.

Stockdill, S. M. J. (1982) Effects of introduced earthworms on the productivity of New Zealand pastures. *Pedobiologia*, **24**, 29–35.

Svendsen, J. A. (1957a) The distribution of Lumbricidae in an area of Pennine moorland (Moor House Nature Reserve). *J. Anim. Ecol.*, **26**, 411–421.

Svendsen, J. A. (1957b) The behaviour of lumbricids under moorland conditions. *J. Anim. Ecol.*, **26**, 423–439.

Swift, M. J., Heal, O. W. and Anderson, J. M. (1979) *Decomposition in Terrestrial Ecosystems*. Blackwell, Oxford. 372 pp.

Syers, J. K., Sharpley, A. N. and Keeney, D. R. (1979) Cycling of nitrogen by surface casting earthworms in a pasture ecosystem. *Soil Biol. Biochem.*, **11**, 181–185.

van Rhee, J. A. (1963) Earthworm activities and the breakdown of organic matter in agricultural soils. In *Soil Organisms* (eds. J. Doeksen and J. van der Drift), North-Holland Publishing Co., Amsterdam, pp. 54–59.

van Rhee, J. A. (1969a) Inoculation of earthworms in a newly drained polder. *Pedobiologia*, **9**, 128–132.

van Rhee, J. A. (1969b) Development of earthworm populations in polder soils. *Pedobiologia*, **9**, 133–140.

van Rhee, J. A. (1977) A study of the effect of earthworms on orchard productivity. *Pedobiologia*, **17**, 107–114.

Vimmerstedt, J. P. and Finney, J. H. (1973) Impact of earthworm introduction on litter burial and nutrient distribution in Ohio strip-mine spoil banks. *Proc. Soil Sci. Soc. Am.*, **37**, 388–391.

Chapter 19

Earthworm ecology in reclaimed opencast coal mining sites in Ohio

J. P. VIMMERSTEDT

19.1 INTRODUCTION

Surface mining for coal disturbs the landscape to an extent approaching glacial action. In Ohio, 4000 ha are affected each year by surface mining. New soil develops on the coal spoils, influenced by soil-forming factors – climate, time, topography, parent material and organisms (Jenny, 1980). Of these factors, topography, parent material and organisms may be controlled during mining and subsequent reclamation.

In the eastern United States, surface-mined areas currently are returned to approximately their original contours, topsoil is stockpiled during mining and graded back on the spoil surface to a depth of 15–20 cm, and grasses, forage legumes, and trees are established using nutrient amendments and other cultural practices required to achieve complete plant cover. The only soil organisms purposely added to reclamation sites are *Rhizobia* inoculants for legumes and, on a limited basis, specific mycorrhizae-forming fungi for some tree species (Marx, 1975; Marx and Artman, 1979). Little, if any, effort is devoted to re-establishing populations of earthworms or other soil fauna, although research results and cultural practices elsewhere indicate significant beneficial effects of earthworm activity on crop yields (Edwards, 1981).

A comprehensive research program on the use of earthworms in mined-area reclamation would include:

1. Studies of earthworm populations, as they exist without deliberate management, in stockpiled topsoil and on reclaimed areas of differing plant cover, age and parent material.
2. Population dynamics of selected species added to revegetated spoils differing in plant cover and parent material.

3. Methods of inoculation – as cocoons, juveniles, adults or as a micro-community of earthworms and associated organisms in a soil block from a populated area.
4. Quantification of effects of specific earthworm species and combinations of species on soil properties and plant growth, especially of effects on soil compaction, root development, water infiltration and organic matter distribution through the soil profile.

In this chapter I report results of research with earthworms on Ohio coal spoils, begun in 1967 and continued to the present. Topics studied have been: alder (*Alnus glutinosa*) and black locust (*Robinia pseudacacia*) litter removal by *Lumbricus terrestris*; growth of populations of *L. terrestris* following inoculation in stands of *A. glutinosa* and *R. pseudacacia*; changes in native earthworm populations during mining and subsequent reclamation; and effects of earthworm activity on soil fertility, tree growth and nutrition.

19.2 MATERIALS AND METHODS

19.2.1 Sites

Studies were conducted on calcareous, clayey spoils from mine no. 9, Meigs Creek, in the Monongahela formation of the Permian system and on dark, friable acid shale spoils from mine no. 6, Middle Kittanning, in the Allegheny formation of the Pennsylvanian system (Brant and DeLong, 1960). These older spoils were graded but not topsoiled. The effect on earthworm populations of stockpiling topsoil and replacing it on the spoil surface was investigated on an active mine in the no. 6 coal seam.

19.2.2 Litter removal

Several approaches were used to determine amount of litter removal from coal spoil surfaces by *L. terrestris* (Vimmerstedt and Finney, 1973; Hamilton and Vimmerstedt, 1980). One approach was to isolate 0.6 × 0.6 m blocks of Middle Kittanning spoil in a 6-year-old stand of *R. pseudacacia* (Fig. 19.1), using fibre glass screening that extended 0.45 m above and below the spoil surface (Fig. 19.2). Furrow grading had created a surface with ridges at 1.5 m intervals; *R. pseudacacia* leaf litter had accumulated in the 0.2 m deep furrows, where the screened blocks were located (Fig. 19.3). Ten living, sexually mature earthworms were placed in six such blocks and ten similar earthworms, killed by freezing, were placed in six other blocks. Two years later, leaflets, leaf rachides, seed pods and twigs were removed by hand, dried at 80°C, and weighed.

Fig. 19.1 *Robinia pseudacacia* plantation on furrow-graded coal spoil 6 years after planting.

A second approach was to remove 15 cm ϕ × 17 cm undisturbed cores of litter and spoil from this same area to a greenhouse, where three living earthworms were placed in 18 cores and three dead ones in another 18 cores. Rate of litter subsidence was observed over 4 months, then litter, humus and wormcastings were removed from the spoil surface by hand, dried (80°C) and weighed.

A third approach was to select 12 paired 0.0625 m² plots in stands of *R. pseudacacia* or *A. glutinosa* growing on calcareous spoils. Earthworms were extracted with formalin from a 0.25 m² area including one member of each pair in early fall, when all litter was removed from each plot. Litter

Fig. 19.2 Isolated block of spoil before installing fibre glass screening and backfilling.

falling on the extracted plots was collected weekly until snowfall, and litter falling on the extracted and unextracted plots was collected in March. The amount of litter removed by worms was determined by difference.

19.2.3 Population growth on revegetated spoils

Ten clitellate *L. terrestris* were released in the middle of 3-year-old *R. pseudacacia* and *A. glutinosa* plantings on the calcareous Meigs Creek spoil on a wet spring day in 1967. Eleven years later, populations were sampled by extraction with dilute formalin from 16–0.25 m² plots in each planting. Worms were preserved in formalin, weighed, and their fresh weights estimated from the regression: fresh wt. (g) = 1.05 (preserved wt.) + 0.099, $r^2 = 0.96$ (Hamilton, 1979).

Five years after living or dead *L. terrestris* were added to the blocks of Middle Kittanning spoil, these were excavated by first digging a trench

Fig. 19.3 Cross-section through furrow and adjacent ridges, showing litter accumulation in the furrow. Light dots mark the litter-spoil boundary.

around the block and then excavating the block itself. Worms were recovered by handsorting. Fourteen years after inoculation, 24–$0.25\,\mathrm{m}^2$ plots were located at random in this same general area and extracted with formalin.

19.2.4 Population studies on topsoiled active mines

Digging and handsorting in $0.1\,\mathrm{m}^2$ plots was used to sample earthworms in undisturbed forests and fields adjacent to active surface mines, in stockpiles of topsoil, in topsoil regraded on the spoil surface, and in topsoiled areas planted to grasses and forage legumes. Worms were anaesthetized in alcohol, preserved in formalin, counted and identified (Reynolds, 1977).

19.2.5 Effect of earthworm activities on minesoil properties

In the acid shale minesoil, six $2\,\mathrm{cm}\ \phi \times 15\,\mathrm{cm}$ cores were removed from each soil block isolated by screening two years previously and combined into two composite samples. Total N, available P, pH and exchangeable cations – K, Ca and Mg, were determined (Black, 1965). Three years later, a further composite sample of 20 randomly chosen 2 cm ϕ × 7.6 cm cores was taken for the same chemical analyses. These chemical properties were also determined on soil particles passing through a 2 mm round hole sieve, taken from the undisturbed cores used in the greenhouse study.

19.2.6 Tree growth and mineral nutrition

Greenhouse studies were conducted on undisturbed cores from the acid shale spoils and on the soil-sized particles derived therefrom. Two months after placing earthworms in the cores, we planted one germinated northern red oak (*Quercus rubra* L.) acorn at the soil surface. After 4 months growth, oak seedlings were harvested, weighed and analysed for content of inorganic nutrients. Then rooted cuttings of an eastern cottonwood (*Populus deltoides* Bartr.) clone were potted in the < 2 mm ϕ particles of minesoil from each core. We measured height growth of the cuttings during one growing season.

19.3 RESULTS AND DISCUSSION

19.3.1 Litter removal

The litter surface in undisturbed cores containing live earthworms moved downward quite rapidly, until after 174 days it averaged 5.5 cm below the original level (Fig. 19.4). Most of the black locust leaf litter had been incorporated into the mineral spoil. The litter surface moved downward in the cores without earthworms too, but at a much slower rate (Fig. 19.5). Similar results were obtained with soil blocks isolated in the field. One year after earthworm introduction, an average of six active middens per block was observed where live earthworms had been added. By the second spring, differences in litter distribution between plots with and without worms were obvious (Fig. 19.6). Black locust leaflets and leaf rachides were removed from most of the plot area and piled around midden entrances. Dry weight differences of litter and humus were large and statistically significant (Table 19.1). The higher values for litter and humus weight in undisturbed cores compared to soil blocks in the field probably result from cores having been taken in the middle of the furrows where litter was deepest while the soil blocks included shallower parts of the furrow cross-section. These two experiments on acid shale spoil indicated how rapidly *L. terrestris* buried layers of litter and humus accumulated over several years in the absence of earthworms. An estimate of the amount of current litter fall removed by *L. terrestris* was obtained in

Live worms   Dead worms

Fig. 19.4 Spoil cores after 174 days in the greenhouse with or without earthworms, showing difference in amount of litter and humus remaining on the spoil surface.

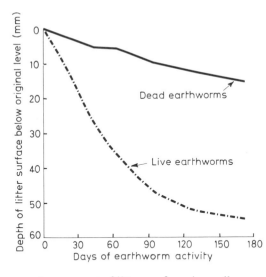

Fig. 19.5 Downward movement of litter surfaces in spoil cores with or without living earthworms during 174 days in the greenhouse.

the *R. pseudacacia* and *A. glutinosa* stands growing on calcareous spoils (Table 19.1). Earthworms removed a greater proportion of *Alnus* than of *Robinia* leaf litter.

Fig. 19.6 Surfaces of field plots in black locust planting showing litter distribution after 2 years with or without live *L. terrestris*.

Table 19.1 Effect of *Lumbricus terrestris* on surface deposits of litter and humus in *Robinia pseudacacia* plantings on acid shale spoil and on leaf litter in *Alnus glutinosa* and *R. pseudacacia* plantings on calcareous spoil.*

	pH†	Oven dry wt. (kg ha^{-1})				Duration of experiment
		Leaflets or leaves		Humus		
		With worms	Without	With worms	Without	
Acid shale spoils						
Soil blocks, field	3.5	1175	6299	31769	53461	2 years
Soil cores, greenhouse		4508	14078	15353	68019	174 days
Calcareous spoils						
A. glutinosa	7.7	107	1209			Oct.–March
R. pseudacacia	8.2	507	890			Oct.–March

* Differences between leaflets, leaf or humus weights between plots with or without earthworms, all significant at $P < 0.01$.
† Measured in 0.01 M-CaCl$_2$.

19.3.2 Population growth on revegetated spoil

On the calcareous spoils, after 12 years, the initial population of 10 clitellate *L. terrestris* had increased to an average of 56 adult worms m^{-2} throughout the alder planting and 28 adults m^{-2} throughout the black locust planting; each planting covered about 0.5 ha. There was no evidence of a population gradient from points of introduction. Besides being more numerous, worms in the alder planting were heavier, 3.03 g fresh wt. under alder, 2.12 g under black locust (Hamilton and Vimmerstedt, 1980). Seven years before this sampling, the *L. terrestris* population in the alder plantation was quite similar with 60 middens m^{-2} up to 15 m from the point of introduction, but then dropped off to 29 middens m^{-2} at 20 m from the point of introduction and 16 middens m^{-2} at 25 m (Vimmerstedt and Finney, 1973).

When the isolated blocks of Middle Kittanning spoil were excavated, five years after inoculation with earthworms, an average of 13.7 sexually mature *L. terrestris* and 9.4 immature earthworms m^{-2} were captured inside the plots where live worms had been introduced, while an average of 12.3 adults and 14.6 juveniles m^{-2} were found while excavating trenches just outside the plots. Most of the earthworm tunnels extended downward just inside the screening. Some of these tunnels went under the bottom of the screening so that many of the earthworms found in the trenches outside the plot had their burrow entrances inside the plot.

In the plots initially without earthworms there were 0.8 adults and 1.5 juveniles m^{-2} inside the plot, and 4.5 adults and 4.3 juveniles m^{-2} outside the plot. All adult earthworms captured were *L. terrestris*. Enough earthworms had escaped through gaps in the screening, or moved through it when newly hatched, to establish a sparse population in the general area. In August, 1981, 14 years after *L. terrestris* was introduced, twenty-four 0.25 m^2 plots scattered in the area had the following mean population density:

	Plots occupied	No. m^{-2}
L. terrestris	17	3.7
L. rubellus	6	1.8
Lumbricus juveniles	18	12.2

19.3.3 Soil properties

Differences in exchangeable cations and available P, between mineral soil from undisturbed cores with or without earthworms, were statistically significant (Table 19.2). In the field study of the same spoil, available P was lower in the plots with earthworms while, in the greenhouse study,

Table 19.2 Nutrient status of spoil samples from field plots and cores with and without earthworms.

Samples	Total N (%)	Available P (mg kg^{-1})	Exchangeable cations			
			K	Ca (meq. 100 g^{-1})	Mg	pH*
1969		Field samples				
With worms	0.12	7.93†	0.28	1.97	1.65	
Without worms	0.11	12.03†	0.24	1.86	1.49	
1972						
With worms	0.13	4.97†	0.32	1.67	0.89	3.5
Without worms	0.11	6.37†	0.36	1.64	0.83	3.4
		Core samples				
With worms	0.06	4.25‡	0.19‡	2.72‡	2.47‡	
Without worms	0.08	1.74‡	0.17‡	1.69‡	1.79‡	

* Measured in 0.01 M-CaCl$_2$.
† Significantly different $P < 0.05$.
‡ Significantly different $P < 0.01$.

soil with earthworms had considerably higher P. Part of the difference in nutrient levels may have resulted from sampling methods used, for in the greenhouse study the total volume of spoil containing earthworms was sampled, while in the field study the spoil near the screening where most earthworm burrows were concentrated was not well represented in the composite samples.

19.3.4 Tree growth and nutrition

Dry weights of northern red oak leaves, stems and roots, their leaf concentrations of N and K, and stem lengths did not differ significantly between seedlings grown in cores with or without *L. terrestris*. Concentrations of P, Mg and Ca were higher in leaves of red oak grown without worms, respectively 0.20, 0.42 and 1.13% compared with 0.17, 0.33 and 0.93% with earthworms. Rooted cuttings of the cottonwood clone grew 29 cm in soil-sized particles of mineral spoil from cores which had previously contained earthworms, while cuttings in the mineral spoil from cores without earthworms grew only 22 cm. These somewhat inconclusive results point up the need for well-designed, replicated field studies of interaction between plants, minesoils and earthworm populations.

19.3.5 Earthworm populations on active mines

At the active surface mine in Middle Kittanning coal, limited sampling indicated that earthworms were sparse in stockpiled topsoil, and absent from newly topsoiled spoils, with or without vegetation (Table 19.3).

Table 19.3 Earthworm populations of stockpiled topsoil, newly topsoiled spoils, topsoiled revegetated spoils and adjacent fields and forests.

No./biomass m^{-2}	Stockpiled topsoil		Freshly topsoiled spoils		Revegetated spoils		Old fields		Forests	
	No.	g	No.	g	No.	g	No.	g	No.	g
Aporrectodea spp.	14	6.9	0	0	0	0	110	31.5	5	0.6
Plots sampled (*n*)	5		4		9		3		2	

19.4 CONCLUSIONS

1. *L. terrestris* can be introduced on revegetated minesoils ranging in pH from quite acid to above neutral.
2. *L. terrestris* will incorporate litter of *A. glutinosa* and *R. pseudacacia* into mineral spoil, thus altering the soil-forming process.

3. On calcareous minesoils, *L. terrestris* populations will be larger and individual worms heavier under *A. glutinosa* than under *R. pseudacacia*.

4. Surface mining reduces earthworm populations to low levels.

19.5 REFERENCES

Black, C. A. (1965) *Methods of Soil Analysis, Part 2. Chemical and Microbiological Properties.* Am. Soc. Agron., Madison, WI. 1572 pp.

Brant, R. A. and DeLong, R. M. (1960) Coal resources of Ohio. *Bull. Geol. Surv. Ohio.*, No. 58, 245 pp.

Edwards, C. A. (1981) Earthworms, soil fertility, and plant growth, In *Workshop on the Role of Earthworms in the Stabilization of Organic Residues* (Compiled by M. Appelhof) Vol. 1. Beech Leaf Press, Kalamazoo, MI, pp. 61–85.

Hamilton, W. E. (1979) *Earthworms on Forested Spoil Banks.* M.S. Thesis, Ohio State University, Columbus.

Hamilton, W. E. and Vimmerstedt, J. P. (1980) Earthworms on forested spoil banks. In *Soil Biology as Related to Land Use Practices* (ed. D. L. Dindal), *Proc. VIIth Int. Soil Zool. Coll., EPA, Washington, D.C.*, pp. 409–417.

Jenny, H. (1980) The soil resource: Origin and behavior. *Ecological Studies*, Vol. 37. Springer-Verlag. 377 pp.

Marx, D. H. (1975) Mycorrhizae and establishment of trees on strip-mined land. *Ohio J. Sci.*, **75**, 288–297.

Marx, D. H. and Artman, J. D. (1979) *Pisolithus tinctorius* ectomycorrhizae improve survival and growth of pine seedlings on acid coal spoils in Kentucky and Virginia. *Reclam. Rev.*, **2**, 23–31.

Reynolds, J. W. (1977) The earthworms (*Lumbricidae* and *Sparganophilidae*) of Ontario. *Life Sci. Misc. Pub., R. Ont. Mus.*, 141 pp.

Vimmerstedt, J. P. and Finney, J. H. (1973) Impact of earthworm introduction on litter burial and nutrient distribution in Ohio strip-mine spoil banks. *Proc. Soil Sci. Soc. Am.*, **37**, 388–391.

Chapter 20

Development of earthworm populations in abandoned arable fields under grazing management.

H. J. P. EIJSACKERS

20.1 INTRODUCTION

An experiment was begun in 1972 on a nature reserve in the southern Netherlands with the object of incorporating adjacent abandoned arable land into the reserve. Icelandic horses were introduced as herbivores additional to the native rabbit population to graze the developing vegetation and so maintain the open character of the landscape and merge the boundaries between the reserve and the fields. This experiment created an opportunity to investigate the development of the earthworm population in relation to the vegetation succession.

The low density of earthworm populations in arable land has been attributed partially to mechanical damage during cultivation but Evans and Guild (1948) considered the main limiting factor to be the low input of suitable plant residues. This study investigated the effect of plant residues from the developing vegetation on the earthworm population and the influence of earthworm activity on the soil conditions and vegetation.

20.2 STUDY AREA AND SAMPLING METHODS

The nature reserve Cranendonck occupies 100 hectares of sandy soil comprising approximately equal areas of coniferous forest, heath and blowing sand, and abandoned arable fields (Fig. 20.1). Ten sampling sites were selected in abandoned fields and four in adjacent fields still under arable cultivation. Earthworms were sampled from 1975 by handsorting duplicate samples (30 × 30 × 30 cm) 3–5 times each year. Soil properties

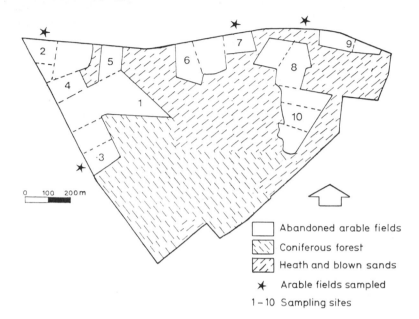

Fig. 20.1 Plan of Cranendonck nature reserve.

were determined in 1973 as follows: pH in KCl; organic matter as loss on ignition; total N by destruction with phenol–H_2SO_4 + Se and distillation of NH_3; total P by extraction with Fleischmann acid; total K, total Fe and Cu by extraction in 0.1 N HCl+0.4 N oxalic acid (1:10). Other workers monitored activities of horses and rabbits, and the development of the vegetation.

20.3 FLUCTUATIONS IN EARTHWORM NUMBERS

Lumbricus rubellus comprised 95 % of all the worms sampled. In four sites (1–4) in the west of the reserve, numbers showed a distinct decline in 1976 followed by an increase up to 1979 and a decrease in 1980. In four sites (6, 7, 8, 10) in the east of the reserve, numbers increased steadily from 1975 to 1979 and fell in 1980 (Fig. 20.2). Two other sites (5, 9) showed irregular fluctuations and were not analysed further.

The decrease in 1976 coincided with a period of very dry hot weather in spring and summer which reduced the populations on the sites with the lowest water tables (Table 20.1). The decrease in 1980 coincided with cold, wet weather conditions in 1979 and 1980. Of the sites in the east of the reserve, the two with the lower initial population densities (8 and 10) had caught up with the remainder. Throughout the whole period

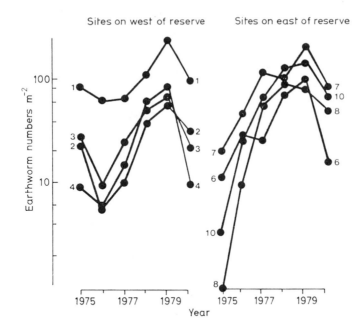

Fig. 20.2 Population densities ($\log n + 1$) of *L. rubellus*.

Table 20.1 Physical and chemical properties of the sites.

Site	Highest ground water level (cm below surface)	pH	Organic matter (g kg⁻¹)	Total N (g kg⁻¹)	Total P (g kg⁻¹)	Total Fe (g kg⁻¹)	K (mg kg⁻¹)	Cu (mg kg⁻¹)
1	< 40/40–80	4.6	33	1.0	0.8	1.1	5	4.2
2	> 80	4.8	33	0.8	0.9	1.2	24	2.6
3	> 80	4.6	36	0.9	0.8	1.4	5	2.4
4	> 80	4.6	34	0.8	0.9	1.4	5	4.0
6	40–80/ > 80	4.6	32	0.8	0.9	2.1	7	1.8
7	40–80/ > 80	4.9	35	0.9	0.9	1.7	6	2.3
8	40–80/ > 80	5.0	25	0.8	1.0	2.1	6	2.5
10	40–80/ > 80	4.8	19	0.6	0.9	3.1	0.6	4.1

1975–80, the population densities in the adjacent fields under arable management remained low. A total of only 22 worms was found in 120 samples.

The proportion of juveniles in the population declined when total numbers fell (Fig. 20.3). *L. rubellus* has only a limited ability to retreat to deeper soil and under adverse weather conditions mortality was greatest amongst the smaller worms.

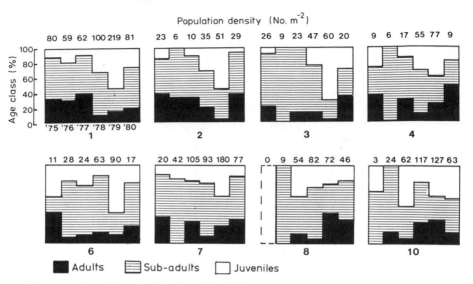

Fig. 20.3 Age class distribution (%) of *L. rubellus*, 1975–80.

20.4 SUCCESSION

During the period of observation, although the vegetation of the abandoned arable fields varied considerably from site to site, arable plants decreased and grasses, typically *Agrostis stolonifera* and *A. tenuis*, increased. This is part of a succession in which arable species are

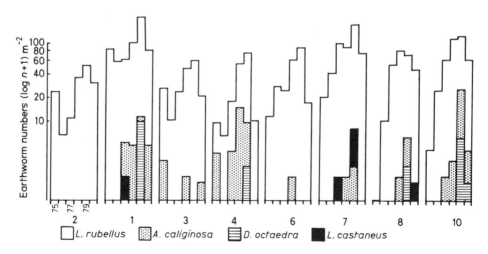

Fig. 20.4 Species composition of sites in supposed order of vegetation succession.

gradually replaced by species typical of dry sandy grasslands (*Koelerio-corynephoreta*) or very poor grasslands tending to heath (*Nardo-calluneta*).

In the first two years of observations, the earthworm populations of all but two of the sites consisted exclusively of *L. rubellus* (Fig. 20.4). By the fourth year *A. caliginosa* had colonized all but one of the sites and between the third and sixth years *Lumbricus castaneus* was recorded in three sites and *Dendrobaena octaedra* in four. Though typically a coprophage, *L. castaneus* showed no relation between the year of colonization and the incidence of horse dung. Invasion by *D. octaedra* also showed no specific relationship with particular plant species but it is typically a species of heath and impoverished sites. The site with only *L. rubellus* was subject to high trampling pressure.

20.5 CONCLUSIONS

Observations have been made over a period of six years in a succession which can be expected to continue developing for several decades. Although the results are therefore preliminary, they nevertheless confirm the results of other studies on colonization. The four earthworm species so far recorded: *L. rubellus*, *A. caliginosa*, *L. castaneus* and *D. octaedra* have been reported as pioneer colonists, although not always in the same combination, in pulverized fuel ash (Satchell and Stone, 1977), municipal refuse (Brockmann *et al.*, 1980), restored opencast coal mines (Dunger, 1969) and pastures (Martin, 1977). These species seem to have little in common, – *L. rubellus* and *L. castaneus* are active surface migrants, *A. caliginosa* is not; *D. octaedra* tolerates sites with very acid organic horizons, *A. caliginosa* does not (Satchell, 1955); *L. rubellus* and *L. castaneus* have high reproductive rates, *A. caliginosa* does not (Evans and Guild, 1948). The success of these species as colonists is clearly not explainable in simple terms of mobility, fecundity or ecological category. Though convergent as pioneers their biology is individual to each species.

20.6 ACKNOWLEDGEMENTS

I thank Mr J. Bodt and Mr D. van den Ham, for help with the earthworm sampling, Dr J. A. van Rhee who started this project and Dr J. E. Satchell for his valuable revision of the manuscript.

20.7 REFERENCES

Brockmann, W., Koehler, M. and Schriefer, T. (1980) Recultivation of refuse tips: soil ecological studies. In *Soil Biology as Related to Land Use Practices* (ed. D. L. Dindal), *Proc. VIIth Int. Soil Zool. Coll.*, *EPA, Washington, D.C.*, pp. 161–168.

Dunger, W. (1969) Fragen der naturlichen und experimentellen Besiedlung kulturfeindlicher Boden durch Lumbriciden. *Pedobiologia*, **9**, 146–151.

Evans, A. and Guild, W. J. McL. (1948) Studies on the relationships between earthworms and soil fertility. IV. On the life cycles of some British Lumbricidae. *Ann. Appl. Biol.*, **35**, 471–484.

Martin, N. A. (1977) Guide to the lumbricid earthworms of New Zealand pastures. *N. Z. J. Exp. Agric.*, **5**, 301–309.

Satchell, J. E. (1955) Some aspects of earthworm ecology. In *Soil Zoology* (ed. D. K. McE. Kevan), Butterworths, London, pp. 356–364.

Satchell, J. E. and Stone, D. A. (1977) Colonization of pulverized fuel ash sites by earthworms. *P. Cent. Pir. Biol. Exp.*, **9**, 59–74.

Chapter 21

Heavy metal uptake and tissue distribution in earthworms

M. P. IRELAND

21.1 INTRODUCTION

Earthworms can contribute to the physical structure and nutritive value of the soil by burrowing and feeding but they can also be a potential pollution hazard. Numerous authors, reviewed by Hughes *et al.* (1980) and Beyer (1981), have reported that earthworms can take up and accumulate in their tissues heavy metals such as cadmium, mercury and gold, when living both in non-contaminated and contaminated environments (Helmke *et al.*, 1979).

Chemical changes which occur in the alimentary tract of earthworms may render various metals more available to plants, and mineralization of dead earthworms will release accumulated heavy metals into the environment (Ireland, 1975a). Earthworms form part of many food chains and carnivorous animals, depending on their assimilation efficiency, may accumulate heavy metals by preying on contaminated earthworms.

The mechanism of detoxification within the oligochaetes varies with the metal concerned but metal accumulation at moderately high levels for reasonably short periods appears to have little deleterious effect on biomass or growth (Williamson and Evans, 1973; Mori and Kurihara, 1979; Hartenstein *et al.*, 1980a). Hartenstein *et al.* (1980b) found that *Eisenia foetida* fed for four weeks on sludge doubled their biomass despite the presence of heavy metals. Long-term sublethal concentrations of heavy metals may however reduce earthworm fecundity.

The object of the present chapter is to review sources of heavy metal contamination that can influence earthworms and to evaluate how earthworms cope with these increased metal levels by accumulation within specific tissues or by efficient excretion.

21.2 HEAVY METAL UPTAKE

Despite the many ways heavy metals can enter the soil, very few contaminated areas, with the exception of some arsenate-impregnated regions around disused copper smelters, are completely devoid of flora and fauna. Grasses such as *Festuca rubra* and *Agrostis tenuis* have evolved a natural tolerance to specific heavy metals and numerous macro-invertebrates, including earthworms, colonize contaminated soil. Heavy metal levels in earthworms and soils from many sites have been reviewed by Hughes *et al.* (1980) and Beyer (1981) and the existing data are divided in Tables 21.1 and 21.2 into non-contaminated and contaminated sites. They confirm previous findings that the ratio of metal concentration in the earthworms divided by the metal concentration in the soil is higher than unity for some metals such as cadmium but not for others, irrespective of the degree of pollution.

The source of the metals, species of earthworm and metal–metal interaction may influence tissue metal concentrations. Several proprietary fungicides used to combat various blights and mildews contain copper. In orchards with a long history of heavy spraying with copper fungicides, the levels of copper in the leaf litter may reach concentrations of $2500\,\mu g$ of copper g^{-1} dry wt. and very few earthworms are present except for a small number of *Lumbricus castaneus* in the surface litter (Raw and Lofty, 1960; Tarman and Raw, 1961). In Holland, van Rhee (1967) reported on a 60-year-old orchard where copper oxychloride winter washes had been applied for 12 years. Mean earthworm density was $70\,m^{-2}$ compared with an untreated orchard where it was $530\,m^{-2}$.

Annelid populations treated with sewage effluent increase in numbers and the treatment selects for *Dendrobaena octaedra* (Dindal, 1975). The heavy metal content of soil in treated and untreated vegetation studied by Dindal *et al.* (1979) showed an increase in copper and zinc at all treated sites. Cadmium increased in the soil in a mixed oak forest and a reed canarygrass monoculture but not in an old field herbaceous community. Earthworms collected at the latter site contained more zinc and cadmium than earthworms at a similar control site.

Accumulation of cadmium, zinc and copper occurs in *Aporrectodea tuberculata* with increasing rates of sludge application (Helmke *et al.*, 1979) while copper-containing pig waste raises the concentration of this metal in the soil and subsequently in earthworms (van Rhee 1975, 1977; Curry and Cotton, 1980). Sewage effluent may be contaminated by heavy metals as the result of industrial effluent discharge and it is difficult to separate the two sources of pollution.

The heavy metal concentration in a simple producer–herbivore food chain in a mixed oak woodland ecosystem 3 km downwind from a primary lead–zinc smelter at Avonmouth, Bristol (Martin and Coughtrey, 1975, 1976) showed that aerial pollution raised the body concentrations of lead and cadmium in earthworms. The metal body burden in earthworms was higher than in gastropods but less than in the isopod *Oniscus asellus*.

Data from studies involving the addition of inorganic heavy metal salts to organic material (Mori and Kurihara, 1979; Hartenstein *et al.*, 1980a) and from studies in which sewage sludge high in heavy metals is applied at varying rates to the soil (Helmke *et al.*, 1979) cannot be compared directly. For instance, 2500 μg copper salt g^{-1} dry wt. added to sewage sludge killed *E. foetida* within one week while non-amended sewage sludge with up to 1500 μg copper g^{-1} dry wt., plus other heavy metals, was non-toxic to earthworms for more than four months of continuous culture (Hartenstein *et al.*, 1980a). *Allolobophora caliginosa* collected in soil containing 10 μg cadmium g^{-1} dry wt. accumulated 63 μg cadmium g^{-1} dry wt. in the tissues (Wright and Stringer, 1980) while the same species subjected to 5 μg cadmium ml^{-1} in solution accumulated 432 μg of the metal g^{-1} dry wt. (Ireland and Richards, 1981). In soil containing 4 μg cadmium g^{-1} dry wt. the total body burden of the metal in *Lumbricus rubellus* was 25 μg g^{-1} dry wt. (Ireland, 1979) but exposure of this species to 5 μg cadmium ml^{-1} in solution for 26 days resulted in a tissue accumulation of 423 μg g^{-1} dry wt. (Ireland and Richards, 1981).

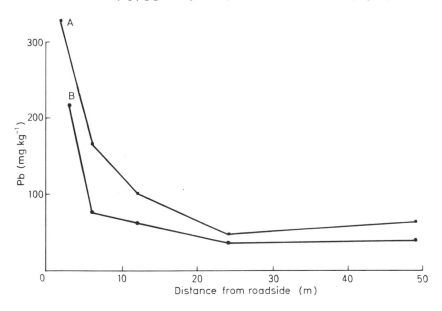

Fig. 21.1 Effect of distance from roadside on the lead concentration in earthworms. Traffic density: A, 46 000/day; B, 25 000/day. (Reproduced with permission from Gish and Christensen (1973).)

All published data on earthworms collected near roadsides show an accumulation of heavy metals above the amounts found in earthworms from non-polluted sites (Andersen, 1979; Wright and Stringer, 1980). The concentration of metal accumulated in the body of earthworms falls off as the distance from the roadside increases (Gish and Christensen, 1973) (Fig. 21.1) and is also dependent on traffic density (Gish and Christensen, 1973; Czarnowska and Jopkiewicz, 1978; Ash and Lee, 1980).

The concentrations of cadmium, lead, copper and iron in *Allolobophora chlorotica* and *Lumbricus terrestris* has been reported for roadsides in Yorkshire (Ash and Lee, 1980). Lead, zinc and cadmium accumulated in both species while iron appeared not to accumulate. Only lead showed significant interspecific variation and was higher in *A. chlorotica*.

One of the earliest reports on earthworms and lead uptake at the site of the Elswick Lead Works was published by Hogg (1895) who studied a small terrestrial oligochaete of undefined species. After removal of the gut contents, the tissues contained 90–180 μg lead g^{-1} dry wt. Similar concentrations of lead were found in *L. terrestris* collected at the site of a derelict lead–zinc mine complex at Minera in North Wales. As expected, the lead and zinc concentrations in the tissues were higher than those in earthworms at control sites and the body burden of metal declined as the distance from the pollution source increased (Roberts and Johnson, 1978).

Studies at a lead–zinc mine spoil heap in the Ystwyth valley near Aberystwyth have revealed very high concentrations of lead in earthworms (Ireland, 1975b). *Dendrobaena rubida* collected at the lead-polluted site and transferred to non-contaminated soil showed after 5 days a significant fall in the tissue lead while calcium and zinc concentrations were not significantly different from control values. Lead and calcium concentrations in *D. rubida* after 20 days were significantly lower than in control tissue, but zinc tissue concentrations remained unchanged. Conversely, control *D. rubida* left in contaminated soil for 20 days showed a significant increase in tissue calcium and lead but not in zinc. There may be a positive correlation between tissue calcium and lead, the higher the concentration of lead the higher the concentration of calcium (Ireland and Wootton, 1976) as reported for *L. terrestris* (Andersen, 1979) and the isopod *Porcellio scaber* (Beeby, 1978).

Ireland (1979) examined the distribution of metals in *L. rubellus* at three sites in Wales containing high concentrations of either lead and manganese, zinc or copper. At the predominantly lead- and manganese-polluted site, the highest concentrations of lead, manganese and calcium were found in the earthworms. Despite the high soil concentrations of zinc at one site (Borth) and copper at the other (Dolgellau), the earthworm body burden of these metals was not significantly different at the three sites and it was suggested that *L. rubellus* may regulate zinc and copper.

Of three species of earthworms found at a highly zinc-polluted site (Ireland, 1979), *L. rubellus* contained the highest concentrations of zinc and manganese, and the lowest concentration of lead. *Dendrobaena veneta* contained the highest concentration of cadmium and *Eiseniella tetraedra* contained the highest concentration of lead. Ireland and Richards (1977) found different lead contents in earthworms taken at Cwmystwyth; *L. rubellus* 2600 μg g^{-1} dry wt., *D. rubida* 7600 μg g^{-1} dry wt.

Differences in the metal content of various species could be the result of differences in uptake mechanisms. Piearce (1972) reported that the intestinal uptake of calcium was much higher in *L. rubellus* than in *A. caliginosa*.

Earthworms that tolerate high environmental concentrations of toxic heavy metals either do not absorb the metal, accumulate it in a non-toxic form or excrete it efficiently. If the concentration of a metal in the body does not exceed the amount found in the pollutant source, it is reasonable to assume that it is either not absorbed or is excreted. This may explain why different species at the same site can contain different concentrations of metal and that certain species have a higher maximum tolerance to a specific metal. This supports the conclusions of Raw and Dobson (1959) who showed that on soil plots treated with lead arsenate to eliminate earthworms, the first species to reappear was *Eiseniella tetraedra*. *E. tetraedra* on a site containing 630 μg lead g^{-1} dry wt. contained more lead in the tissues than two other species (Ireland, 1979).

Interaction between metals can also influence the amount of a specific metal assimilated (Kirchgessner, 1978), notably, high dietary copper and calcium can reduce lead uptake (Shields and Mitchell, 1941; Petering, 1974; Baltrop and Khoo, 1976). In the sites studied by Ireland (1979), the lead concentration in *L. rubellus* at Borth was approximately 400 times less than the lead tissue level in the same species at Cwmystwyth, while the metal content of the soil differed by a factor of approximately two. Both the soil calcium and the copper concentrations are much higher at Borth than at Cwmystwyth.

The word 'accumulation' is used in different senses by different authors. According to Van Hook (1974), accumulation occurs when the ratio of the concentration of metal in the tissue over the concentration in the environment exceeds 1.0. This is disputed by Hartenstein *et al.* (1980a) who point out that anomalously high levels of metal may be present in the environment which could not exist chemically or physically within a living organism. Accumulation is said to occur when animals exposed to a single level of heavy metal in the environment over a period of time show increasingly high levels in the tissues or, for a given time, the animals' tissue shows increasing levels related to the concentrations in the environment. They note that their definition is not totally adequate since for numerous metals a normal physiological level within an organism is not known.

Of all the metals studied, cadmium appears to accumulate (Van Hook's definition) in most species of earthworms at greater levels than any other metal (Tables 21.1 and 21.2). It does not increase in the tissues indefinitely and the concentration ratio decreases with increasing cadmium concentrations in the environment (Hartenstein *et al.*, 1980a). This decrease in the concentration ratio was not observed after 60 days in the same species (*E. foetida*) studied by Mori and Kurihara (1979). The other toxic heavy metal extensively studied in earthworms is lead, the concentration of which increases in the tissues with increasing concentrations in the environment (Hughes *et al.*, 1980; Beyer, 1981). Cwmystwyth (Ireland, 1975b) appears to be unique in that the concentration ratio exceeds unity when the concentration of the metal is high in the soil. Whether this ecosystem contains earthworms with a particularly high inherited tolerance for lead, as in several species of plants (Antonovics

Table 21.1 Concentration factors* for cadmium, lead, zinc and copper at non-contaminated sites.

Site	Species	Cd	Pb	Zn	Cu	Reference
Soil	Mixed	4.6	0.8	5.3	—	Gish and Christensen (1973)
Soil	Mixed	16	0.2	7.4	—	Van Hook (1974)
Compost	*Dendrobaena rubida*	—	0.8	0.7	—	Ireland (1975b)
Soil	Mixed	12	1.3	1.8	—	Martin and Coughtrey (1976)
Soil	Mixed	—	—	—	1.5	van Rhee (1977)
Soil	*Lumbricus terrestris*	27	0.5	18	0.8	Czarnowska and Jopkiewicz (1978)
Soil	*L. terrestris*	—	0.04	0.7	—	Roberts and Johnson (1978)
Soil (Plot 1)	*L. terrestris*	58	0.04	—	—	Andersen (1979)
	2 *Allolobophora* spp.	40–90	0.2–0.3	—	—	
Soil (Plot 7)	2 *Allolobophora* spp.	71–151	0.2–0.4	—	—	
Soil	*Aporrectodea tuberculata*	32	—	2.9	—	Helmke *et al.* (1979)
Compost	*Eisenia foetida*	3.8	—	0.5	—	Mori and Kurihara (1979)
Soil	5 *Allolobophora* spp.	16–33	0.3–0.5	3–10	—	Wright and Stringer (1980)
	L. terrestris	15	0.3	5.2	—	

* The ratio of metal in earthworms to that in soil.

et al., 1971), remains to be seen. Both *L. rubellus* and *D. rubida* present in the contaminated soil are smaller in the adult stage than control earthworms but they breed and cocoons are frequently found in the lead- and zinc-polluted soil. Lead tissue concentration ratios > 1 have been reported (Tables 21.1 and 21.2) by Martin and Coughtrey (1976) and Gish and Christensen (1973) but in both cases the soil concentration of lead was low. It is clear that numerous parameters have to be considered when using the concentration ratio as an indication of metal accumulation (Beyer, 1981).

Table 21.2 Concentration factors for cadmium, lead, zinc and copper at contaminated sites. (*Contd overleaf*)

Site	Species	Cd	Pb	Zn	Cu	Reference
3–7 m from roadside	Mixed (degutted)	—	0.1	—	—	Williamson and Evans (1972)
3 m from highway	Mixed	10	0.6	3.2	—	Gish and Christensen (1973)
40 m from highway		11.5	1.1	6.4	—	
Acid mine spoil	*Dendrobaena rubida*	—	2.4	0.3	—	Ireland (1975b)
Near zinc smelter	Mixed	7.6	0.3	0.5	—	Martin and Coughtrey (1976)
Soil amended with pig waste	Mixed	—	—	—	0.4–0.8	van Rhee (1977)
Near roadside	*Lumbricus terrestris* (degutted)	17	0.4	7.3	0.6	Czarnowska and Jopkiewicz (1978)
200 m from roadside		31	0.6	17	0.8	
Lead mine complex	*L. terrestris*	—	0.03	0.02	—	Roberts and Johnson (1978)
Soil amended with sludge (Plot 5)	*L. terrestris* 4 *Allolobophora* spp.	26 / 9–17	0.4 / 0.1–0.2	—	—	Andersen (1979)
Soil amended with sludge (Plot 6)	*L. terrestris* 4 *Allolobophora* spp.	9 / 9–20	0.1 / 0.1–0.2	—	—	
10–50 m from roadside	*L. terrestris*	20–100	0.1	—	—	
Soil amended with sludge	*Aporrectodea tuberculata*	236	—	4.5	—	Helmke *et al.* (1979)
Lead mine complex	*Lumbricus rubellus*	7.5	2.7	5.4	0.7	Ireland (1979)
Copper mine complex		6.3	0.7	4.2	0.03	

Table 21.2 (*Contd.*)

Site	Species	Cd	Pb	Zn	Cu	Reference
Soil amended with sludge	*L. rubellus*	1.0	0.01	0.7	0.04	
	Dendrobaena veneta	1.8	0.03	0.2	0.06	
	Eiseniella tetraedra	0.8	0.03	0.4	0.03	
Soils amended with metals	*Eisenia foetida*	4.7	—	0.3	—	Mori and Kurihara (1979)
Soil amended with pig waste	Mixed	—	—	—	0.1– 0.8	Curry and Cotton (1980)
Sludge amended with metals	*E. foetida*	1.7	0.01	0.07	0.2	Hartenstein *et al.* (1980a)
Sludge	*E. foetida*	0.8	0.6	0.6	0.1	Hartenstein *et al.* (1980b)
Near zinc smelter	5 *Allolobophora* spp.	4–6	0.4– 0.6	1–2	—	Wright and Stringer (1980)
	L. terrestris	5.6	0.3	2.3	—	

21.3 CASTS

The concentrations of total cadmium in earthworm casts are higher than the concentrations in the respective soils (Table 21.3). The concentration ratios of zinc, lead and copper are variable but not directly related to the soil concentration. Casts produced by earthworms living in contaminated environments may contain metals which by passing through the alimentary tract become more available to plants. Decay of dead earthworms may also release heavy metals in plant-available forms. Ireland (1975a) concluded that *D. rubida* living in heavy-metal-contaminated acid soil can increase the amounts of lead, zinc and calcium made available to plants. The method used to estimate the availability of the metals was based on the technique of Alloway and Davies (1971) using 2.5 % acetic acid. Neuhauser and Hartenstein (1980), comparing various methods of extracting heavy metals to estimate the availability to living plants, consider 0.1 N-HCl to be preferable although they admit that the acetic acid concentration used gives a pH value more realistic for the plant root environment. Contrary to the findings of Ireland (1975a), Hartenstein *et al.* (1980) reported that passage of sludge through the gut of *E. foetida* did not increase the 0.1 N-HCl-extractable cadmium, copper, nickel, lead and zinc. This difference is unlikely to result from the method of extraction since 0.1 N-HCl will extract more of these five metals than 2.5 % acetic acid. A difference in the physiology or uptake mechanism of the two species cannot be ruled out and warrants further investigation.

Table 21.3 Concentration of metals in soil and earthworm casts (μg g^{-1} dry wt.).

Site	Species	Cd Soil	Cd Casts	Pb Soil	Pb Casts	Zn Soil	Zn Casts	Cu Soil	Cu Casts	Reference
Lead mine complex 1*	Dendrobaena rubida	—	—	1810	2070	154	152	—	—	Ireland (1975a)
Near roadside	Lumbricus terrestris	1.1	1.0	170	185	275	354	52	58	Czarnowska and Jopkiewicz (1978)
200 m from roadside		0.3	0.5	64	45	103	132	27	27	
2 near roadside		0.9	0.9	130	118	170	300	55	32	
200 m from roadside		0.3	0.3	30	21	57	88	15	12	
3 Near roadside		0.6	0.9	170	123	225	140	26	22	
200 m from roadside		0.3	0.3	39	24	105	66	18	10	
Control soil		0.1	0.3	20	22	40	66	9	9	
Control soil	Aporrectodea tuberculata	0.5	0.7	—	—	100	38	—	—	Helmke et al. (1979)
Soil amended with pig waste	Mixed	—	—	—	—	—	—	483	460	Curry and Cotton 1980
Soil amended with sludge	Dendrobaena veneta	—	—	629	315	992	1277	252	275	Ireland (unpublished)
	Eiseniella tetraedra	—	—	629	449	992	1171	252	281	

* Traffic density 24 h^{-1}: 1, 27 000; 2, 15 270; 3, 19 640.

21.4 PREDATOR–PREY RELATIONSHIPS

Earthworms are an important link in the food chain of numerous birds, mammals and amphibia (Williamson and Evans, 1972; Martin and Coughtrey, 1975; Ireland, 1977) as well as several species of beetles and carnivorous slugs (Edwards and Lofty, 1977). Martin and Coughtrey (1975) showed that thrushes living near a primary lead–zinc smelter accumulated cadmium in kidney tissue. The highest concentrations of cadmium in the area were found in isopods and earthworms. Cadmium accumulation in the carnivorous common shrew (*Sorex araneus*) was high compared with the omnivorous field mouse (*Apodemus sylvaticus*) at the site of a derelict lead–zinc mine complex at Minera, North Wales (Roberts and Johnson, 1978). Lead accumulates in small mammals living on roadside verges (Jefferies and French, 1972; Mierau and Favara, 1975; Welch and Dick, 1975). Roberts and Johnson found that cadmium concentration in the food of *S. araneus* exceeded that in the diet of *A. sylvaticus* and of the field vole (*Microtus agrestis*) irrespective of the degree of site contamination. The concentration ratio for cadmium in *S. araneus* always exceeded unity but the ratios for lead and zinc, except for zinc at the control site, did not exceed unity. Lead and zinc concentrations were higher for *M. agrestis* than for the other two species which may reflect differences in diet.

Ireland (1977) compared the lead concentration ratios in bone and soft tissues of the clawed toad *Xenopus laevis*, fed on lead-contaminated earthworms, and the small rodent *Peromyscus maniculatus*. Although the two animals had similar bone lead concentrations, the ratios of bone lead to lead in the kidney, liver and muscle were considerably lower in the toad than in the rodent, indicating relatively more lead in the soft tissues of the toad. Lead situated in bone is relatively unavailable to digestion by predators and *Xenopus*, and possibly other amphibia, may therefore present a greater lead pollution hazard in the food chain than small mammals. This may not apply however to cadmium which in mammals accumulates mainly in the soft tissues (Friberg *et al.*, 1974).

21.5 DEPOSITION OF METALS IN TISSUES

There are few studies of the distribution of heavy metals in earthworm tissues but histochemical research by Wielgus-Serafinska and Kawka (1976) on *E. foetida* showed that lead accumulated in the glandular epithelial cells of the epidermis and intestine. In *D. rubida* from a lead–zinc-contaminated site (Ireland, 1975c) most of the lead and zinc was found in the posterior alimentary tract. Calcium was located mainly in the anterior alimentary tract which in *D. rubida* is the site of active calciferous glands containing a relatively high concentration of calcium (Piearce, 1972). The lowest concentrations of lead, zinc and calcium were in the body wall and small amounts of lead and zinc were also detected in the urine, mucus and coelomic fluid (Ireland, 1976). It is difficult to assess the importance of these sites in excretion of heavy metals since the rate of secretion of urine, mucus and coelomic fluid is not known. Mucus probably has an excretory function since it contains about half the excreted nitrogen (Needham, 1957). Many species of earthworms eject coelomic fluid through the dorsal pores in response to mechanical or chemical irritation. The dorsal pores are involved in the regulation of intracoelomic pressure and are also an important part of the excretory system. Foreign substances such as injected iron, Indian ink and bacteria are discharged in the coelomic fluid through the dorsal pores (Cameron, 1932).

Cadmium can profoundly effect the metabolism of calcium, copper, zinc, iron and manganese in vertebrates (Kirchgessner, 1978). Experimental exposure of *L. rubellus* and *A. caliginosa* to 5 μg cadmium g^{-1} dry wt. for 26 days resulted in accumulation of the metal but did not affect the concentrations of calcium, copper, zinc, iron or manganese. The post-clitellar region contained the highest cadmium concentrations which were higher in the intestine than in the body wall. Fleming and Richards (1981) subjected cadmium-treated earthworms to activated Sepharose 6B to distinguish between surface-associated and internalized cadmium, and observed that after this treatment 68 % of the cadmium associated with the body wall was removed and was therefore associated with the surface mucus.

21.5.1 Intracellular bodies

The main storage organ in oligochaetes is the chloragogenous tissue which consists of modified epithelial cells disposed as a thick layer around the gut. Electron-microscope study of the chloragocytes shows the presence of predominantly two types of cytoplasmic organelles, electron-dense, roughly spherical, chloragosomes (van Gansen and Vandermeerssche, 1958; Lindner, 1965) and irregularly shaped debris vesicles (Ireland and Richards, 1977). Chloragosomes are easy to isolate (Roots, 1957; Urich, 1959) and contain numerous organic compounds such as carbohydrates (van Gansen, 1956; Roots, 1960; Fischer, 1971a; Richards and Ireland, 1978), proteins (Fischer, 1971b), lipids, phospholipids and pigments (Urich, 1959; Roots and Johnston, 1966; Fischer, 1973). The chloragosomes also contain metals such as calcium, iron, zinc and magnesium together with phosphate which have been identified chemically, histochemically and by electron microscope X-ray microprobe analysis (Lindner, 1965; Prento, 1979; Wroblewski *et al.*, 1979; Ireland, 1978).

If lead salts are treated with potassium chromate under controlled conditions, insoluble lead chromate is formed. Sections cut through the intestinal region of *D. rubida* collected near a derelict lead–zinc mine showed the presence of lead chromate crystals in the chloragogenous tissue. X-ray microanalysis of contaminated *L. rubellus* and *D. rubida* (Ireland and Richards, 1977) revealed the lead to be deposited as electron-dense flecks associated with the chloragosomes and within the debris vesicles. Studies on isolated chloragosomes from non-contaminated *L. terrestris* have shown that these intracellular organelles bind lead and copper to a greater extent than body wall homogenates and that the chloragosomes have a higher affinity for lead than copper. This binding appears to be unrelated to the protein and phospholipid content (Ireland, 1978). *L. rubellus* and *A. caliginosa* exposed to cadmium for 26 days showed deposition of electron-dense flecks mainly within the debris vesicles, and over 70 % of the chloragocyte cadmium found by chemical analysis was not in the chloragosomes (Ireland and Richards, 1981). Since none of the five other metals analysed in the tissues increased after cadmium treatments, the electron-dense flecks observed with the electron microscope were assumed to be cadmium.

Fischer (1975) reported that isolated chloragosomes of *L. terrestris* contain redox pigments such as riboflavin, flavin nucleotides, thiamin, carotenoids and metalloporphyrins. He concludes that chloragosomes act as specific electron acceptors, substituting molecular oxygen in the metabolism when the environment becomes hypoxic.

A more important function of the chloragosomes as regards heavy metals is their cation-exchange properties (Fischer, 1973). Isolated chloragosomes also contain phosphoric acid, carboxyl, phenolic hydroxyl and sulphonic acid groups, all of which are constituents of ion-exchange compounds (Kunin and Myers, 1952). The addition of lead to a chloragosomal suspension caused a drop in pH and release of mainly calcium, indicating that cation exchange was taking place (Ireland, 1978). Earthworm chloragosomes may therefore function as a cation-exchange system capable of taking up and retaining heavy metals, thus reducing their toxic effects.

21.5.2 Metal–protein complexes

Metals associated with a protein moiety may be involved in respiration. These associations include the cuproprotein haemocyanin and the iron-containing respiratory pigments including the haemoglobins, erythrocruorins and haemerythrins. Metals are also an integral part of numerous enzymes such as amylases (calcium), succinate dehydrogenase (iron), cytochrome oxidase (copper) and carbonic anhydrase (zinc) (Vallee and Wacker, 1970). Non-enzymic metalloproteins function in storage, transport or detoxification systems. Ferritins are proteins with a capacity for storing iron while transferritins are glycoproteins responsible for the transport of iron among sites of absorption, storage and excretion.

Metalloproteins are associated with the cytoplasmic fraction of animal tissues exposed to high levels of the heavy metals bismuth, cadmium, copper, gold, mercury, silver and zinc (Cherian and Goyer, 1978). These proteins are characterized by low molecular weight, a peak absorption of 250 nm, high cysteine content and absence of aromatic amino acids. Metallothioneins are normally found only in traces but exposure to sublethal amounts of heavy metals induces increased synthesis.

Fresh starved *D. rubida*, collected from a lead–zinc mine site, were homogenized at different pH values and after centrifugation the residue and supernatant were analysed for lead, zinc and calcium (Ireland, 1975c). The pH of whole earthworms homogenized in distilled water was 6.9 and at this pH less than 20% of the lead and zinc was found in the soluble fraction but over 80% of the protein. Most of these metals are therefore not present in an aqueous soluble form. Over 90% of the calcium was located in the aqueous fraction which may be explained by the presence of high levels of calcium carbonate in earthworms. At lower pHs of 1.0 and 3.1, more lead and zinc was found in the soluble fraction. This suggests that the metals were present as soluble chlorides and acetates because of the buffers used, or were released from an organic ligand.

In subsequent experiments using gel filtration on Sephadex G-25 at pH 7.0, < 3% of the lead was eluted in the fast moving fraction or associated with high molecular weight substances. Baltrop and Smith (1971), using radioactive lead, reported that lead in red blood corpuscles is bound to low molecular weight material in addition to haemoglobin. Gel filtration of earthworm tissue homogenates at pH 3.1, 5.7 and 7.0 produced a gradual decrease in the amounts of zinc and calcium in the high molecular weight fraction. This can be accounted for by the reduction in pH which can result in release of metallic ions from metalloenzymes (Vallee and Wacker, 1970). Numerous zinc and calcium metalloenzymes have been identified from earthworm tissue (Florkin and Scheer, 1969). The results for lead are more difficult to explain and may be attributable to an interaction with the Sephadex (Tarutani and Watanabe, 1973) but at the pH of aqueous tissue homogenates, very little lead is associated with high molecular weight substances.

The ubiquitous cadmium-binding protein is probably present in oligochaetes. Gel filtration on Sephadex G-75 of whole tissue homogenate from *E. foetida* exposed to cadmium showed that three different molecular weight proteins were associated with this metal (Suzuki *et al.*, 1980). The three proteins were resistant to heat denaturation, their absorption spectra suggesting that they are rich in cysteinyl residues and poor in aromatic amino acids. A close similarity to a metallothionein was found in at least one of the protein fractions (Cherian and Goyer, 1978). As in the case of the slug *Arion ater* (Ireland, 1981) and vertebrates (Nordberg, 1978), the cadmium-binding proteins were not accompanied by increased zinc and copper contents despite high concentrations of these metals in the compost on which the earthworms were fed.

Recent work (Ireland, unpublished) on *L. terrestris* treated with 5 µg cadmium g^{-1} dry wt. for 14 days confirms Ireland and Richards' (1981) conclusion that very little of the cadmium is associated with the chloragosomes. The supernatant after disruption of isolated chlorago-cytes was placed on a Sephadex G-50 column at pH 8.0 and the absorption monitored at 250 nm (Fig. 21.2). The cadmium-treated earthworm fraction showed a peak of absorption in the mol. wt. range 9000, coinciding with a second peak of cadmium activity (A). This peak was not present in a control earthworm fraction, but the comparable elution volume was collected and freeze-dried (B). Cadmium-treated (A) and control (B) protein fractions were subsequently passed through a small

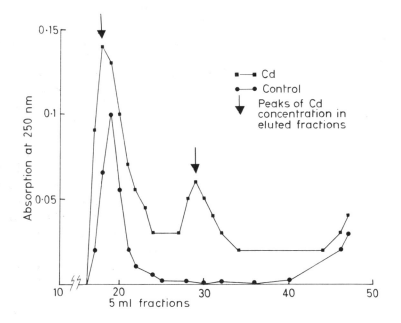

Fig. 21.2 Sephadex G-50 chromatography of *L. terrestris* particle-free chlora-gocyte fraction in cadmium-treated and control earthworms.

Sephadex G-15 column at pH 8.0 and with cadmium added to B. All of the cadmium in A was eluted in the protein fractions while in the B sample all the cadmium was eluted in the low molecular weight fraction and was not bound to the protein. The protein isolated from the control earthworms was not capable of binding cadmium at pH 8.0, suggesting that the cadmium-binding protein is not normally present at any appreciable concentration in earthworms from unpolluted sites.

21.6 CONCLUSION

It is now recognized that the extent to which earthworms accumulate heavy metals differs between species and that various physiological factors are involved. There is perhaps much now to be learned by recognizing that earthworms are not a form of standard laboratory material but that ecological factors such as diet and season can also influence their body burden of heavy metals.

Heavy metals enter primary producers not only by root uptake but by absorption from aerial deposits on leaves and stems (Hughes *et al.*, 1980). Since earthworms show marked preferences for certain types of leaves (Satchell and Lowe, 1967; Piearce, 1978), their feeding behaviour may result in tissue metal levels bearing little relation to concentrations in the soil.

In temperate climates, maximum activity of soil-dwelling earthworms occurs when the soil temperature is between 4 and 11°C and when the moisture content is high. It is perhaps no coincidence that the highest temporal body burden of lead in earthworms reported by Ireland and Wootton (1976) was obtained when the temperature was within this range and moisture levels were conducive to maximum activity.

These two examples illustrate the need in future investigations to take account of seasonal factors, behavioural characteristics and food sources in assessing the relationships between heavy metals in the environment and the body burden in earthworms.

21.7 REFERENCES

Alloway, B. J. and Davies, B. E. (1971) Trace element content of soils affected by base metal mining in Wales. *Geoderma*, **5**, 197–208.

Andersen, C. (1979) Cadmium, lead and calcium content, number and biomass, in earthworms (Lumbricidae) from sewage sludge treated soil. *Pedobiologia*, **19**, 309–319.

Antonovics, J., Bradshaw, A. D. and Turner, R. G. (1971) Heavy metal tolerance in plants. *Adv. Ecol. Res.*, **7**, 1–85.

Ash, C. P. J. and Lee, D. L. (1980) Lead, cadmium, copper and iron in earthworms from roadside sites. *Environ. Pollut.*, **22A**, 59–67.

Baltrop, D. and Khoo, H. E. (1976) The influence of dietary minerals and fat on the absorption of lead. *Sci. Total Environ.*, **6**, 265–273.

Baltrop, D. and Smith, A. (1971) Interaction of lead with erythrocytes. *Experientia*, **27**, 92–93.

Beeby, A. (1978) Interaction of lead and calcium uptake by the woodlouse, *Porcellio scaber* (Isopoda, Porcellionidae). *Oecologia (Berl.)*, **32**, 255–262.

Beyer, W. H. (1981) Metals and terrestrial earthworms (Annelids: Oligochaeta). In *Workshop on the Role of Earthworms in the Stabilization of Organic Residues* (M. Appelhof, Compiler), Vol. 1, Proc., Beech Leaf Press, Kalamazoo, MI, pp. 137–150.

Cameron, G. R. (1932) Inflammation in earthworms. *J. Pathol. Bacteriol.*, **35**, 933–972.

Cherian, M. G. and Goyer, R. A. (1978) Metallothioneins and their role in the metabolism and toxicity of metals. *Life Sci.*, **23**, 1–9.

Curry, J. P. and Cotton, D. C. F. (1980) Effects of heavy pig slurry contamination on earthworms in grassland. In *Soil Biology as Related to Land Use Practices* (ed. D. L. Dindal), *Proc. VIIth Int. Soil Zool. Coll., EPA, Washington D.C.*, pp. 336–343.

Czarnowska, K. and Jopkiewicz, K. (1978) Heavy metals in earthworms as an index of soil contamination. *Pol. J. Soil Sci.*, **11**, 57–62.

Dindal, D. L. (1975) Effects of sewage effluent disposal on community structure of soil invertebrates. In *Progress in Soil Zoology* (ed. J. Vanek), *Proc. Vth Int. Soil Zool. Coll., Academia, Prague*, pp. 419–427.

Dindal, D. L., Newell, L. T. and Moreau, J. (1979) Municipal wastewater irrigation: Effects on Community Ecology of Soil Invertebrates. In *Municipal Wastewater and Sludge Recycling on Forest and Disturbed Land*. Pennsylvania State Univ. Press, University Park. pp. 197–205.

Edwards, C. A. and Lofty, J. R. (1977) *Biology of Earthworms*. 2nd edn. Chapman & Hall, London. 333 pp.

Fischer, E. (1971a) The carbohydrate content of chloragosomes in *Lumbricus terrestris*. *Acta Biol. Acad. Sci. Hung.*, **22**, 343–350.

Fischer, E. (1971b) Amino acids and proteins in the chloragosomes of *Lumbricus terrestris*. *Acta Biol. Acad. Sci. Hung.*, **22**, 365–368.

Fischer, E. (1973) The chloragosomes of Lumbricidae as cation exchangers. *Acta Biol. Acad. Sci. Hung.*, **24**, 157–163.

Fischer, E. (1975) Chloragosomes of Lumbricidae as specific electron acceptors. *Acta Biol. Acad. Sci. Hung.*, **26**, 135–140.

Fleming, T. P. and Richards, K. S. (1981) A technique to quantify surface adsorption of heavy metals by soft-bodied invertebrates. *Comp. Biochem. Physiol.*, **69C**, 391–394.

Florkin, M. and Scheer, B. T. (1969) *Chemical Zoology, Vol IV. Annelida, Echiura and Sipuncula*. Academic Press, London.

Friberg, L., Piscator, M., Nordberg, G. G. and Kjellstrom, T. (1974) *Cadmium in the Environment*. CRC Press, Cleveland.

Gish, C. D. and Christensen, R. E. (1973) Cadmium, nickel, lead and zinc in earthworms from roadside soil. *Environ. Sci. Technol.*, **7**, 1060–1062.

Hartenstein, R., Neuhauser, E. F. and Collier, J. (1980a) Accumulation of heavy metals in the earthworm *Eisenia foetida*. *J. Environ. Qual.*, **9**, 23–26.

Hartenstein, R., Leaf, A. L., Neuhauser, E. F. and Bickelhaupt, D. H. (1980b) Composition of the earthworm *Eisenia foetida* (Savigny) and assimilation of 15 elements from sludge during growth. *Comp. Biochem. Physiol.*, **66C**, 187–192.

Helmke, P. A., Robarge, W. P., Korotev, R. L. and Schomberg, P. J. (1979) Effects of soil-applied sewage sludge on concentrations of elements in earthworms. *J. Environ. Qual.*, **8**, 322–327.

Hogg, T. A. (1895) Immunity of some low forms of life from lead poisoning. *Chem. News*, **71**, 223–224.

Hughes, M. K., Lepp, N. W. and Phipps, D. A. (1980) Aerial heavy metal pollution and terrestrial ecosystems. In *Advances in Ecological Research* (ed. A. Macfadyen), Academic Press, London, pp. 218–327.

Ireland, M. P. (1975a) The effect of the earthworm *Dendrobaena rubida* on the solubility of lead, zinc and calcium in heavy metal contaminated soil in Wales. *J. Soil Sci.*, **26**, 313–318.

Ireland, M. P. (1975b) Metal content of *Dendrobaena rubida* (Oligochaeta) in a base metal mining area. *Oikos*, **26**, 74–79.

Ireland, M. P. (1975c) Distribution of lead, zinc and calcium in *Dendrobaena rubida* (Oligochaeta) living in soil contaminated by base metal mining in Wales. *Comp. Biochem. Physiol.*, **52B**, 551–555.

Ireland, M. P. (1976) Excretion of lead, zinc and calcium by the earthworm *Dendrobaena rubida* living in soil contaminated with zinc and lead. *Soil Biol. Biochem.*, **8**, 347–350.

Ireland, M. P. (1977) Lead retention in toads *Xenopus laevis* fed increasing levels of lead-contaminated earthworms. *Environ. Pollut.*, **12**, 85–92.

Ireland, M. P. (1978) Heavy metal binding properties of earthworm chloragosomes. *Acta Biol. Acad. Sci. Hung.*, **29**, 385–394.

Ireland, M. P. (1979) Metal accumulation by the earthworms *Lumbricus rubellus*, *Dendrobaena veneta* and *Eiseniella tetraedra* living in heavy metal polluted sites. *Environ. Pollut.*, **19**, 201–206.

Ireland, M. P. (1981) Uptake and distribution of cadmium in the terrestrial slug *Arion ater* (L.). *Comp. Biochem. Physiol.*, **68A**, 37–41.

Ireland, M. P. and Richards, K. S. (1977) The occurrence and localisation of heavy metals and glycogen in the earthworms *Lumbricus rubellus* and *Dendrobaena rubida* from a heavy metal site. *Histochemistry*, **51**, 153–166.

Ireland, M. P. and Richards, K. S. (1981) Metal content after exposure to cadmium of two species of earthworms of known differing calcium metabolic activity. *Environ. Pollut.*, **26A**, 69–78.

Ireland, M. P. and Wootton, R. J. (1976) Variations in the lead, zinc and calcium content of *Dendrobaena rubida* (Oligochaeta) in a base metal mining area. *Environ. Pollut.*, **10**, 201–208.

Jefferies, D. J. and French, M. C. (1972) Lead concentrations in small mammals trapped on roadside verges and field sites. *Environ. Pollut.*, **3**, 147–156.

Kirchgessner, M. (1978) *Trace Element Metabolism in Man and Animals*. 3, Freising-Weihenstephan, Technische Universitat Munchen.

Kunin, R. and Myers, R. J. (1952) *Ion Exchange Resins*. Wiley, New York.

Lindner, E. (1965) Ferritin und Hämoglobin im Chloragog von Lumbriciden (Oligochaeta). *Z. Zellforsch Mikrosk. Anat.*, **66**, 891–913.

Martin, M. H. and Coughtrey, P. J. (1975) Preliminary observations on the levels of cadmium in a contaminated environment. *Chemosphere*, **3**, 155–160.

Martin, M. H. and Coughtrey, P. J. (1976) Comparisons between the levels of lead, zinc and cadmium within a contaminated environment. *Chemosphere*, **5**, 15–20.

Mierau, C. A. and Favara, B. E. (1975) Lead poisoning in roadside populations of deer mice. *Environ. Pollut.*, **8**, 55–64.

Mori, T. and Kurihara, Y. (1979) Accumulation of heavy metals in earthworms (*Eisenia foetida*) grown in composted sewage sludge. *Sci. Rep. Tohoku Univ., Ser. IV (Biol.)*, **37**, 289–297.

Needham, A. E. (1957) Components of nitrogenous excreta in the earthworms *Lumbricus terrestris* L. and *Eisenia foetida* (Savigny). *J. Exp. Biol.*, **34**, 425–446.

Neuhauser, E. F. and Hartenstein R. (1980) Efficiencies of extractants used in analyses of heavy metals in sludges. *J. Environ. Qual.*, **9**, 21–22.

Nordberg, M. (1978) Studies on metallothionein and cadmium. *Environ. Res.*, **15**, 381–404.

Petering, H. G. (1974) The effect of cadmium and lead on copper and zinc metabolism. In *Trace Element Metabolism in Animals* (eds. W. G. Hoekstra, J. W. Suttie, H. E. Ganther and W. Mertz), Butterworths, London, pp. 311–325.

Piearce, T. G. (1972) The calcium relations of selected Lumbricidae. *J. Anim. Ecol.*, **41**, 167–188.

Piearce, T. G. (1978) Gut contents of some lumbricid earthworms. *Pedobiologia*, **18**, 153–157.

Prentø, P. (1979) Metals and phosphate in the chloragosomes of *Lumbricus terrestris* and their possible physiological significance. *Cell Tissue Res.*, **196**, 123–134.

Raw, F. and Dobson, R. M. (1959) Rehabilitation of marginal grassland. *Rep. Rothamsted Exp. Stn. 1958*, 141–143.

Raw, F. and Lofty, J. R. (1960) Earthworm populations in orchards. *Rep. Rothamsted Exp. Stn. 1959*, 134–135.

Richards, K. S. and Ireland, M. P. (1978) Glycogen–lead relationship in the earthworm *Dendrobaena rubida* from a heavy metal site. *Histochemistry*, **56**, 55–64.

Roberts, R. D. and Johnson, M. S. (1978) Dispersal of heavy metals from abandoned mine workings and their transference through terrestrial food chains. *Environ. Pollut.*, **16**, 293–310.

Roots, B. I. (1957) Nature of chloragogen granules. *Nature (London)*, **179**, 679–680.

Roots, B. I. (1960) Some observations on the chloragogenous tissue of earthworms. *Comp. Biochem. Physiol.*, **1**, 218–226.

Roots, B. I. and Johnston, P. V. (1966) The lipids and pigments of the chloragosomes of the earthworm *Lumbricus terrestris* L. *Comp. Biochem. Physiol.*, **17**, 285–288.

Satchell, J. E. and Lowe, D. G. (1967) Selection of leaf litter by *Lumbricus terrestris*. In *Progress in Soil Biology* (eds. O. Graff and J. E. Satchell), North-Holland Publ. Co., Amsterdam, pp. 102–119.

Shields, J. B. and Mitchell, H. H. (1941) The effect of calcium and phosphorus on the metabolism of lead. *J. Nutr.*, **21**, 541–552.

Suzuki, K. T., Yamamura, M. and Mori, T. (1980) Cadmium-binding proteins induced in the earthworm. *Arch. Environ. Contam. Toxicol.*, **9**, 415–424.

Tarman, C. and Raw, F. (1961) Earthworm populations in orchards. *Rep. Rothamsted Exp. Stn. 1960*, 157.

Tarutani, T. and Watanabe, M. (1973) Chromatographic behaviour of some divalent metal ions on sodium chloride, sulphate and nitrate solutions on a Sephadex G-10 column. *J. Chromatogr.*, **75**, 169–289.

Urich, K. (1959) Über den Stoffbestand der Chloragosomen von *Lumbricus terrestris*. *Zool. Beit.*, **5**, 281–289.

Vallee, B. L. and Wacker, E. C. (1970) Metalloproteins. In *The Proteins* (ed. H. Neurath), Vol. V., Academic Press, New York, pp. 30–60.

van Gansen, P. S. (1956) Les cellules chloragogenes des Lombriciens. *Bull. Biol. Fr. Belg.*, **90**, 335–356.

van Gansen, P. S. and Vandermeerssche, G. (1958) L'ultrastructure des cellules chloragogenes. *Bull Microsc. Appl.*, **8**, 7–13.

Van Rhee, J. A. (1967) Development of earthworm populations in orchard soils. In (eds. O. Graff and J. E. Satchell) *Progress in Soil Biology*. North-Holland Publishing Co., Amsterdam. pp. 360–371.

Van Rhee, J. A. (1975) Copper contamination effects on earthworms by disposal of pig waste in pastures. In *Progress in Soil Zoology* (ed. J. Vanek), *Proc. Vth Int. Soil Zool. Coll., Academia, Prague*, pp. 451–456.

Van Rhee, J. A. (1977) Effects of soil pollution on earthworms. *Pedobiologia*, **17**, 201–208.

Welch, W. R. and Dick, D. L. (1975) Lead concentrations in tissues of roadside mice. *Environ. Pollut.*, **8**, 15–21.

Wielgus-Serafinska, E. and Kawka, E. (1976) Accumulation and localization of lead in *Eisenia foetida* (Oligochaeta) tissues. *Folia Histochem. Cytochem.*, **14**, 315–320.

Williamson, P. and Evans, P. R. (1972) Lead levels in roadside invertebrates and small mammals. *Bull. Environ. Contam. Toxicol.*, **8**, 280–288.

Williamson, P. and Evans. P. R. (1973) A preliminary study of the effects of high levels of inorganic lead on soil fauna. *Pedobiologia*, **13**, 16–21.

Wright, M. A. and Stringer, A. (1980) Lead, zinc and cadmium content of earthworms from pasture in the vicinity of an industrial smelting complex. *Environ. Pollut.*, **23A**, 313–321.

Wroblewski, R., Roomans, G. M., Ruusa, J. and Hedberg, B. (1979) Elemental analysis of histochemically defined cells in the earthworm *Lumbricus terrestris*. *Histochemistry*, **61**, 167–176.

Chapter 22

Heavy metals in earthworms in non-contaminated and contaminated agricultural soil from near Vancouver, Canada

A. CARTER, E. A. KENNEY,
T. F. GUTHRIE and H. TIMMENGA

22.1 INTRODUCTION

The objectives of this study were (1) to describe the concentrations and amounts of cadmium, zinc and copper in earthworms, their food and predators and (2) to evaluate the effects of applications of sewage sludge with high levels of these metals on earthworms and their food. To achieve this, quantitative sampling was combined with laboratory experiments.

22.2 METHODS

22.2.1 Field sampling

A non-contaminated soil on Westham Island, approximately 40 km from Vancouver, British Columbia, was sampled. The soil was a silty clay loam derived from deltaic deposits and was of pH 4.1 (1:2 in 0.01 M $CaCl_2$).

At each sampling, six to eight random samples were taken along a line transect in a plot of 2500 m^2 in a field of red clover. Two soil blocks, 0.063 m^2 × 0.15 m, were taken from each sampling position. One set was handsorted for earthworms in the laboratory. Arthropods were extracted from the other with large controlled-draught Tullgren funnels.

22.2.2 Egestion experiments

Chromic oxide powder was used as an indicator. The powder was thoroughly mixed with wetted soil and air-dried until moist. Fifteen Petri dishes were filled with the coloured soil and two mature *Lumbricus rubellus* Hoffm. were added to each and kept at 15°C. After 24 hours, the worms were transferred to clean soil and after a further 24 hours were returned to the chromic oxide–soil mixture. Faeces were collected daily using a microscope. Faeces and earthworms were dried at 65°C for four days and weighed on a micro-balance. The experiments were carried out in March 1980 and July 1981.

22.2.3 Sewage sludge experiments

Milorganite sewage sludge from Milwaukee, US was analysed for heavy metals. The mean concentrations and standard deviations in mg kg^{-1} were: Cd 99 (31), Cu 320 (36), Ni 67 (7), Pb 573 (78) and Zn 792 (31).

Soil collected from Westham Island and sewage sludge was air-dried for eight days and mixed in various proportions to give 0, 5, 10, 25, 50 and 100 g sludge kg^{-1} soil. The control and sludge–soil mixtures were added to two litre flower pots with six replicates of each treatment and were watered to field capacity. Four mature *L. rubellus* were added to each container, the tops and bottoms of which were covered with fine cheesecloth to prevent the worms from escaping and to reduce evaporation. The cultures were then kept in a growth chamber at 15°C for ten days.

22.2.4 Sample preparation and chemical analyses

Earthworms and beetles were washed with distilled water and their guts voided over four days. They were then dried at 65°C for four days and weighed on a micro-balance. Litter, soil and sewage sludge–soil mixtures were also dried at 65°C and weighed.

Samples were digested as described in Carter *et al.* (1980). Animals were digested with nitric acid in small glass vials. Litter and soil were digested in 100 ml glass beakers (after Van Loon and Lichwa, 1973) on a hot plate. Digests were analysed using an atomic absorption spectrophotometer, with deuterium arc background compensation. National Bureau of Standards bovine liver, orchard leaves and soil were used as standards.

22.3 RESULTS

22.3.1 Non-contaminated soil

Heavy metal concentrations in earthworms, their general food items and predators are shown in Fig. 22.1. The concentrations of metals in the organic and mineral matter consumed by the worms were similar. Zn and Cd were assimilated and concentrated by *L. rubellus* and *Allolobophora*

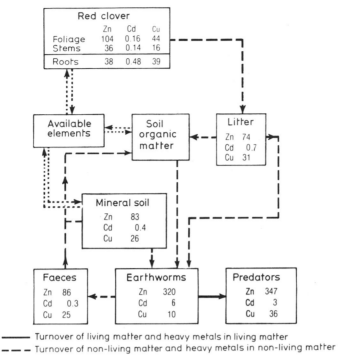

Turnover of living matter and heavy metals in living matter
Turnover of non-living matter and heavy metals in non-living matter
Chemical transformations in relation to root uptake and release

Fig. 22.1 Heavy metal concentrations (mg kg^{-1} dry wt.) in earthworms, their surroundings and predators.

Table 22.1 Amounts of heavy metals in earthworms, their food and egesta.

	No. m^{-2} ($\bar{x} \pm$ SD)	Biomass (g dry wt. m^{-2})	Cd (mg m^{-2})	Cu (mg m^{-2})	Zn (mg m^{-2})
Contaminated soil					
Sewage sludge	—	707.9	70.1	226.5	560.6
Mineral soil	—	156.3 × 10^3	63	4064	12973
Earthworms	—	13.23	0.423	0.238	3.929
Egestion rate (g dry wt. 2 weeks^{-1})	—	170.4	1.08	9.88	15.32
Non-contaminated soil					
Surface litter	—	225.7	0.065	3.160	8.35
Mineral soil	—	115.0 × 10^3	46	2990	9545
Earthworms	640 ± 184	13.23	0.079	0.132	4.23
Egestion rate (g dry wt. 2 weeks^{-1})	—	170.4	0.063	4.43	14.14
Arthropod predators	11 ± 13	0.28	0.000	0.004	0.032

chlorotica Sav. as found in other earthworms by Van Hook (1974), Ireland (1979) and Helmke *et al.* (1979). Predatory carabid larvae and spiders also had high Zn and Cd contents.

Earthworms were abundant and because they concentrated Zn and Cd strongly in their tissues, the amounts of these metals in their biomasses were extremely high. The amount of Cd exceeded that in the surface litter (Table 22.1). Amounts of Cd and Cu egested by the earthworm population in two weeks at 15 °C were similar to those in litter but low in comparison with mineral soil.

22.3.2 Contaminated soil

The amounts of heavy metals in earthworms kept in contaminated soil in the laboratory are presented in Table 22.1. Large amounts of organic matter with high levels of metals were introduced in the sewage sludge and the amounts of metals in this material were much higher than in surface litter from the non-contaminated soil.

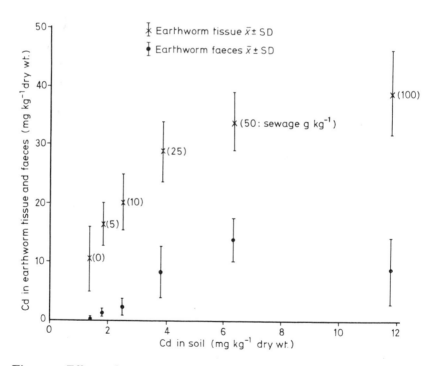

Fig. 22.2 Effects of sewage sludge applications on cadmium levels in sexually mature *Lumbricus rubellus*. Rates of sludge application (g kg^{-1} of air-dried soil) in parentheses.

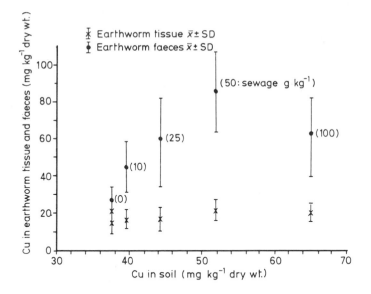

Fig. 22.3 Effects of sewage sludge applications on copper levels in sexually mature *L. rubellus*. Rates of sludge applications in parentheses.

The effects of sewage sludge applications on concentrations of heavy metals in the tissues and faeces of *L. rubellus* are shown in Figs 22.2–22.4. Cd levels in earthworm tissue and faeces increased steadily as those in the soil–sludge mixture rose to approximately 6 mg kg^{-1} dry wt., after which concentrations levelled off (Fig. 22.2). By contrast, Cu levels in earthworm tissue did not change with increasing level in the substrate (Fig. 22.3). Concentrations in the faeces had high variances but showed a definite pattern of increase until substrate levels exceeded 50 mg kg^{-1}. Zn levels in worm tissue and faeces were not significantly different in the different substrates although the concentration in the faeces of worms kept in 50 g sludge kg^{-1} soil was significantly different ($P < 0.05$, Duncan's Multiple Range Test) from preceding treatments (Fig. 22.4).

The results suggest that there was regulation of Zn and Cu which are essential elements for many organisms, but not of the non-essential Cd. This metal is very mobile and may be readily incorporated into soft and non-calcareous tissues.

22.4 DISCUSSION

Earthworms recycle elements by egestion, excretion, mortality and cocoon production. In our study, the amounts of heavy metals in the

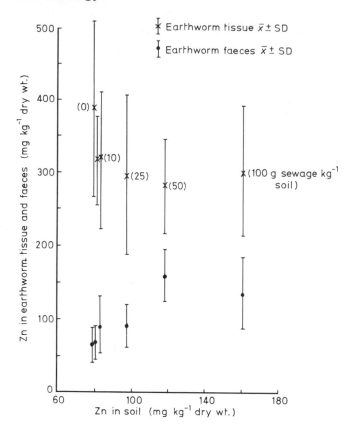

Fig. 22.4 Effects of sewage sludge applications on zinc levels in sexually mature *L. rubellus*. Rates of sludge applications in parentheses.

earthworm biomass were low compared with those in the sewage sludge. In the non-contaminated soil, Cd in the earthworm biomass was high in relation to that in the surface litter. Earthworm mortality would therefore recycle large amounts of this metal. Surface litter is only one component of the pool of large organic matter which also includes dead roots. The amount of Cd egested by the earthworm population was much higher in the contaminated soil but there was little difference between the two soils in the amounts of Cu and Zn egested (Table 22.1). Whereas the Cd levels in the faeces increased as the sludge concentration increased up to 25 g sludge kg^{-1} soil, the Cu and Zn levels showed no relationship with sludge concentration. Helmke *et al.* (1979) studied the effects of sewage sludge applications on earthworms in field plots and found Cd levels up to 100 mg kg^{-1} dry wt. in tissues. Mean concentrations in our worms did not exceed 40 mg kg^{-1} but our study ran for ten days while that of Helmke *et al.* continued for three years.

Mori and Kurihara (1979) added large amounts of Cd and Zn salts to manure and found extremely high concentrations of these metals, up to 400 and 1400 mg kg^{-1} dry wt. for Cd and Zn respectively. These levels are much higher than any observed in worms living in soil with heavy applications of sewage sludge. The metals in the sludge are probably less available because they are complexed with organic matter (Hartenstein *et al.*, 1980). Heavy metals in particulate matter from automobile exhaust, coal burning or metal-refining plants fall on to forests and agricultural land. In this form, as with inorganic salts, they may be more available to earthworms than are metals in sewage sludge. The high levels of Zn in tissues of earthworms collected near US highways (Gish and Christensen, 1973) may originate in particulate matter.

22.5 ACKNOWLEDGEMENTS

We thank Professor E. Steen (Swedish University of Agricultural Sciences), Carina Lindström and B. von Spindler, for their assistance. The study was funded by Agriculture Canada, the University of British Columbia and the Natural Sciences and Engineering Research Council of Canada.

22.6 REFERENCES

Carter, A., Kenney, E. A. and Guthrie, T. F. (1980) Earthworms as biological monitors of changes in heavy metal levels in an agricultural soil in British Columbia. In *Soil Biology as Related to Land Use Practices* (ed. D. L. Dindal), *Proc. VIIth Int. Soil Zool. Coll., EPA, Washington D.C.*, pp. 344–367.

Gish, C. D. and Christensen, R. E. (1973) Cadmium, nickel, lead and zinc in earthworms from roadside soil. *Environ. Sci. Technol.*, **7**, 1060–1062.

Hartenstein, R., Neuhauser, E. F. and Collier, J. (1980) Accumulation of heavy metals in the earthworm *Eisenia foetida. J. Environ. Qual.*, **9**, 23–26.

Helmke, P. A., Robarge, W. P., Korotev, R. L. and Schomberg, P. J. (1979) Effects of soil-applied sewage sludge on concentrations of elements in earthworms. *J. Environ. Qual.*, **8**, 322–327.

Ireland, M. P. (1979) Metal accumulation in the earthworms *Lumbricus rubellus, Dendrobaena veneta* and *Eiseniella tetraedra* living in heavy metal polluted sites. *Environ. Pollut.*, **19**, 201–206.

Mori, T. and Kurihara, Y. (1979) Accumulation of heavy metals in earthworms (*Eisenia foetida*) grown in composted sewage sludge. *Sci. Rep. Tohoku Univ., Ser. IV (Biol.)*, **37**, 289–297.

Van Hook, R. I. (1974) Cadmium, lead and zinc distributions between earthworms and soils: potentials for biological accumulation. *Bull. Environ. Contam. Toxicol.*, **12**, 509–511.

Van Loon, J. C. and Lichwa, J. (1973) A study of the atomic absorption determination of some important heavy metals in fertilizers and domestic sewage plant sludges. *Environ. Lett.*, **4**, 1–8.

Chapter 23

Earthworms and TCDD (2,3,7,8-tetrachlorodibenzo-*p*-dioxin) in Seveso

G. B. MARTINUCCI, P. CRESPI, P. OMODEO,
G. OSELLA and G. TRALDI

23.1 INTRODUCTION

On 10th July 1976 after an explosion in a chemical plant near Seveso a vast inhabited area was contaminated by certain chemicals including the extremely toxic 2,3,7,8-tetrachlorodibenzo-*p*-dioxin (TCDD).

At the request of the Lombardia Region we investigated the earthworm fauna in zones with different degrees of contamination from November 1979 to March 1981 (Fig. 23.1). Our aims were:

1. To evaluate the species in the fauna of a highly polluted area and to compare them with those of similar adjoining areas with little or no pollution
2. To establish the contamination ratio in the soil, earthworms and casts of the three commonest species in the most contaminated soil horizon
3. To evaluate the extent of soil mixing by earthworm burrowing and consequently the amount of TCDD transported
4. To assess the role of earthworms in passing TCDD along food chains.

23.2 MATERIALS AND METHODS

23.2.1 Study sites

In 1976 the polluted area of the communes involved in the accident at the ICMESA chemical plant was divided into zones A, B and R, where A was the most contaminated and R the least. Table 23.1 describes the size,

275

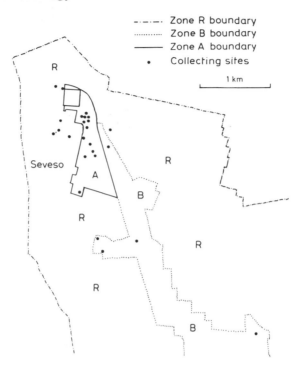

Fig. 23.1 Map of the contaminated zones showing sampling sites for earthworms, wormcasts and small mammals.

Table 23.1 Size, population and TCDD concentration in soil of contaminated zones. (Taken from Fanelli *et al.* (1980b).)

				TCDD level in soil			
				$(\mu g\ m^{-2})$		$(\mu g\ kg^{-1})$	
Zone	Area (ha)	Perimeter (km)	Human population	$\bar{x}$	range	$\bar{x}$	range
A	108	6	739	240	n.d.*–5 477	1.6	n.d.*–36.51
B	269.4	16.5	4 699	3	n.d.*–43.83	0.02	n.d.–0.29
R	1430	26	31 800	—	n.d.–5	—	n.d.–0.03

* n.d. (not determinable) $< 0.750\ \mu g\ m^{-2}$ or $0.005\ \mu g\ kg^{-1}$.

population and degree of TCDD contamination of the top 7 cm of soil in these three zones (Regione Lombardia, 1977; Adamoli *et al.*, 1978). A fourth, clear zone, C, was added as a control, representative of the uncontaminated environment of Brianza, with apparently similar pedological features and vegetation: i.e. fields long left uncultivated and meadows with Graminaceae. Samples of earthworms, worm casts and small mammals and numerous soil samples were collected from each of the four zones.

23.2.2 Collection and analysis of earthworms

Earthworms were collected in zones A and C in autumn 1979, spring–summer 1980 and autumn 1980; in B and R, earthworms were also collected in the spring of 1981. These were periods of maximum earthworm activity. The populations were sampled by handsorting after digging and chemical extraction using formalin solution (0.55%). Specimens reserved for taxonomic examination were fixed *in situ* in 5% formalin (alcohol dissolves lipids and may also dissolve TCDD).

Specimens of the three commonest species were sent alive for analysis of the TCDD content of the body tissues to the Institute of Pharmacological Research 'M. Negri' at Milan, where they were anaesthetized, classified and the earth content of the intestine removed. Samples were then deep-frozen for later analysis. The analysts were not told where the animals came from nor did the collectors know the exact concentration of TCDD at the collection sites. TCDD content was determined by the method of Fanelli *et al.* (1980a) for tests on rabbit liver. The limit of sensitivity is 0.25 μg kg^{-1}; results were expressed in μg kg^{-1} wet wt.

23.2.3 Collection and analysis of wormcasts

Wormcasts evacuated on the surface were collected from April to June 1980, one of the periods of greatest earthworm activity in Lombardy. We carefully protected six areas of 2 m × 1 m in zone A, two in zone R and one in zone C. Casts were collected fortnightly, air-dried, and weighed. Their TCDD content was determined at the Milan Public Health Laboratory.

23.2.4 Collection and analysis of small mammals

In autumn 1980, we trapped a number of small wild rodents in zones A and C. In zone A we caught one specimen of *Pitymys savii*, which feeds entirely on vegetable matter, four specimens of *Apodemus sylvaticus*, which are mainly phytophagous, and one dying specimen of *Rattus norvegicus*, an omnivore. None of the specimens was more than a year old. In zone C (Seveso – Altopiano and Via Marsala) we caught one specimen of *Sorex araneus*, an insectivore which also feeds on earthworms and slugs, two specimens of *A. sylvaticus* and one specimen of *Clethrionomys glareolus*, exclusively vegetarian.

The livers and the skins (including fur) of all these mammals were analysed separately. TCDD levels were calculated at the 'M. Negri' Institute as before.

23.3 RESULTS AND DISCUSSION

In all, 3433 earthworms were collected and examined (Table 23.2), 37% from zone A, 24% from zone B, 21% from zone R and 18% from zone C. They comprised 14 species and were subdivided into soil species, litter species and species from particular environments. Of the eight soil species, the most numerous were *Allolobophora chlorotica* and *A. rosea* which live among grass roots and in the topsoil between 2 and 10 cm, *A. caliginosa* which occurs mainly in the layers rich in humus (2 to 30 cm) and *Lumbricus terrestris* which burrows down to 30 cm but feeds in or on the topsoil.

We are concerned with the species inhabiting the most contaminated horizons, 2 to 20 cm, containing 90% of the toxin. *A. caliginosa*, *A. chlorotica* and *A. rosea* together made up 96% of the specimens of the soil species in zone A, 92% in zone B, 95% in zone R and 93% in zone C. The frequency of *L. terrestris*, found in every zone, was probably underestimated.

A. caliginosa was everywhere the most abundant species, never less than 50%. The ratio of *A. chlorotica* to *A. rosea*, which share the same soil horizon, varied from zone to zone: 1.5:1 in zone A, 2.5:1 in zone B, 1:7.5 in zone R, and 1:6.4 in zone C. This seems to indicate that *A. rosea* was more sensitive to dioxin. Of the three litter species, *Lumbricus castaneus* was most abundant, particularly in zones R and B which were similar to the other zones except that there was more leaf litter. The presence of *Eisenia foetida* and *Dendrobaena hortensis* was associated with compost heaps and anthropogenic environments; *Eiseniella tetraedra* was found in semiaquatic habitats. The soil species constituted about 79% of the total specimens collected, litter species 19%, and species of particular environments about 2%.

Cast production was estimated as $5.9 \, \mathrm{g \, m^{-2} \, day^{-1}}$ in zone A, $7 \, \mathrm{g \, m^{-2} \, day^{-1}}$ in zone R and $7.7 \, \mathrm{g \, m^{-2} \, day^{-1}}$ in zone C. Seasonal averages for the three zones were $530 \, \mathrm{g \, m^{-2}}$, $630 \, \mathrm{g \, m^{-2}}$ and $690 \, \mathrm{g \, m^{-2}}$. Extrapolating these estimates on the assumption that an earthworm is active 200 days a year, the amount of earth moved by earthworms was calculated as $11.8 \, \mathrm{t \, ha^{-1}}$ in zone A, $14 \, \mathrm{t \, ha^{-1}}$ in zone R and $15.4 \, \mathrm{t \, ha^{-1}}$ in zone C. These values are reasonably close and are similar to estimates from analogous soils in our latitudes (cf. Edwards and Lofty, 1977). They indicate that in the most contaminated zones earthworm activity was more or less normal.

Table 23.2 Earthworm numbers in the four zones of Seveso, autumn 1979 to spring 1981.

	Zone A		Zone B		Zone R		Zone C		Total	
	No.	%	No.	%	No.	%	No.	%	No.	%
Soil species										
A. caliginosa	597	56.2	398	66.9	233	49.8	288	49.6	1516	56.0
A. chlorotica	253	23.8	93	15.6	25	5.4	34	5.9	405	15.0
A. rosea	170	16.0	38	6.4	188	40.2	217	37.4	613	22.6
A. georgii	1	<1	—	—	—	—	6	1.0	7	<1
A. antipai	7	<1	—	—	—	—	17	2.9	24	<1
Allolobophora sp.	—	—	12	2.0	—	—	—	—	12	<1
Octolasion lacteum	1	<1	1	<1	3	<1	3	<1	8	<1
L. terrestris	18	1.7	44	7.4	10	2.1	10	1.7	82	3.0
L. rubellus	16	1.5	9	1.5	9	1.9	6	1.0	40	1.5
	1063		595		468		581		2707	
Litter species										
L. castaneus	144	93.5	203	96.2	232	92.1	41	100.0	620	94.2
Lumbricus sp.	1	<1	5	2.4	—	—	—	—	6	<1
D. octaedra	2	1.3	—	—	20	7.9	—	—	22	3.4
D. rubida	7	4.6	3	1.4	—	—	—	—	10	1.5
	154		211		252		41		658	
Stenotopic species										
E. foetida	64		—		—		—		64	
D. hortensis	1		—		—		—		1	
E. tetraedra	1		1		—		1		3	
	66		1		—		1		68	
Total	1283		807		720		623		3433	

Table 23.3 Relationship between TCDD content of earthworm tissues and soil.

Zones	No. specimens	Species	TCDD in earthworms (μg kg^{-1})	TCDD in soil (μg m^{-2})	TCDD in soil (μg kg^{-1})	TCDD in earthworms /TCDD in soil	TCDD in casts (μg kg^{-1})
A 1	15	*A. caliginosa*	196.8	1388	9.3	21.3	0.60
	10	*A. caliginosa*	104.9			11.3	
	7	*A. rosea*	73.1	884	5.9	12.4	0.86
	11	*A. caliginosa*	248.2			42.1	
	7	*A. rosea*	102.4	1388	9.3	11.1	
	15	*A. caliginosa*	107.6			11.6	
	16	*A. rosea*	9.5	102	0.7	14.0	
		A. rosea	n.d.			—	
A 3	5	*A. caliginosa*	7.0			5.0	
	16	*A. caliginosa*	9.3	209	1.4	6.7	
	16	*A. chlorotica*	7.6			5.5	
	8	*A. caliginosa*	1.7	38	0.25	6.8	
B	10	*A. caliginosa*	0.76	~10	0.06	12.6	
	9	*A. caliginosa*	0.40			6.6	
	10	*A. caliginosa*	0.83	5.7	0.04	21.8	
	11	*A. caliginosa*	1.06			27.9	
R	10	*A. caliginosa*	n.d.*	n.d.	n.d.	—	
C	9	*A. caliginosa*	n.d.	n.d.	n.d.	—	

* not determinable $< 0.005\ \mu$g kg^{-1} or $0.750\ \mu$g m^{-2}

TCDD content of two samples of wormcasts from part of zone A was determined as 0.6 and 0.86 $\mu g\ kg^{-1}$ (Table 23.3). This indicates that part of the toxin ingested by earthworms comes back to the surface, even if less concentrated. Considering the mass of earth transported by earthworms, it is quite certain that they are continually remixing dioxin into the soil.

All the species removed from the polluted zones concentrated dioxin in their tissues (Table 23.3), except for one sample out of 16 where dioxin was not measurable. Possibly these earthworms fed under some impermeable natural shelter. If this sample is excluded, the coefficient of concentration varied between 5 and 42 with an average about 14.5. Amounts of TCDD in the soil and in the earthworms show a more or less linear relationship; $y' = 2.38 + 0.96x'$, $r = 0.97$ (Fig. 23.2). Earthworms may thus be useful in demonstrating the presence of very low or immeasurable quantities of TCDD in the soil. We have no information about the accumulation of TCDD in earthworms in relation to its concentration in the soil; Gish (1970) however has shown some relationship between the total amount of organochlorine pesticide residues in the soil and the amount in earthworms, with an average concentration factor of about nine for all compounds and doses tested. Other authors

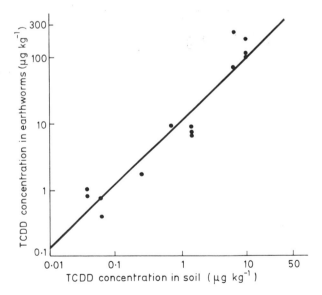

Fig. 23.2 TCDD level in earthworms as function of TCDD concentration in soil. *Soil*: points represent averages of nine determinations made in 1978 on an area of 150 m × 150 m. *Earthworms*: points represent analyses of 5–16 specimens of *A. caliginosa*, *A. rosea* or *A. chlorotica*.

(cf. Edwards and Lofty, 1977) indicate that not all earthworm species concentrate insecticides in the same way.

No earthworm predators were caught in the polluted zone but small rodents from zone C showed no traces of dioxin, whereas most of those from zone A had traces in the skin and more in the liver (Table 23.4). In zone A (45 ha), which is fenced, there are no recent traces of moles although there are signs that they were once present in the area. Mole burrows that are in use are present in large numbers in the unpolluted zones even quite close to zone A. Since earthworms are an important part of the diet of moles, one must presume that these predators died from TCDD poisoning.

Table 23.4 TCDD concentrations in the skin including fur and in the liver of some wild rodents, and average TCDD concentrations in the soil.

Zones	No. specimens	Species	TCDD in mammals		TCDD in soil	
			Skin (μg kg^{-1})	Liver (μg kg^{-1})	(μg m^{-2})	(μg kg^{-1})
	1	*A. sylvaticus*	0.4	7		
A 1	1	*A. sylvaticus*	0.7	31.4	102	0.68
	1	*A. sylvaticus*	0.7	0.4		
	1	*A. sylvaticus*	n.d.*	n.d.		
A 3	1	*P. savii*	n.d.	n.d.	38	0.25
A 5	1	*R. norvegicus*	n.d.	6.8	~10	0.06
	1	*S. araneus*	n.d.	n.d.	n.d.	n.d.
C	1	*A. sylvaticus*	n.d.	n.d.	n.d.	n.d.
	1	*A. sylvaticus*	n.d.	n.d.	n.d.	n.d.
	1	*C. glareolus*	n.d.	n.d.	n.d.	n.d.

* n.d. $< 0.005\ \mu$g kg^{-1} or $0.750\ \mu$g m^{-2}.

23.4 SUMMARY

We investigated the Seveso area involved in the accident at the ICMESA chemical plant, sampling earthworms, casts and small animals. Collections were made in three zones with different degrees of TCDD contamination and in a fourth unpolluted zone. Our results indicated that:

1. Earthworms which ingest contaminated soil accumulate dioxin in their tissues and concentrate it on average 14.5-fold
2. The three earthworm species analysed do not concentrate TCDD with any significant difference

3. The burrowing and casting activity of earthworms tends to bring TCDD back to the surface
4. Except for *A. rosea*, earthworms do not appear to suffer from an accumulation of the toxin
5. In zone A, which is fenced, moles have completely disappeared.

23.5 ACKNOWLEDGEMENTS

Financial support for this project from Regione Lombardia, Ufficio Speciale di Seveso and by grant no. 80.01747.90 from C.N.R. is acknowledged. We thank Professor R. Mitchell and Mrs Miranda Nardo Mitchell for help with the manuscript.

23.6 REFERENCES

Adamoli, P., Angeli, E., Bandi, G., Bertolotti, A., Bianchi, E. and Boniforti, L. (1978) Analysis of 2,3,7,8-tetrachlorodibenzo-paradioxin in the Seveso area. *Ecol. Bull. (Stockholm)*, **27**, 31.

Edwards, C. A. and Lofty, J. R. (1977) *Biology of Earthworms.* 2nd edn. Chapman & Hall, London. 333 pp.

Fanelli, R., Bertoni, M. P., Bonfanti, M., Castelli, M. G., Chiabrando, C., Martelli, G. P., Noè, M. A., Noseda, A. and Sbarra, C. (1980a) Routine analysis of 2,3,7,8-tetrachlorodibenzo-p-dioxin in biological samples from the contaminated area of Seveso, Italy. *Bull. Environ. Contam. Toxicol.*, **24**, 818–823.

Fanelli, R., Bertoni, M. P., Castelli, M. G., Chiabrando, C., Martelli, G. P., Noseda, A., Garattini, S., Binaghi, C., Marazza, V. and Pezza, F. (1980b) 2,3,7,8-Tetrachlorodibenzo-p-dioxin toxic effects and tissue levels in animals from the contaminated area of Seveso, Italy. *Arch. Environ. Contam. Toxicol.*, **9**, 569–577.

Gish, C. D. (1970) Organochlorine insecticide residues in soils and soil invertebrates from agricultural land. *Pest. Mon. J.*, **3**, 241–252.

Regione Lombardia (1977) Provvedimenti per Seveso, *Bollettino Ufficiale della Regione Lombardia*, **7**, Suppl. 28.

Earthworms as a source of food and drugs

J. R. SABINE

24.1 INTRODUCTION

In recent years considerable attention has been focused upon the potential role of intensive earthworm culture, or vermiculture, in the amelioration of severe problems associated with the disposal of large quantities of organic wastes (Hartenstein, 1978; Appelhof, 1981). It is now accepted that the economic value of vermiculture lies in (i) diminution of noxious qualities associated with organic wastes, e.g. elimination of smell; (ii) generation of a useful compost; and (iii) production of earthworm biomass. This review covers the value of the worms produced as a source of food, primarily proteins, and of drugs.

24.2 EARTHWORMS AS A SOURCE OF FOOD

Now that various vermiculture systems, which have been designed primarily for biological waste control, are producing worms in large quantities the question arises of the possible value of these worms as food for domestic animals, or even for direct human consumption. This question demands consideration of five aspects: chemical composition, practical value in animal feeds, potential hazards, production economics and legal constraints.

24.2.1 Chemical composition of earthworms

The chemical composition of any potential animal foodstuff should be considered in a number of ways – the content of macro- and micro-nutrients, the content of specific organic compounds and the content of any possible hazardous materials, either natural constituents or con-

Table 24.1 Gross nutrient content of earthworms (*Eisenia foetida*).

	McInroy (1971)	Fosgate and Babb (1972)	Schulz and Graff (1977)	Sabine (1978)	Taboga (1980)	Hartenstein (1981)
Dry matter (%)	12.9	22.9		20–25	15–20	18
As % of dry matter:						
Protein (N × 6.25)	68.1	58.2	66.3	62–64	62–71	65
Fat (ether extract)	6.4	2.8	7.9	7–10	2.3–4.5	9
Fibre		3.3				
Carbohydrate			14.2			21
Ash	5.2		11.6	8–10		5–8
Calcium		0.54		0.55		0.3–0.8
Phosphorous		0.90		1.0		0.7–1.0
Gross energy (kJ kg⁻¹)				16 380–17 220		

taminants. Table 24.1 shows some published values for the gross nutrient content of earthworms.

The most striking feature of these data is the high protein content, 58–71 % on a dry-weight basis, and most of the interest in the potential value of worms for livestock feeding has concentrated upon this aspect, particularly for the intensive pig and poultry industries. The actual nutritive value of a feed protein, however, depends upon its specific amino acid composition. Table 24.2 shows the amino acid composition of earthworm protein, as reported by several laboratories and as compared with two regular sources of high-protein supplements, meat meal and fish meal. Earthworm protein is high in the essential amino acids, including those containing sulphur, and should be of high biological quality.

Tables 24.1 and 24.2 also show considerable variability, some of which could be due to variations in analytical techniques between different laboratories. Graff's work (1981) is so far the only report of a direct comparison between the composition of various worm species (*Eisenia foetida* and *Eudrilis eugeniae*), and it is particularly interesting that in these species, when raised under comparable conditions, the content of individual amino acids differed by no more than 17 % and usually by considerably less. No work has yet been reported on the composition of the same species grown under different conditions, nor for comparisons between field and vermiculture samples, but considerable differences would be expected.

Recent papers on the presence or absence of specific nutritive compounds in earthworm tissue include studies of triglycerides (Hansen and Czochanska, 1974), squalene (Voogt *et al.*, 1975) and free fatty acids (Naya and Kotake, 1967).

Table 24.2 Amino acid composition of earthworm protein (g 100 g^{-1} of protein).

	McInroy (1971)[a]	Taboga (1980)[b]	Sabine (1981)[a]	Graff (1981)[a]	Graff (1981)[c]	FM[d]	MM[d]
Ala		5.4		6.0	5.2		
Arg*	6.1	7.3	6.8	6.1	6.1	6.7	6.5
Asp		10.5		11.0	10.3		
Cys	1.8	1.8	3.8	1.4	1.6	1.1	1.3
Glu		13.2		15.4	13.8	14.8	13.8
Gly		4.3	4.8			4.0	7.2
His*	2.2	3.8	2.6	2.3	2.6	2.0	2.5
Ile*	4.6	5.3	4.2	4.7	4.5	3.5	6.0
Leu*	8.1	6.2	7.9	8.2	7.9	6.4	8.4
Lys*	6.6	7.3	7.1	7.5	7.1	6.9	10.4
Met*	1.5	2.0	3.6	1.8	2.0	1.5	3.0
Phe*	4.0	5.1	3.7	3.5	4.1	3.5	4.2
Pro		5.3					
Ser		5.8	4.7	4.8	4.8		
Thr*	5.3	6.0	4.8	4.7	4.8	3.3	4.6
Try*		2.1				0.5	1.1
Tyr		4.6	2.2	3.0	3.4	1.6	3.0
Val*	5.1	4.4	4.9	5.2	5.0	4.7	5.7

* Essential amino acids.
[a] *Eisenia foetida.*
[b] Mixture of *E. foetida* and *Lumbricus rubellus.*
[c] *Eudrilus eugeniae.*
[d] MM, Meat meal; FM, Fish meal (Morrison, 1957).

24.2.2   The quality of earthworm protein

Pellett and Young's recent (1980) review, *Nutritional Evaluation of Protein Foods*, has highlighted the deficiencies and limitations of any method so far proposed to measure protein 'quality'. Table 24.3 lists assays that have been suggested, based on chemical analysis, enzymic digestion, microbiological growth and a variety of animal feeding trials. In many of these a score is measured relative to some theoretical or actual 'ideal' protein, e.g. milk casein or egg albumin. Some idea of the complexities involved in such experimentation can be gained from the reviewers' comment that the 'conditions under which a particular measurement has been made should be stated, e.g. the age or weight, sex and species of the animal used, its energy intake, the composition of the diet fed (especially the protein and energy contents), details of the experiment, including the period of measurement, the environmental temperature, and the previous nutrition of the animals. Other factors may relate to cage size and whether the animals are housed singly or together.'

Table 24.3 Assays suggested for protein quality (Pellett and Young, 1980).

A. Based on chemical analyses:
    Amino acid score/Chemical score/Protein score
    Available amino acids
    Calculated protein efficiency ratio
    Dye-binding capacity
    Essential amino acid index
    Pepsin digest residue
    Protein quality index

B. Based on microbiological growth:
    Relative nutritional value
    Relative nutritive value/Relative growth index
    Tetrahymena protein efficiency ratio

C. Based on animal feeding:

Biological value	Nitrogen growth index
Digestibility	Protein efficiency ratio
Gross protein value	Protein rating
Net protein ratio/Relative net protein ratio	Protein retention efficiency
	Protein value/Relative protein value
Net protein utilization	Relative nitrogen retention
Net protein value	Repletion
Nitrogen balance index	Slope ratio
Nitrogen efficiency ratio	

With these limitations it is not surprising that little work of this nature has been done on earthworm protein. From the results of a rat-growth assay Schulz and Graff (1977) reported for earthworm protein (*E. foetida*) a Biological Value of 84% and a Net Protein Utilization of 79%, both excellent results relative to the fish meal protein tested at the same time. Cieslak and Benevega, at the University of Wisconsin, have developed computer programs to evaluate the Amino-Acid Scores of protein and protein mixtures and to calculate optimum complementary protein mixtures relative to the known amino acid requirements of various domestic animals. Table 24.4 (private communication) shows that earthworm meal is clearly superior to meat meal in this type of analysis.

The alternative approach to determining the quality of earthworm meal is to try it in some specific production system. Taboga (1980) in the United States, Yoshida and Hoshii (1978) and Mekada *et al.* (1979) in Japan and Harwood (1976) in Australia have all fed earthworm meal to chickens, and Harwood and Sabine (1978) have fed it to growing pigs. In earlier studies in the United States (McInroy, 1971) and Germany (Schulz and Graff, 1977) earthworm meal was tested in the diets of mice and rats respectively. In all of these trials the results have been

Table 24.4 Relative values of worm meal and meat meal in the diets of broiler chickens. (From Cieslak and Benevega (pers. comm.)

| Proportions of protein from | | | | | |
Worm meal	Meat meal	Cereal*	Requirement (g kg^{-1} body wt.)	LAA†	Chemical score (%)
50 :	0 :	50	21.8	Arg	92
60 :	0 :	40	20.9	Arg	96
70 :	0 :	30	20.1	Arg	100
80 :	0 :	20	19.3	Arg	103
90 :	0 :	10	20.5	Try	98
0 :	50 :	50	23.6	Lys	85
0 :	60 :	40	22.6	His	89
0 :	70 :	30	22.8	Ile	88
0 :	80 :	20	24.5	Ile	82
0 :	90 :	10	26.4	Ile	76

* A mixture of wheat and barley.
† Limiting amino acid.

consistent – those animals fed on earthworm meal as the major source of protein in their diets have all grown at rates equal to or better than those displayed by animals fed conventional protein meals. Moreover, both Harwood (1976) and Mekada et al. (1979) reported that chickens fed on earthworm meal had better feed-conversion ratios than control birds, i.e. the same weight gain could be obtained from less food consumed. This saving, if confirmed in more extensive trials, would represent a considerable benefit to producers since the cost of food is generally more than half the cost of intensive pig and poultry production.

There is a belief widespread in the earthworm industry that chickens fed on earthworms will lay low-cholesterol eggs. Although the cholesterol content of eggs can indeed be lowered by dietary means (Clarenburge et al., 1971; Godfrey et al., 1976), there is no published evidence that earthworms can do this. Preliminary studies in my laboratory (unpublished) showed no effect on egg cholesterol of the inclusion of earthworm meal in the diet of laying birds.

Finally, a role for earthworms has been suggested in the commercial feeding of fish, eels, frogs, gamecocks, domestic pets and even humans. From the results already obtained with poultry, worm meal seems likely to become an important ingredient in commercial fish diets, but despite reports of earthworms in the diets of two native peoples, the Aborigines of southern Australia (Smyth, 1878) and the Maoris of New Zealand (McInroy, 1971), earthworms are unlikely to be significant in human nutrition for many years to come.

24.2.3 Biohazards of earthworm meal

Before widespread use of earthworms in animal or human nutrition could be contemplated, work will be needed to allay fears that they might carry disease or contain toxic residues.

Although the life cycles of a number of organisms that cause human and animal diseases have stages involving worms, little work has yet been done on the likely extent of this hazard in commercial systems. Brown and Mitchell (1981) have shown that the concentration of *Salmonella* spp., a major disease organism of poultry, can be significantly reduced in laboratory culture by the presence of *E. foetida*. Earthworms are not ordinarily significant in the natural distribution of two important parasitic nematodes, *Ascaridia galli* in chickens (Augustine and Lund, 1974) and *Ascaris suum* in pigs (Jakovljevic, 1975) but Lund (quoted by Augustine and Lund, 1974) has suggested that earthworms may be particularly important in the transmission of *Heterakis gallinarum*. Studies on the persistence of disease organisms in earthworm meal that is prepared for inclusion in livestock rations will be essential, especially where the worms are cultured on animal manures.

Presumptive evidence from the successful poultry- and pig-feeding trials referred to above, and from numerous food-chain studies, is that earthworms are unlikely to contain any naturally occurring toxins. A possible exception is the giant earthworm of Gippsland (*Megascolides australis*) which is said to contain a creosote-like substance (McCoy, 1878). Such materials, however, seem more likely to inhibit consumption than to cause toxicity problems and various unpublished reports suggest that *E. foetida* is distasteful to fish.

Toxic residues accumulated by earthworms include particularly metals and agrochemicals. Since at least 1895 (Hogg, 1895) we have known that earthworms can accumulate lead, and more recently Gish and Christensen (1973), Hartenstein *et al.* (1980), Beyer (1981) and Ireland (Chapter 21) have reported accumulation by earthworms of lead, cadmium, chromium, copper, nickel, mercury and zinc. Worms seem to tolerate large concentrations of these metals in their tissues and continuous monitoring will be necessary if earthworms are introduced in large quantities into other food chains. An expanding area of interest in this context is the induction in worms of specific metal-binding proteins and of the fate of these specific metal–protein adjuncts when fed to other organisms. High levels of cadmium induce in the earthworm relatively large quantities of at least three cadmium-binding proteins (Suzuki *et al.*, 1980a) and in the rat the metabolic fate of each of these is distinct. (Suzuki *et al.*, 1980b).

Recent work on earthworms and agrochemicals seems to follow three lines – the effects of the chemicals upon the earthworms themselves (e.g. Thompson, 1971), the role of earthworms in biodegradation (e.g. Cooke, 1979) and the extent to which worms accumulate the primary chemicals or their degradation products (e.g. Bailey *et al.*, 1974). Little work has yet appeared on concentrations of agrochemicals in earthworm meal, but since modern feed formulations for the intensive animal industries include a large range of organic and inorganic chemicals, extensive monitoring of meals derived from worms grown on animal manures will be necessary.

24.2.4 Economic potential

The economic return possible from earthworm production depends upon the rate and cost of production and the price available for the product.

Given that the rate of reproduction and growth of earthworms can be particularly high (Graff, 1974; Hartenstein *et al.*, 1979), at least in the laboratory, several authors have suggested that the rate of protein production that could be attained with earthworms is also particularly high. Fosgate and Babb (1972) fed dairy manure to worms (reputedly *Lumbricus terrestris*, but there are some doubts about this) and obtained a rate of production of worm protein of $0.42 \, kg \, cow^{-1} \, day^{-1}$, a rate higher than that of the South Australian average production of milk protein, approximately $0.38 \, kg \, cow^{-1} \, day^{-1}$ (Sabine, 1975). McInroy (1971) calculated, by extrapolation from small-scale experiments, a similar production figure of $190 \, kg$ of worm protein $1000 \, lb \, cow^{-1} \, y^{-1}$.

Hartenstein (1981) has suggested that a rate of protein production of $6685 \, kg \, ha^{-1} \, y^{-1}$ should be attainable, but again this estimate is based on an extrapolation from very small laboratory-scale tests. Figures derived from actual large-scale operations and calculated as yield per unit of food input have, however, not yet been published.

Nor have any accurate estimates of production costs been published, but if food for the worms could be obtained at no cost, or at negative cost – which could be the case with a biological waste that would otherwise be a financial burden to dispose of – costs of production of earthworm protein would be quite low, certainly relative to the intrinsic value of the product.

In many countries, commonly-cultured earthworms such as *E. foetida* can be sold for fishing bait for $US 1–2 per 100. Night crawlers (*Lumbricus terrestris*) might bring roughly the same return to the picker, but cost perhaps 5–10 times as much to the angler (Tomlin, Chapter 29). This market, however, is strictly limited, and any realistic assessment of the economic potential for large-scale earthworm production must consider that the financial return will be related rather to the value of

earthworm protein to the intensive livestock industries. In most indus-
trialized countries the market price that can be obtained for any novel
protein depends upon the current price of meat meal, the most commonly
used protein supplement in animal feeds. Mohr and Littleton (1978) have
computed 'shadow prices' on the Australian market for a number of exotic
foodstuffs, using a least-cost linear program, and have calculated that if
meat meal costs $177 t^{-1} to the feed manufacturer, then earthworm meal
is an economic substitute for poultry rations up to $236 t^{-1}. This figure,
however, was based on a low protein value for earthworm meal and could
realistically be higher. Nevertheless, if an average mature specimen of *E.
foetida* weighs about 1 g, then a value of over $300 t^{-1} for earthworm
protein is equivalent to less than 1 cent per 100 worms, a figure much
smaller than that currently attainable on the bait market.

   Large-scale vermiculture operations, however, produce two saleable
products, worms and vermicompost. The economic viability of any
vermiculture system, at least for a long time to come, will probably
depend upon the financial return from compost production rather than
from worm protein.

24.2.5 Legal constraints

Many food laws distinguish between an 'ingredient' and an 'additive',
with different regulations and restrictions applying to each. Earthworm
growers should concentrate, at least initially, upon classification of their
product as a regular high-protein ingredient. Most food laws also contain
a blanket prescription that food should be 'wholesome'. If some
ingredient or additive can be demonstrated to have caused ill health to the
consumer, then *de facto* it was not wholesome and therefore contravened
the law. Because of reluctance to risk this outcome the adoption of
earthworms in animal feeds is likely to be a slow process and extensive
safety trials will be necessary. Feed manufacturers are more likely to
accept unusual ingredients if they come from a usual source, e.g. triticale
before single-cell protein, or are prepared by a usual process, e.g. the
cooking and drying that converts abattoir waste to meat meal. Earthworm
meal, if prepared by a standard meat or fish meal process, should meet
little opposition, but again continual safety monitoring will be necessary.

24.3 EARTHWORMS AS A SOURCE OF DRUGS

Folk history abounds with stories of the medicinal value of earthworms,
dating from at least as far back as 1340 A.D. and covering a range of
diseases from pyorrhea to postpartal weakness, from smallpox to jaundice
to rheumatism (Reynolds and Reynolds, 1972). While the absolute value

of many traditional remedies cannot be denied, the amount of scientific evidence for earthworms as a potent source of drugs is meagre.

24.3.1 Pharmacologically active materials derived from worms

Since the addition of earthworm meal to the diet of chickens may promote improved feed conversion, worms may contain one or more 'pharmacologically active' substances. Hori *et al.* (1974) have reported significant anti-pyretic activity derived from earthworms (*Lumbricus spencer, Perichaeta communissima*) when tested in rabbits in which fever had been induced by injection of *Escherichia coli* pyrogen. The authors suggest that this activity is due to *all cis*-5,8,11,14-icosatetraenoic acid (archidonic acid) and *all cis*-5,8,11,14,17-icosapentaenoic acids. Since such long-chain fatty acids are readily available from other natural sources their presence in earthworms is unlikely to affect the economics of vermiculture. This work excepted, there is no well-documented evidence for any drug-like activity of earthworm extracts in any organism other than in the earthworm itself. Many workers have sought in worms physiologically important compounds similar to those found in higher organisms, e.g. β-endorphin and enkephalin (Alumets *et al.*, 1979), hyperglycaemic factor (Lawrence *et al.*, 1972) and various monoamines (Gardner and Cashin, 1975). While all these compounds have undoubted roles to play in earthworm physiology, it seems most unlikely that they would be present in sufficient concentration or activity to play a role in human or animal medicine. There is consistent historical evidence of the value of earthworms in alleviating rheumatism (Reynolds and Reynolds, 1972) and the search for a possible pharmacologically active ingredient here may be justifiable.

24.4 CONCLUSIONS

Earthworms contain high levels of protein (58–71 % dry weight) and this protein is high in the amino acids considered essential for animal diets. Thus earthworm meal should be a valuable ingredient for livestock rations and this has been demonstrated in practical feeding trials with chickens and pigs, as well as in laboratory tests with rats and mice. Earthworms can be cultured in large numbers on a variety of substrates, including many biological materials at present regarded solely as wastes. Since worms grown on such wastes, and the worm meal derived from them, might harbour disease organisms or contain unacceptable quantities of heavy metals and/or agrochemical residues, various safety trials will be necessary. There is, nevertheless, good reason to believe that worm meal will become an important high-protein component in commercial feed rations for the intensive livestock industries, including fish culture.

There is little scientific evidence for the belief, common in the folklore of many countries, that earthworms cure a wide range of common diseases.

24.5 REFERENCES

Alumets, J., Hakanson, R., Sundler, F. and Thorell, J. (1979) Neuronal localisation of immunoreactive enkephalin and β-endorphin in the earthworm. *Nature (London)*, **279**, 805–806.

Appelhof, M. (compiler) (1981) *Workshop on the Role of Earthworms in the Stabilization of Organic Residues*, Vol. 1, Proc. Beech Leaf Press, Kalamazoo, MI. 315 pp.

Augustine, P. C. and Lund, E. E. (1974) The fate of eggs and larvae of *Ascaridia galli* in earthworms. *Avian Dis.*, **18**, 394–398.

Bailey, S., Tunyan, P. J., Jennings, D. M., Norris, J. D., Stanley, P. I. and Williams, J. H. (1974) Hazards to wildlife from the use of DDT in orchards: II. A further study. *Agro-Ecosystems*, **1**, 323–338.

Beyer, W. N. (1981) Metals and terrestrial earthworms (*Annelida: Oligochaeta*). In *Workshop on the Role of Earthworms in the Stabilization of Organic Residues* (M. Appelhof, compiler), Vol. 1, Proc. Beech Leaf Press, Kalamazoo, MI, pp. 137–150.

Brown, B. A. and Mitchell, M. J. (1981) Role of the earthworm, *Eisenia foetida*, in affecting survival of *Salmonella enteritidis* ser. *typhimurium*. *Pedobiologia*, **21**, 434–438.

Clarenburge, R., Kim Chung, I. A. and Wakefield, L. M. (1971) Reducing the egg cholesterol level by including emulsified sitosterol in standard chicken diet. *J. Nutr.*, **101**, 289–298.

Cooke, B. K. (1979) DDT residues in soil and earthworms. *Report. Long Ashton Res. Stn.*, University of Bristol, p. 113.

Fosgate, O. T. and Babb, M. R. (1972) Biodegradation of animal wastes by *Lumbricus terrestris*. *J. Dairy Sci.*, **55**, 870–872.

Gardner, C. R. and Cashin, C. H. (1975) Some aspects of monoamine function in the earthworm (*Lumbricus terrestris*). *Neuropharmacology*, **14**, 493–500.

Gish, C. D. and Christensen, R. E. (1973) Cadmium, nickel, lead and zinc in earthworms from roadside soil. *Environ. Sci. Technol.*, **7**, 1060–1062.

Godfrey, J. C., Luttinger, J. R., Taylor, H. D. and Sanheuza, G. M. (1976) Dietary plant sterol-induced reduction of egg yolk cholesterol in the chicken. *Nutr. Rep. Int.*, **13**, 263–271.

Graff, O. (1974) Gewinnung von Biomasse aus Abfallstoffen durch Kultur des Kompostregenwurms *Eisenia foetida* (Savigny 1826). *LandbForsch.-Völkenrode*, **24**, 137–142.

Graff, O. (1981) Preliminary experiments of vermicomposting of different waste material using *Eudrilus eugeniae* Kinberg. In *Workshop on the Role of Earthworms in the Stabilization of Organic Residues* (M. Appelhof, compiler), Vol. 1, Proc. Beech Leaf Press, Kalamazoo, MI, pp. 179–191.

Hansen, R. P. and Czochanska, Z. (1974) Occurrence of triglycerides in earthworms. *Lipids*, **9**, 363–364.

Hartenstein, R. (ed.) (1978) *Proc. Utilization of Soil Organisms in Sludge Management*, Syracuse, NY. 171 pp.

Hartenstein, R. (1981) Use of *Eisenia foetida* in organic recycling based on laboratory experiments. In *Workshop on the Role of Earthworms in the Stabilization of Organic Residues* (M. Appelhof, compiler), Vol. 1, Proc. Beech Leaf Press, Kalamazoo, MI, pp. 155–165.

Hartenstein, R., Neuhauser, E. F. and Kaplan, D. L. (1979) Reproductive potential of the earthworm *Eisenia foetida*. *Oecologia (Berl.)*, **43**, 329–340.

Hartenstein, R., Neuhauser, E. F. and Collier, J. (1980) Accumulation of heavy metals in the earthworm *Eisenia foetida*. *J. Environ. Qual.*, **9**, 23–26.

Harwood, M. (1976) Recovery of protein from poultry waste by earthworms. *Proc. 1st Austr. Poult. Stockfeed Conv. Melbourne*, pp. 138–143.

Harwood, M. and Sabine, J. R. (1978) The nutritive value of worm meal. *Proc. 2nd Austr. Poult. Stockfeed Conv., Sydney*, pp. 164–171.

Hogg, T. W. (1895) Immunity of some low forms of life from lead poisoning. *Chem. News*, **71**, 223–224.

Hori, M., Kondon, K., Yoshida, T., Konishi, E. and Minami, S. (1974) Studies of antipyretic components in the Japanese earthworm *Biochem. Pharmacol.*, **23**, 1583–1590.

Jakovljevic, D. D. (1975) Some aspects of the epizootology and economical significance of ascariasis in swine. *Acta Vet., Beogr.*, **25**, 315–325.

Lawrence, J. McV., Craig, J. V. and Clough, D. (1972) The presence of a hyperglycaemic factor in the suprapharyngeal ganglia of *Lumbricus terrestris*. *Gen. Comp. Endocrinol.*, **18**, 260–267.

McCoy, F. (1878) *Megascolides australis* (McCoy). The giant earthworm. *Natural History of Victoria – Prodromus of the Zoology of Victoria*, **1**, 21–24.

McInroy, D. M. (1971) Evaluation of the earthworm *Eisenia foetida* as a food for man and domestic animals. *Feedstuffs*, Feb. 20th, 37–46.

Mekada, H., Hayashi, N., Yokota, H. and Okumura, J. (1979) Performance of growing and laying chickens fed diets containing earthworms (*Eisenia foetida*). *Jpn. Poult. Sci.*, **16**, 293–297.

Mohr, G. and Littleton, I. (1978) Preliminary economic evaluation of some exotic feedstuffs. *Proc. 2nd Austr. Poult. Stockfeed Conv., Sydney*, pp. 226–231.

Morrison, F. B. (1957) *Feeds and Feeding*. 22nd edn. Morrison Publishing Co., Ithaca, NY.

Naya, Y. and Kotake, M. (1967) Untersuchung der freien Fettsauren, der freien Sterine, der Sterinester und der Glyzeride im Regenwurm (*Lumbricus spencer*). *Bull. Chem. Soc. Jpn.*, **40**, 880–884.

Pellett, P. L. and Young, V. R. (eds.) (1980) *Nutritional Evaluation of Protein Foods*. (WHTR-3/UNUP-129). The United Nations University, Tokyo.

Reynolds, J. W. and Reynolds, W. M. (1972) Earthworms in medicine. *Am. J. Nurs.*, **72**, 1273.

Sabine, J. R. (1975) Where has all the protein gone? (or Rediscovering animal wastes). *Proc. 3rd Combined Conf. Austr. Chicken Meat Fed. Austr. Stockfeed Manuf. Assoc., Adelaide*, pp. 181–189.

Sabine, J. R. (1978) The nutritive value of earthworm meal. In *Utilization of Soil Organisms in Sludge Management* (ed. R. Hartenstein), State Univ. N.Y., Syracuse. pp. 122–130.

Sabine, J. R. (1981) Vermiculture as an option for resource recovery in the intensive animal industries. In *Workshop on the Role of Earthworms in the Stabilization of Organic Residues* (M. Appelhof, compiler), Vol. 1, Proc. Beech leaf Press Kalamazoo, MI, pp. 241–252.

Schulz, E. and Graff, O. (1977) Zur Bewertung von Regenwurmmehl aus *Eisenia foetida* (Savigny 1826) als Eiweissfuttermittel. *LandbForsch-Völkenrode*, **27**, 216–218.

Smyth, R. B. (1878) *The Aborigines of Victoria*, Vol. 1, Melbourne.

Suzuki, K. T., Yamamura, M. and Mori, T. (1980a) Cadmium-binding proteins induced in the earthworm. *Arch. Environ. Contam. Toxicol.*, **9**, 415–424.

Suzuki, K. T., Yamamura, M. and Mori, T. (1980b) Metabolic fate of earthworm cadmium-binding proteins in rats. *Arch. Environ. Contam. Toxicol.*, **9**, 519–531.

Taboga, L. (1980) The nutritional value of earthworms for chickens. *Br. Poult. Sci.*, **21**, 405–410.

Thompson, A. R. (1971) Effects of nine insecticides on the numbers and biomass of earthworms in pasture. *Bull. Environ. Contam. Toxicol.*, **5**, 577–586.

Voogt, P. A., van Rheenen, J. W. A. and Zandee, D. I. (1975) What about squalene in the earthworm *Lumbricus terrestris*? *Comp. Biochem. Physiol.*, **50B**, 511–513.

Yoshida, M. and Hoshii, H. (1978) Nutritional value of earthworms for poultry feed. *Jpn. Poult. Sci.*, **15**, 308–311.

Chapter 25

Assimilation by the earthworm *Eisenia fetida*

R. HARTENSTEIN

25.1 INTRODUCTION

Eisenia fetida (Savigny) is an earthworm which grows rapidly (Neuhauser *et al.*, 1980), reproduces prodigiously (Hartenstein *et al.*, 1979), is potentially deployable for management of wastes rich in microbial biomass (Hartenstein, 1981a, b), and is possibly more efficient on a time–space scale for production of animal (meat, in contrast to eggs and milk) protein than any other organism husbanded at present (Hartenstein, 1981b).

Proximate analytical data are available on the elemental composition of this earthworm (Hartenstein *et al.*, 1980). Information is available on some of its physicochemical environmental requirements (Kaplan *et al.*, 1980a), its response to various salt and oxide additives to activated sludges derived from domestic wastes (Hartenstein *et al.*, 1981), and certain physical and chemical effects which it imposes on activated sludge in converting the sludge into wormcasts (Hartenstein and Hartenstein, 1981).

Intensive production of *E. fetida* to satisfy a large demand for bait suggests that it may be suitable in aquaculture, and it has already been shown to be an organism which could replace fish meal or soybean in the diets of domestic animals such as chickens and pigs (Sabine, 1978; Abe *et al.*, 1979).

The main objective of the present study was to determine the efficiency of *E. fetida* in converting several types of substrates into biomass. Two equally important derivative objectives were also sought: to obtain some insight into earthworm–microbial interactions and to determine whether *E. fetida* is able to derive energy from cellulose and from lignocellulose complexes.

25.2 METHODS AND MATERIALS

Stock cultures of *E. fetida* were maintained on mixtures of 1:4 wet wt./wet wt. cellulose and activated sludge. The cellulose was obtained from Sigma Chemical Co., St. Louis, Missouri, USA and wetted with three parts of water by weight. The sludge was obtained from the Meadowbrook-Limestone Wastewater Treatment Plant, Onondaga, NY, USA, and dehydrated to about 12 % solids as described elsewhere (Hartenstein and Hartenstein, 1981).

For experiments, cocoons were removed within a week of production and placed on wet paper, from which hatchlings were obtained three weeks later. The hatchlings weighed 5 to 10 mg and were not weighed, since the values would be within the experimental error of values expected at the conclusion of an experiment.

A sample of activated sludge with 13 % solids (wet wt.) was lyophilized (freeze-dried) and is called lyoaer. Another sample was placed with *E. fetida* into a wooden box with screen floor as described earlier (Hartenstein and Hartenstein, 1981). Following its conversion to castings, this sample was lyophilized and is called lyocasts. A third sample was sealed into a polythylene bag and stored in a sealed glass jar at room temperature for a year; it was then lyophilized and is referred to as lyoaeranaer. A sample of anaerobic sludge was obtained directly from a digester at the Metropolitan Wastewater Treatment Plant, Syracuse, New York, USA; it was centrifuged at 1000 *g* 10 min, lyophilized and is called lyoanaer. Horse manure, uncontaminated with urine, was obtained from a farm in Tully, New York, USA, allowed to age at room temperature for 5 days, lyophilized and called lyoman. Groundwood pulp, derived from firs and spruces, was obtained from the College's pulp mill. Teel silt loam, described by Mitchell *et al.* (1978), was purchased from Action Topsoil, Minoa, New York, USA. Elemental analyses were performed according to procedures reported previously (Hartenstein and Hartenstein, 1981).

Four experiments were run, all at 24 ± 1 °C. In the first, 30 g silt loam and 12 ml distilled water were placed into 100 × 20 mm Petri dishes, a mass of sludge and a number of hatchlings were added as indicated in Table 25.1, and worms were weighed after 4 and 8 weeks. Each variable was replicated twice.

The second experiment differed from the first in (1) use of a wider range of both earthworm populations and sludge rations (Table 25.3), (2) use of three replicates, and (3) weighing the earthworms after 4 and 7 weeks. Larger dishes (100 mm ϕ × 5 cm) were used where more than 50 g wet wt. sludge was tested.

In the third experiment, either moist cellulose (25 % solids, w/w) or moist groundwood pulp (25 % solids, w/w) was mixed with sludge (12 %

solids) to give w/w ratios of 20, 40, 60 and 80% of these test materials; these values corresponded to 34, 58, 76 and 89% cellulose (or pulp) on a dry wt. basis, and are referred to as such hereafter. Sludge not amended with fibre, or one of the foregoing mixtures, was placed together with either two or four hatchlings in a single dish. Each variable was replicated twice. Worms were weighed after five weeks.

In the fourth experiment, 5 g silt loam and 1.5 ml water were placed into a 60 × 15 mm Petri dish. A mixture of 1.25 g dry cellulose, 0.1 to 1.0 g lyoaer, lyoanaer, lyoaer-anaer, lyocasts or lyoman, and 3.75 ml water was added. Two hatchlings were placed into a single dish and weighed four weeks later. Each variable was replicated twice.

Weights in all experiments were obtained after rinsing and blotting. Dry weight of *E. fetida* was assumed to be 18% of live weight (Hartenstein *et al.*, 1980).

25.3 RESULTS

25.3.1 Experiment 1

When one or two *E. fetida* were placed into 20 to 100 g wet wt. sludge (about 11.3% solids, w/w), a greater earthworm biomass was always present at 8 weeks than at 4 weeks. When four hatchlings were added instead, a larger biomass occurred after 4 weeks than after 8 weeks in some of the dishes which contained less than 4.5 g dry wt. sludge. The lower weight in these cases signified an exhaustion of food supply.

On the basis only of maximum biomass produced in each dish, either at 4 or 8 weeks (Table 25.1), regressions of total mg dry wt. *E. fetida* in relation to initial dry wt. ration of sludge for one, two or four earthworms were calculated (Table 25.2). These show that use of two worms instead of

Table 25.1 Weight achieved by *E. fetida* hatchlings in relation to sludge ration. Means of two replicates.

Ration (mg dry wt.)	Total biomass (mg dry wt.)		
	One worm	Two worms	Four worms
1690	51	53	79
2260	61	63	114
2820	80	69	121
3390	88	145	157
3950	110	132	188
4520	118	172	216
5080	114	179	229
5650	124	199	269

Table 25.2 Regression of dry wt. (g) of *E. fetida* in relation to initial dry wt. (g) of sludge.

Number of earthworms	Regression	r
1	$Y = 0.019\,X + 21.8$	0.98
2	$Y = 0.037\,X - 8.1$	0.96
4	$Y = 0.060\,X - 60.8$	0.94

one yielded twice as much biomass per initial ration, while the use of four worms instead of two did not result in a further doubling of yield. The slope of the regression for four worms suggests production of about 60 mg dry wt. of *E. fetida* per 1 g dry wt. sludge, a 6% yield.

25.3.2 Experiment 2

In this experiment, as in experiment 1 – where more than two *E. fetida* were added to a culture dish – a greater biomass of *E. fetida* was present in certain dishes at 4 weeks than at 7 weeks. Values which are underscored in Table 25.3 are the higher weights achieved at 4 weeks.

Table 25.3 *E. fetida* biomass produced in relation to activated sludge ration. Means of three determinations.

Population density	Activated sludge (mg dry wt.)								
	2260	3390	4520	5650	5780	7910	9040	10 170	11 300
1	58	101	82	80	81	104	104	131	140
2	96	120	158	134	119	145	135	149	200
4	97	162	183	233	179	230	276	263	309
8	126	189	243	263	292	252	408	440	560
12	129	195	257	294	268	258	390	522	676
16	140	159	231	283	290	351	487	642	742

Underscoring based on weight at 4 weeks; weight was lost by these earthworms between 4 and 7 weeks.

On the basis of the maximum weight achieved by any population of *E. fetida* at each weight-level of sludge, a regression of $Y = 0.063X - 48$, $r = 0.98$, was calculated. Thus, about 63 mg of dry biomass was produced from 1 g of dry sludge. As shown in Fig. 25.1, this is equivalent to the carrying capacity of the system and represents a yield of about 6%

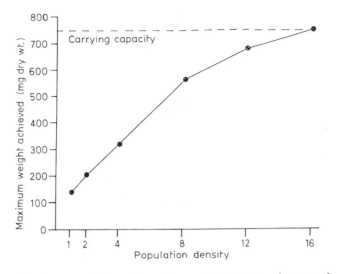

Fig. 25.1 Maximum weight achieved on 2.26 to 11.3 g dry wt. sludge by individuals or populations of *E. fetida*.

(dividing the highest ordinate, 750 mg by 11.3 g, which corresponds to the ration on which the 16 worms achieved 750 mg, giving a yield of 6.6%).

25.3.3 Experiment 3

A greater amount of biomass was produced in each culture into which four hatchlings, rather than two, were introduced. Accordingly, only the data for the former are reported (Table 25.4).

Calculations from the data of Table 25.4 show that in dishes with the two lowest rations, significantly more biomass was produced per unit weight of substrate which contained 34, 58 or 76% cellulose than per unit weight of activated sludge alone or sludge which contained 89% cellulose (Table 25.5). In contrast, equal yields were obtained on the lowest ration of 34 and 58% pulp in comparison to lowest ration of sludge alone, and lower yields were obtained for all the remaining rations and pulp mixtures (Table 25.5).

Regressions of mg dry wt. *E. fetida* in relation to mg dry wt. substrate show that approximately 4 to 5% of the substrates with 0 to 76% cellulose were converted into earthworm biomass (Table 25.6). Significantly lower yields were obtained for the corresponding range of pulp, except for the mixture which contained 58% pulp. Here a yield of 3.9% was obtained. Mixtures with 89% cellulose or pulp yielded significantly less biomass than all other mixtures.

Table 25.4 Production of *E. fetida* from five hatchlings on mixtures of cellulose (C) or groundwood pulp (P) and activated sludge.

Concentration of P or C in mixture (%)	0		34			58			76			89		
Concentration of nitrogen (%)	5		3.3			2.1			1.2			0.55		
Ration (R) (g dry wt.) and biomass (mg dry wt.) produced on C or P	R	C and P¹	R	C	P	R	C	P	R	C	P	R	C	P
	1.2	54	1.5	141	66	1.7	145	76	2.0	252	72	2.2	59	30
	2.4	109	2.9	299	109	3.4	314	130	4.0	324	153	4.5	100	74
	3.6	180	4.4	324	122	5.2	268	219	6.0	410	166	6.7	153	79
	4.8	271	5.8	389	128	6.9	376	272	8.0	415	147	9.0	112	49
	6.0	346	7.3	367	175	8.6	409	339	10.0	419	194	11.2	168	99

¹ Means based on four replicates; all other means based on two replicates.

Table 25.5 Ratio of biomass production on mixtures of cellulose or groundwood pulp and activated sludge, to production on activated sludge alone. (Based on data in Table 25.4, corrected for differences in dry wt. of rations; rations numbered 1 to 5 correspond to quantities ranging from low to high in Table 25.4.)

Ration	34%/0%	58%/0%	76%/0%	89%/0%
Cellulose mixtures				
1	2.1	1.9	2.8	0.5
2	2.2	2.0	1.7	0.5
3	1.4	1.0	1.4	0.4
4	1.1	1.0	0.9	0.2
5	0.9	0.8	0.7	0.3
Pulp mixtures				
1	1.0	1.0	0.7	0.3
2	0.8	0.8	0.8	0.4
3	0.6	0.8	0.5	0.2
4	0.4	0.7	0.3	0.1
5	0.4	0.7	0.4	0.2

Table 25.6 Production of *E. fetida* (mg dry wt.) per mg dry wt. activated sludge mixed with cellulose or groundwood pulp. Regressions are based on data in Table 25.4.

	Regression equation	r
% Cellulose		
0	$Y = 0.044X + 7.1$	0.95
34	$Y = 0.051X + 66.8$	0.92
58	$Y = 0.044X + 60.7$	0.93
76	$Y = 0.038X + 112.6$	0.88
89	$Y = 0.013X + 23.6$	0.91
% Pulp		
34	$Y = 0.021X + 22.9$	0.95
58	$Y = 0.039X + 3.7$	0.99
76	$Y = 0.017X + 38.3$	0.86
89	$Y = 0.007X + 15.3$	0.82

25.3.4 Experiment 4

After 4 weeks, 45%, 57% and 67% of 40 test specimens on lyoanaer, lyocasts and lyoaer-anaer respectively were dead. In contrast, 92% of 40 hatchlings placed into cultures with either lyoaer or lyoman were alive. Growth on these latter substrates was approximately proportional to

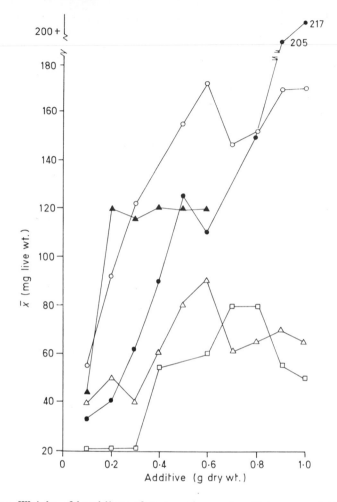

Fig. 25.2 Weight of hatchlings after 4 weeks on 5 g silt loam overlaid with a mixture of 1.25 g cellulose, 3.75 ml water and 0.1 to 1.0 g lyoaer (●), lyoaer-anaer (□), lyoanaer (Δ), lyocasts (▲) or lyoman (○).

concentration of test material, while on the substrates which caused high mortality, growth of the survivors was stunted (Fig. 25.2).

Considering the similarity in elemental analyses among the five lyophilized materials (Table 25.7), a modified experiment was run. This time the test substances were allowed to interact with the cellulose for 4 weeks before introduction of hatchlings, one hatchling was added per dish, and only one level of test substance, 1.0 g, was tested. Ten replicates were run.

Table 25.7 Composition of lyophilized test substances.

Substance	Nitrogen (%)	Phosphorus (%)	Potassium (%)	Sulphur (%)	Calcium (%)	Magnesium (%)
Lyoaer	4.5	2.4	0.17	0.72	4.7	2.0
Lyoaer-anaer	2.4	2.6	0.19	0.75	5.8	2.8
Lyoanaer	4.2	2.2	0.10	1.23	5.3	0.6
Lyocasts	4.2	2.3	0.12	0.66	4.5	1.8
Lyoman	1.8	0.4	0.59	0.22	1.3	0.4

Survival on lyoanaer was not improved, and growth was significantly inhibited on lyoaer-anaer (Table 25.8). Growth was statistically similar on the other three substances, though the greatest yield per mg of N was derived from the lyophilized manure.

Table 25.8 Survival and growth of hatchlings on lyophilized microbial additives to a cellulose–soil substrate.

Material	Survival (%)	Growth $\bar{x}$ mg live wt. $\pm$ SE	Dry wt. of worm (mg mg^{-1} N)
Lyoaer	95	578 ± 94	0.34
Lyoaer-anaer	95	106 ± 16	0.10
Lyoanaer	45	54 ± 21	0.03
Lyocasts	83	366 ± 106	0.23
Lyoman	83	437 ± 53	0.66

25.4 DISCUSSION

On a dry weight basis *E. fetida* contains about 10 % nitrogen (Abe *et al.*, 1979; Hartenstein *et al.*, 1980). The yield of about 4–6 % *E. fetida* biomass (experiments 1–3) from sludge samples which contained about 5 % nitrogen suggests that, assuming denitrification was not occurring, only 10 % of the sludge nitrogen was incorporated into the earthworm. This suggestion is supported by data reported previously (Hartenstein and Hartenstein, 1981).

When cellulose was mixed into sludge to a point where the nitrogen concentration was reduced to about 1.2 %, a yield of *E. fetida* similar to that which occurs on sludge with a fourfold greater concentration of nitrogen was obtained (Table 25.4). This suggests that *E. fetida* may derive a considerable amount of energy from cellulose and that it may selectively remove a larger proportion of microbes from the ambient cellulosic matrix. Both of these suggestions are valid, the latter because of the large amount of nitrogen that was required to achieve the growth reported in Table 25.4, and the former because of the twofold greater yields obtained on limited food supplies which contained 34, 58 or 76 % cellulose, relative to yield on sludge which was not amended with cellulose (Table 25.5).

In contrast to cellulose, wood did not lend itself to production of *E. fetida*. This suggests that the ligneous components of wood must be microbially or enzymically separated from the cellulose if the latter is to support weight gain.

In experiments independent of this study it was learned that, regardless of which of 13 different combinations of five wood-decay fungi (selected from a total of 40 species) and three concentrations of mineral nutrients was used, less than 1 % of the groundwood pulp used in this experiment decayed in ten weeks at 25°C. Failure to achieve rapid decay occurred regardless of whether the groundwood pulp/water ratio was 1:2, 1:3 or 1:4 (w/w). The groundwood may have contained preservatives or natural fungistatic or fungicidal material and it is essential to realize that, while some species of finely ground wood may provide a nutrient (cellulose to *E. fetida* in the presence of wood-decay microbes), the spruce and fir pulpwood used here did not.

The yield of only about 5 % may be related to castings toxicity (Kaplan *et al.*, 1980b). It appears that a certain amount of space is required if *E. fetida* is to sustain itself. In general, my experiences lead me to believe that *E. fetida* processes about 20 times its own volume or weight of material (assuming density (m/v) is about 1) in 10 days and then may begin to search for fresh substrate. When the processed material is quite moist (*ca.* 12 % solids) searching is very apt to proceed. In considerably drier surroundings (*ca.* 20 %), *E. fetida* may elect to remain sedentary instead, and to thrive on its own energy reserves, which diminish at a rate equivalent to a loss of about 1 % body weight per day at 25°C (unpublished data). (This information on space requirement should be viewed as tentative and in need of further research.)

The low yield is also partly due to the unavailability of some of the starting material to the earthworms with time. Exactly how much material decayed through microbial action was not determined. Assuming as much as 25 %, however, the yield of *E. fetida* biomass would still not exceed about 7 %.

Not all microbes, at least not the secretory or other derivative products of microbes, are suitable for production of earthworm biomass (Fig. 25.2, Table 25.8). The data of Fig. 25.2 show that, despite the similarity in elemental composition of freeze-dried preparations of sludge, certain sludges were toxic. These include samples which were derived anaerobically (lyoanaer), samples which were derived aerobically and then allowed to become anaerobic (lyoaer-anaer), and samples which had passed through the gut of *E. fetida* (lyocasts). It is known that earthworms do not practice coprophagy (Kaplan *et al.*, 1980b) and the anaerobic sludge digests are toxic to earthworms (Kaplan *et al.*, 1980a). What is of interest is that, despite a 4 week incubation with cellulose, the anaerobic sludge remained morbidly toxic. Also, the aerobic sludge which was allowed to become anaerobic, though less toxic than initially, nonetheless inhibited growth (Table 25.8). Only castings, of the three previously toxic materials, had lost toxicity significantly.

Three comments are required on Table 25.5. First, when small rations (corresponding to the three lowest levels in Table 25.4) were offered, and 34, 58 or 76 % of these rations was cellulose, more *E. fetida* was produced than on the same weight of ration consisting entirely of sludge. This suggests that cellulose is required for rapid growth of *E. fetida*.

Second, when a critical ration was offered during a unit of time (the second highest ration in Table 25.5; 5 weeks), growth occurred at the same rate, independently of whether the ration contained exogenous cellulose or sludge alone. This suggests that there was sufficient endogenous cellulose in the sludge itself to enable *E. fetida* to grow as rapidly as it could under the constraints of the kinds and quantities of microbes available in the food, moisture content, temperature, and perhaps other ambient factors and conditions. (The endogenous source of cellulose in the sludge is undecomposed toilet paper, undigested fibres from the diet of the populace, and possibly cellulose present in certain micro-organisms).

Third, there was a trend, heading in the direction of small to large rations, of obtaining smaller differences in quantity of *E. fetida* produced in the presence compared with the absence of exogenous cellulose. In fact, where the largest ration was offered, growth occurred more slowly on sludge with exogenous cellulose. Since each of the five levels of ration for each variable tested had been obtained from a common mix, an explanation for this observation must lie in differences between the compositions of the diets with and without exogenous cellulose. The most plausible explanation would be an extension and expansion of the argument advanced in the second comment. That is, *E. fetida* grew faster on the highest ration of sludge with exogenous cellulose because the quantity of cellulose present in the former was not a limiting factor, and had a much higher population density of microbes which might be selectively ingested by *E. fetida*. Some support for this is given by the data for groundwood pulp in Table 25.5, where the same trend is shown. Additional support is provided in Neuhauser *et al.* (1980), where it was shown that much larger *E. fetida* can be produced on unrestricted rations of activated sludge than on unrestricted rations of manures.

The data of this paper suggest a need for fundamental work with axenic cultures of microbes, combinations of axenic cultures, and combinations of microbes, soil minerals and humus, if we are to obtain a basic understanding of the ecological roles of earthworms.

25.5 ACKNOWLEDGEMENT

Research was supported by a grant from National Science Foundation. Mr M. Malecki and Ms K. A. Starwood provided technical assistance.

25.6 REFERENCES

Abe, R. K., Braman, W. L. and Simpson, O. (1979) Producing earthworms. USDA Report, SEA/CR 616-15-12.

Hartenstein, R. (1981a) Position paper on use of *Eisenia foetida* in organic recycling based on laboratory experiments. In *Workshop on the Role of Earthworms in the Stabilization of Organic Residues* (M. Appelhof, compiler), Vol. 1. Proc. Beach Leaf Press, Kalamazoo, MI, pp. 155–165.

Hartenstein, R. (1981b) Production of earthworms as a potentially economical source of protein. *Bioeng. Biotech.*, **23**, 1797–1811.

Hartenstein, R. and Hartenstein, F. (1981) Physicochemical changes effected in activated sludge by the earthworm *Eisenia foetida*. *J. Environ. Qual.*, **10**, 377–382.

Hartenstein, R., Neuhauser, E. F. and Kaplan, D. L. (1979) Reproductive potential of the earthworm *Eisenia foetida*. *Oecologia (Berl.)*, **43**, 329–340.

Hartenstein, R., Leaf, A. L., Neuhauser, E. F. and Bickelhaupt, D. (1980) Composition of the earthworm *Eisenia foetida* (Savigny) and assimilation of 15 elements from sludge during growth. *Comp. Biochem. Physiol.*, **66**C, 187–192.

Hartenstein, R., Neuhauser, E. F. and Narahara, A. (1981) Effects of heavy metal and other elemental additives to activated sludge on growth of *Eisenia foetida*. *J. Environ. Qual.*, **10**, 372–376.

Kaplan, D. L., Neuhauser, E. F., Hartenstein, R. and Malecki, M. R. (1980a) Physicochemical requirements in the environment of the earthworm *Eisenia foetida*. *Soil Biol. Biochem.*, **12**, 347–352.

Kaplan, D. L., Hartenstein, R. and Neuhauser, E. F. (1980b) Coprophagic relations among the earthworms *Eisenia foetida, Eudrilus eugeniae,* and *Amynthas* spp. *Pedobiologia*, **20**, 74–84.

Mitchell, M. J., Hartenstein, R., Swift, B. L., Neuhauser, E. F., Abrams, B. I., Mulligan, R. M., Brown, B. A., Craig, D. and Kaplan, D. (1978) Effects of different sewage sludges on some chemical and biological characteristics of soil. *J. Environ. Qual.*, **7**, 551–559.

Neuhauser, E. F., Hartenstein, R. and Kaplan, D. L. (1980) Growth of the earthworm *Eisenia foetida* in relation to population density and food rationing. *Oikos*, **35**, 93–98.

Sabine, J. R. (1978) The nutritive value of earthworm meal. In *Utilization of Soil Organisms in Sludge Management* (ed. R. Hartenstein), Conf. Proc. Natn. Tech. Info. Svces., Springfield, VA, PB 286932, pp. 122–130.

Chapter 26

The culture and use of *Perionyx excavatus* as a protein resource in the Philippines

R. D. GUERRERO

26.1 INTRODUCTION

The production of animal protein in the Philippines is insufficient to supply the requirements of its animal industry, much less its human population. In 1978, we imported $26 million worth of fish meal, meat and bone meal to augment local supplies. Several earthworm species are presently being cultured experimentally or on a limited commercial scale as a possible substitute for these imports. One that has shown promise is *Perionyx excavatus* (Perrier 1872) which is widely distributed in Luzon Island, the Philippines and elsewhere in Asia. The growth of *Tilapia nilotica* fingerlings and of Japanese Quail (*Coturnix coturnix*) fed on diets containing meal prepared from *P. excavatus* has been studied.

26.2 MATERIALS AND METHODS

26.2.1 Preparation of *P. excavatus* meal

P. excavatus weighing 0.3–0.5 g produced at the Central Luzon State University were used for propagation. Cultures were set up in three concrete circular tanks, 100 cm ϕ × 45 cm stocked with 200 worms each. The tanks were kept outdoors and covered with galvanized iron sheets. Dried and fermented Murrah buffalo manure (20 kg) was placed in each tank and the substrate was overturned weekly to improve aeration and distribute the moisture which was maintained at 40–60%, by watering twice a week.

The original worms were transferred after 4 weeks to new breeding tanks, and dried *Leucaena leucocephala* leaves were added to the original culture tanks as food for the newly hatched juveniles which then weighed 0.1–0.29 g. These were collected by hand with the use of wire screens, washed and sun-dried or oven-dried at 50°C for 6 h and analysed. The buffalo manure had a N content of 1.9% and the manure/leaf mixture at the end of the experiment, 2.4%.

26.2.2 *Tilapia* studies

Two feeding trials were conducted on *Tilapia nilotica* fingerlings in floating cages in a pond. Trial I compared controls, which were not fed, with groups fed on four different diets: A, 25% fish meal (approx. 50% crude protein), 75% fine rice bran; B, 25% earthworm meal, 75% fine rice bran; C, 10% earthworm meal, 15% fish meal, 75% fine rice bran; D, 15% earthworm meal, 10% fish meal, 75% fine rice bran. In Trial II, three diets were tested: A, 25% fish meal, 75% fine rice bran; B, 10% fish meal, 15% earthworm meal, 75% fine rice bran; C, 15% earthworm meal, 85% fine rice bran.

For Trial I, fifteen 1 m³ floating cages with mesh size of 19 mm were each stocked with 200 fingerlings of *T. nilotica*, 2–5 g each. Three replicate cages were used per treatment in a randomized design. The trial was continued for 60 days from November 1979 to January 1980.

For Trial II, nine 1 m³ floating cages similar to those in Trial I were utilized with 100 fingerlings of *T. nilotica*, 11–12 g each, per cage. Three replicates per treatment were used as before and the trial was conducted for 30 days from September to October 1980.

The diets were fed to the fish as mash at the rate of 5% of fish body weight per day, six days a week. Half of the ration for the day was given in the morning and half in the afternoon. Every two weeks, 10% of the fish in each cage were weighed and the feeding rate adjusted. The weight gain, feed conversion and survival of the fish were determined at the end of the trials. Feed conversion ratio was calculated as total weight of feed given/total weight gain of fish per cage, using the F-test to determine significant differences at the 1% level.

26.2.3 Quail studies

Five diets were tested (Table 26.1). Five batches of *Coturnix coturnix*, each of 20 birds, with body weights of 17.3–17.5 g were used and each batch was assigned a diet randomly. The birds were kept in coops and had

Table 26.1 Composition of diets used in Japanese Quail studies.

	Ingredients (%)				
	A	B	C	D	E
Yellow corn	38	38	38	38	38
Rice bran	9.4	9.4	9.4	9.4	9.4
Soybean meal	25	25	25	25	25
Copra meal	14	14	14	14	14
Fish meal	10	7.5	5	2.5	—
Earthworm meal	—	2.5	5	7.5	10
Bone meal	3	3	3	3	3
Common table salt	0.3	0.3	0.3	0.3	0.3
Afsillin*	0.3	0.3	0.3	0.3	0.3

* Vitamin–mineral premix.

access to feed and water *ad libitum*. After 28 days, the final weight was measured and the gain in weight, average daily gain, total feed consumed and feed efficiency were calculated.

26.3 RESULTS AND DISCUSSION

The breeding tanks for *P. excavatus* yielded an average of 2386 young $tank^{-1} month^{-1}$. Mean survival and weight of the introduced worms after four weeks were 97% and 0.66 g and an average of 1420 juveniles weighing 270 g was collected from each rearing tank after another month.

Fresh *P. excavatus* contained 85.2% moisture after being washed and drained; 5 kg of live earthworms produced approximately 1 kg of

Table 26.2 Chemical analysis of earthworm meal.

Proximate analysis (%)		*Partial amino acid analysis* (%)	
Crude protein (N × 6.25)	69.8	Lysine	4.0
Moisture	4.0	Methionine	2.4
Crude fat	5.8	Cystine	0.28
Crude fibre	1.5		
Ash	5.4	*Calorific value* (cal g^{-1})	6400
Non-fat extract	13.5		
		Mineral analysis (mg 100 g^{-1})	
Vitamin assay ($100 g^{-1}$)			
		Calcium	286
Vitamin A	733 iu	Phosphorus	1004
Thiamin	0.12 mg	Iron	13.5
Riboflavin	2.86 mg	Sodium	400
Niacin	5.0 mg	Potassium	830

earthworm meal. The chemical composition of the meal is shown in Table 26.2.

26.3.1 *Tilapia* studies

Trial I showed that the weight gain of fish fed with diet D was significantly higher than those of the other treatments ($P < 0.01$) (Table 26.3). The control fish gained weight slightly as a result of natural food in the pond water, but the weight gains of the fish fed with diets A, B and C were significantly higher than those of the unfed controls ($P < 0.01$). The survival rates of fish fed with diets B, C and D were significantly better than those fed on diet A and the controls. The best feed conversion ratio was also obtained with diet D. These findings indicate that the diet in which fish meal had been partly replaced with earthworm meal was more efficient compared with the diet with fish meal as the only animal protein source.

Trial II gave similar results. Weight gains, survival rates and feed conversions were better for fish fed with diets containing earthworm meal than with the standard diet (Table 26.4).

Table 26.3 Effects on *Tilapia* of diets including earthworm meal (Trial I).

$\bar{x}(n = 3)$	Control	Diet A	Diet B	Diet C	Diet D
Initial weight (g)	3.15	3.05	3.18	3.14	3.17
Final weight (g)	8.70	12.05	15.07	14.40	22.74
Weight gain (g)	5.55	9.0	11.89	11.26	19.57
Feed conversion ratio	—	2.12	1.76	1.96	1.42
Survival (%)	81.66	89.0	96.3	94.6	98.16

Table 26.4 Effects on *Tilapia* of diets containing earthworm meal (Trial II).

$\bar{x}(n = 3)$	Diet A	Diet B	Diet C
Initial weight (g)	12.7	11.9	11.8
Final weight (g)	22.7	24.73	24.10
Weight gain (g)	10.0	12.83	12.3
Feed conversion ratio	1.58	1.34	1.31
Survival (%)	95.7	97.3	96.3

26.3.2 Quail studies

Weight gain and feed conversion for *C. coturnix* improved with increasing levels of earthworm meal in the diet (Table 26.5). The best results

Table 26.5 Effects on quail of diets containing earthworm meal.

$\bar{x}$ ($n = 20$)	Diet A	Diet B	Diet C	Diet D	Diet E
Initial weight (g)	17.50	17.47	17.49	17.32	17.32
Final weight (g)	99.37	102.58	106.27	107.75	111.22
Weight gain (g)	81.87	85.11	88.78	90.43	93.90
Feed conversion ratio	4.26	4.14	3.95	3.92	3.82

were obtained with diet E in which fish meal was totally replaced with earthworm meal as 10% of the diet. No mortality of birds was recorded during the study.

The results of both studies support the conclusion of Sabine (1978) that earthworm meal is nutritionally satisfactory as a protein supplement in animal feedstuffs.

26.4 REFERENCE

Sabine, J. R. (1978) The nutritive value of earthworm meal. In *Utilisation of Soil Organisms in Sludge Management* (ed. R. Hartenstein), Conference Proc., Syracuse, N.Y., pp. 122–130.

Chapter 27

Utilization of *Eudrilus eugeniae* for disposal of cassava peel

C. C. MBA

27.1 INTRODUCTION

The bitter cassava, *Manihot utilissima*, is widely grown in the tropics for its large tuberous roots, a major source of food carbohydrate. The rind of the roots, though rich in nutrients, has a high cyanide content and, as cassava peel, forms a toxic waste which kills soil invertebrates and can inhibit root respiration at very low cyanide (KCN) concentrations (Tanaka and Tadano, 1972). The toxicity of cassava peel limits its use as an organic manure and the peel is therefore dumped. The dumps are commonly on the most fertile land in the villages, presenting problems of soil conservation.

Earlier investigations (Mba, 1978) showed that the earthworm *Eudrilus eugeniae* (Kinberg) is capable of ingesting and excreting similar organic materials at a high rate. The present study concerns its possible utilization in the disposal of cassava peel, production of which is very high in Nigeria.

27.2 MATERIALS AND METHODS

27.2.1 Growth experiments

In September 1980, 200 g air-dried cassava peel was introduced into each of eight plastic pots (20 cm ϕ × 10 cm) with perforated bottoms and watered to field capacity. The pots were enclosed in nylon mesh bags and eight sub-adult *E. eugeniae* were introduced to each after 2 days. The worm cultures were kept in a thatched shelter and were watered to field capacity every other day. Air-dried cassava peel (100 g) was added to each at monthly intervals. After 4 months the worms and cocoons were

315

separated by hand from the residues and a residue sample was taken from each pot for chemical analysis. At this time the worm biomass was on average 6.32 g 100 g^{-1} of cassava peel used. In February 1981, eight of the cocoons obtained were placed in each of eight pots prepared as before. These were examined daily until the cocoons hatched and the young worms were then weighed singly and returned to the pots. Thereafter the worms were weighed individually at monthly intervals and after 2 months they were examined fortnightly and morphological changes recorded. During this period the temperature in the shelter ranged from 22°C min. to 34°C max. and the relative humidity, measured at 10:00 h and 16:00 h, ranged from 34 % min. to 81 % max.

In May 1981 another set of eight pots was similarly set up, but with the addition of 50 g of cow dung, air-dried and ground, to each pot. Eight *E. eugeniae* from the previous experiment, 6–7 weeks old and each weighing about 0.5 g, were introduced to each pot and the cultures were examined after 1 month. The temperature range in the shelter was 22°C min.–31°C max. and the relative humidity 72 % min.–83 % max.

27.2.2 Analysis of residues

The residues taken from the worm cultures after 4 months were compared with air-dried cassava peel, and peel allowed to decompose without earthworms for 8 months. Four pots were taken at random from each treatment and duplicate analyses were made on each pot. Microbial biomass was determined by the method of Anderson and Domsch (1978) using NaOH and BaCl$_2$ titration for CO$_2$ estimation. Dehydrogenase activity was estimated by a modification of Thalmann (1967) using Na-dithionate to calibrate the solution of 2,3,5-triphenyltetrazolium chloride. The method of Hoffmann and Hoffmann (1966) was used to estimate acid phosphatase activity with K$_3$Fe(CN)$_6$ and 4-aminophe-nazone used as indicator to determine the phenol concentration. Total N was determined by the Kjeldahl method, organic C by Walkley and Black's method, K, Ca, Mg, Na and cation-exchange capacity after Jackson (1958), P by Bray's method and HCN by that of the Association of Agricultural Chemists (1965). Water-holding capacity was determined with a tension table constructed according to Hartge (1971).

27.3 RESULTS AND DISCUSSION

A growth curve for *E. eugeniae* based on 64 worms on a diet of cassava peel is shown as drawn by eye in Fig. 27.1. The curve shows an acceleration of growth up to about 7 weeks when clitellar development was complete and a decline during cocoon production which continued from about 11

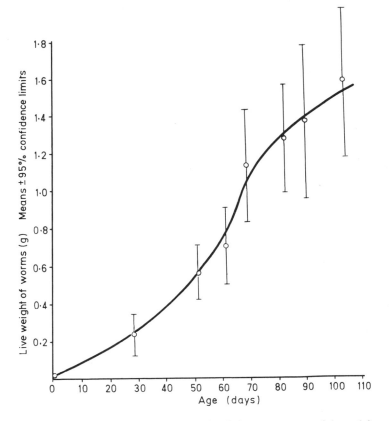

Fig. 27.1 Growth curve of *E. eugeniae* fed on cassava peel ($n = 64$).

Table 27.1 Growth and cocoon production of *E. eugeniae* after 1 month in cassava peel with and without cow dung.

	Survival (No. worms pot^{-1})		Growth (weight (g) worm^{-1})			Cocoon production (No. pot^{-1})
	Initial*	Final	Initial*	Final	Increase	
Cassava peel	8	6	0.57	1.64	1.07	6
SE	—	0.5	0.13	0.08	0.19	1
Peel + cow dung	8	8	1.52	2.16	1.64	24
SE	—	—	0.07	0.09	0.11	5

* At age 6–7 weeks.

Table 27.2 Chemical composition of cassava peel and of cassava peel worm-culture residues. Means ± SE, $n = 8$.

	C (%)	Total N (%)	C/N	Total P (%)	C/P	Total K (%)	Total Ca (%)	Total Mg (%)	HCN (mg kg^{-1})
Air-dried peel									
$\bar{x}$	39.0	0.96	40.6	0.19	205	1.10	1.20	1.20	344.2
SE	0.6	0.16	—	—	—	—	—	—	9.5
Culture residue									
$\bar{x}$	27.0	1.98	13.6	0.14	171	1.80	—	—	182.2
SE	0.3	0.37	—	0.05	—	0.10	—	—	9.9

weeks. Table 27.1 indicates that maximum growth was not attained on a diet of cassava peel alone and that the worms grew much larger and produced more cocoons when cow dung was added. Cocoons hatched, under the conditions of the experiment, in approximately 21 days.

Comparison of analyses of air-dried cassava peel and the 4-month worm-culture residues (Table 27.2) showed a marked loss of carbon and corresponding decreases in C/N ratio and C/P ratio and a relative increase in total N. The cyanide (HCN) concentration fell by almost half.

The worm-culture residues after 4 months compared with decomposed cassava peel after 8 months (Table 27.3) contained substantially higher concentrations of exchangeable K but very much lower concentrations of exchangeable Na. Concentrations of HCN were similar in the two materials. Base saturation of 80.3% and a mean cation-exchange capacity of 81 meq. 100 g^{-1} indicated the high availability of base nutrients and the colloidal nature of the worm-worked residues. The decomposed peel not worked by worms was markedly less colloidal and had a cation-exchange capacity of 46 meq. 100 g^{-1}. The loss of exchangeable Na from the worm-worked residues is at present unexplained but may in part result from active uptake by the worms.

Table 27.3 Chemical composition of decomposed cassava peel and of cassava peel worm-culture residues. Means $\pm$ SE, $n = 8$.

	Water-holding capacity (%)	pH	Cation-exchange capacity	Exchangeable bases				
				K	Mg	Ca	Na	HCN (mg kg^{-1})
				(meq. 100 g^{-1})				
Decomposed peel								
$\bar{x}$	279.4	7.8	46.0	11.1	10.1	18.3	4.30	168.5
SE	5.9	0.1	5.9	4.0	5.2	0.6	0.40	6.0
Culture residues								
$\bar{x}$	634.9	7.2	81.0	32.0	12.1	20.3	0.63	182.2
SE	6.6	0.1	9.7	2.0	4.3	0.9	0.11	9.9

The microbial contents of the four-month worm-culture residues and the eight-month decomposed peel (Table 27.4) were 6.5 and 5.5 times that of the air-dried peel and contained a slightly higher proportion of bacterial to fungal biomass. The ratios of fungi to bacteria in the decomposed peel and the culture residues, respectively 66:34 and 61:39, were significantly different at the 1% level as tested by LSD analysis (Steel and Torrie, 1960). Phosphatase activity was very much greater in

Table 27.4 Microbial biomass and enzyme activity in cassava peel, decomposed cassava peel and cassava peel worm-culture residues. Means $\pm$ SE, $n = 8$.

	Air-dried peel		Decomposed peel		Culture residues	
	$\bar{x}$	SE	x	SE	$\bar{x}$	SE
Microbial biomass ($\mu g\,C\,g^{-1}$ dry wt.)						
Bacterial	52.8	4.6	819.5	58.0	1068.6	82.0
Fungal	382.4	20.0	1593.6	120.0	1767.3	134.0
Total	435.2	16.9	2413.1	124.5	2835.9	102.0
Acid phosphatase						
(mg phenol g^{-1} dry wt.)	4.9	0.0	24.6	2.8	182.0	10.4
Dehydrogenase (mg TTX*						
100 g^{-1} dry wt.)	14.0	0.9	59.3	1.9	127.3	3.4

* 2,3,5-Triphenyltetrazolium chloride.

the culture residues than in the decomposed peel. Dehydrogenase activity, a general indicator of high mineralization potential, was markedly higher in the worm-culture residues than in the decomposed peel.

The effects of *Eudrilus eugeniae* on cassava peel appear to be beneficial, increasing the availability of nutrient elements, reducing Na availability and producing a biologically active humified material with high colloid content, high exchange capacity and high water-holding capacity. The cyanide toxicity of cassava peel was reduced by worm action in 4 months to a level attained by decomposition in the absence of worms in 8 months.

27.4 REFERENCES

Anderson, J. P. E. and Domsch, K. H. (1978) A physiological method for the quantitative measurement of microbial biomass in soils. *Soil Biol. Biochem.*, **10**, 215–221.

Association of Agricultural Chemists (1965) *Official Methods of Analysis of the Association of Official Agricultural Chemists*, 10th edn, A.O.A.C., Washington D.C.

Hartge, K. H. (1971) *Die physikalische Untersuchung von Böden*. F. Enke Verlag, Stuttgart, 71 pp.

Hoffmann, E. and Hoffmann, G. (1966) Die Bestimmung der biologischen Tatigkeit in Boden mit Enzymmethoden. *Adv. Enzymol. Relat. Subj. Biochem.*, **28**, 365–390.

Jackson, M. L. (1958) *Soil Chemical Analysis*. Prentice Hall Inc., Englewood Cliffs, New Jersey. 498 pp.

Mba, C. C. (1978) Influence of different mulch treatments on the growth rate and activity of the earthworm *Eudrilus eugeniae* (Kinberg). *Z. Pflanzenernahr-Bodenk.*, **141**, 453–468.

Steel, R. G. D. and Torrie, J. H. (1960) *Principles and Procedures of Statistics.* McGraw Hill, London and New York, 481 pp.

Tanaka, A. and Tadano (1972) Der Einfluss des Kaliums auf die Eisentoxizitat in der Reispflanze. *Kali-Brief,* **9/21,** 1–12.

Thalmann, A. (1967) Über die mikrobielle Aktivität und ihre Beziehung zu Fruchtbarkeitsmerkmalen einiger Böden unter Berücksichtigung der Dehydrogenaseaktivitat. Dissertation, University of Gissen.

Chapter 28

Cultivation of *Eisenia fetida* using dairy waste sludge cake

K. HATANAKA, Y. ISHIOKA and E. FURUICHI

28.1 INTRODUCTION

Treatment of dairy waste water by the active sludge process always produces an excess of active sludge. This is usually dehydrated to cake and then incinerated or discarded. This chapter describes experiments in which earthworms were reared in sludge cake derived from dairy waste and presents data on their growth and reproductive rates and the change in composition of the food.

28.2 MATERIALS

Adult worms of *Eisenia fetida*, purchased from the Agricultural Co-operative of Kawakami Village, Nagano Prefecture, were fed on dehydrated sludge cake, the composition of which was 85% water, 7.3% protein, 0.46% lipid, 4.5% carbohydrate, 0.71% cellulose, 2.1% inorganic materials, 0.64% P_2O_5, 0.08% K_2O. Since the sludge contained a minimal amount of cellulose, which earthworms are able to digest by secretion of cellulase, dry rice straw was added to some of the feeds. Early in the study when the surface of the feed in the cultivation boxes was covered with paper waste to control evaporation, it was noticed that the worms fed on this preferentially and waste paper was therefore added to some of the feeds in small quantities as an additional source of cellulose. Cow manure, known to be one of the best natural feeds for earthworms, was used as a control feed and in one sludge feed after it had been left outdoors for one month. A parallel series of feeds supplemented with a commercial animal feed was also set up. The compositions of the feeds are shown in Table 28.1.

Table 28.1 Composition of earthworm feeds.

Feed	Sludge cake	Rice straw	Cow manure	Newspaper	Commercial feed
A	95.8%	3.6%		0.6%	
B	63.9%	3.6%	31.9%	0.6%	
C		3.6%	95.8%	0.6%	
D	99.4%			0.6%	
A′		Feed A 85%			15%
B′		Feed B 85%			15%
C′		Feed C 85%			15%
D′		Feed D 85%			15%

The growth chambers comprised plastic boxes (230 × 135 × 60 mm) with perforated lids.

28.3 METHODS

28.3.1 Preference tests

The experimental feeds A–D were weighed, well mixed and placed with 20 worms of average weight 1.33 g in each of four culture boxes. The four boxes were placed in a large container with their lids slightly open to permit movement of the earthworms between them. At intervals, the adult worms, hatchlings and cocoons were handsorted from the feed, counted and weighed.

28.3.2 Growth rate

Five newly hatched worms were placed in a culture box containing feed A and kept at room temperature ($21 \pm 2°C$). At intervals the worms were removed and weighed.

28.3.3 Hatchling production

Sixteen cocoons were incubated on wet filter paper in a Petri dish at room temperature.

28.3.4 Tissue analysis

Lipid assay was by the ether extraction method and protein was estimated as $N \times 6.38$.

28.3.5 Analysis of food and faeces

Faeces present on top of the feed could be easily separated for analysis. The remaining mixture of food and faeces was thoroughly mixed for analysis. Water content was determined by drying at 110°C to constant weight, ash content by incineration at 550°C, and organic matter content was estimated by difference. N was assayed by the Kjeldahl method, P by the ammonium molybdate method and K by atomic absorption.

28.4 RESULTS

28.4.1 Preference tests

The number and weight of worms in the culture boxes was determined by reproduction as well as by movement in and out of them. After 70 days (Fig. 28.1), feed D' contained the least worms, and feed B the most. Feed A was preferred less than B but produced the highest weight gains. In the series A–D, after 45 days, the weight of earthworms (Fig. 28.2) was lowest in feed D (sludge cake + newspaper) and highest in feed A (sludge cake, newspaper and rice straw). In the series A'–D', the weight of worms was in the same relative order. The addition of commercial feed appeared to have increased the weight of worms in feed D' but not in the others.

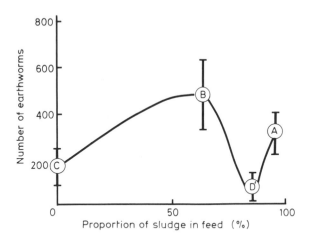

Fig. 28.1 Numbers of worms after 70 days in relation to proportion of sludge in feed.

The results for feed A are shown in Table 28.2. The worms laid cocoons after 10 days and hatchlings appeared after 30 days. Thereafter the worms increased to 10 times the original number at 65 days and 50

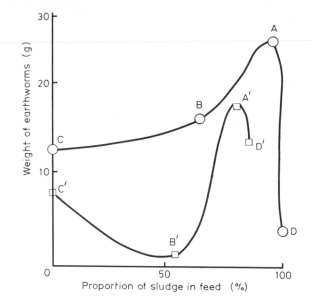

Fig. 28.2 Weight of worms after 45 days in relation to proportion of sludge in feed.

Table 28.2 Increase in number and biomass of earthworms in feed A including immigrant worms.

Day	Number of worms present	Increase ratio of number	Average weight (g)	Total biomass (g)	Increase ratio of biomass
0	20	1	1.33	26.6	1
30	20	1	1.33	26.6	1
70	321	16	0.115	61.1	2.3
90	919	46	0.100	117.0	4.4
120	1522	76	0.152	255.3	9.6

times at 90 days. The total earthworm biomass increased to tenfold at 120 days.

28.4.2 Growth rate

The weight at hatching, 0.003 g worm^{-1} (Fig. 28.3), rose tenfold at 10 days and 100-fold to 0.3 g at 40 days. After 80 days the rate of weight increase became much slower. The rates of increase in numbers and biomass are given in Table 28.3.

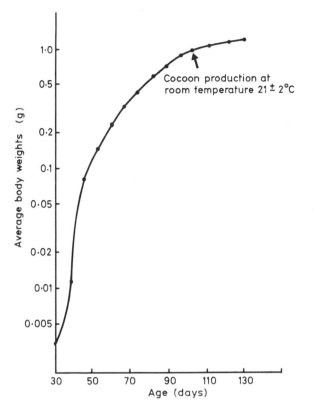

Fig. 28.3 Age/body weight curve in feed A.

Table 28.3 Increase in number and biomass of earthworms in closed cultures.

Day	Number of worms	Increase ratio of number	Average weight (g)	Total biomass (g)	Increase ratio of biomass
0	5	I	1.33	6.7	I
30	5	I	1.33	6.7	I
70	165	33	0.18	35.5	5.3
90	240	49	0.35	90.7	13.6
120	365	73	0.60	222.7	33.4

28.4.3 Cocoon and hatchling production

After 70 days all cocoons, including evacuated shells, were removed and counted. The results showed that one earthworm laid on average 14

Table 28.4 Average number of progeny cocoon^{-1}.

Progeny cocoon^{-1}	Cocoons		Worms hatched		Mean worms cocoon^{-1}
	Total	%	Total	%	
1	1	6.2	1	1.6	
2	2	12.5	4	6.4	
3	2	12.5	6	9.5	
4	6	37.6	24	38.1	3.9
5	3	18.8	15	23.8	
6	1	6.2	6	9.5	
7	1	6.2	7	11.1	

cocoons in 70 days or one cocoon every 5 days. The number of hatchlings per cocoon varied from one to seven with an average of 3.9 (Table 28.4).

28.4.4 Tissue analysis

A sample of the experimental earthworms had the following percentage composition: water, 83.3; protein, 9.74; lipid, 2.11; ash, 1.08.

Changes in food and faeces

The organic matter, N and P contents of feed A fell during the experiment. The organic matter/ash ratio, useful as an index of transformation to compost, fell from 4.54 to 2.15 at 120 days (Table 28.5). The water content was 84% both initially and after 120 days.

Table 28.5 Composition of feed A before and after the experiment (% dry wt.).

Day	Organic matter	Ash	Total N	Total P	Total K	Organic matter/ash
0	82.00	18.00	5.40	9.10	1.00	4.54
120	67.72	32.28	3.47	6.14	1.76	2.15

28.5 DISCUSSION

The observation that weight increase was higher if rice straw was added to the food indicates the importance of cellulose in the earthworm diet. The high reproductive rate in the protein-rich sludge cake feed could be accounted for in two ways. It may be that protein in the sludge cake, more particularly in cells of the dead micro-organisms of which it is composed, is easily digested by earthworms. Alternatively the protein in the sludge cake may be already decomposed to simpler compounds in the presence of various enzymes before it is ingested by earthworms. The direct usability of the sludge cake as feed suggests that organic materials have already been digested by microbial activity during treatment of the sludge as industrial waste water.

After 120 days, 11 % of the solid matter in the feed had been converted into the body weight of earthworms, 69 % remained as residual solid matter and 20 % was lost as CO_2, H_2O etc., presumably by respiration.

The organic matter/ash ratio is usually about 2.0 in natural feeds. The high ratio of 4.54 in the sludge cake feed fell as the experiment proceeded, reaching 2.15 at 120 days, close to the level observed in natural feeds. It is interesting to note that this is the time when the reproductive rate starts to fall and the earthworms should be transferred to fresh feed.

28.6 SUMMARY

The authors attempted to cultivate earthworms in sludge cake from two dairy plants. The sludge cake could support the growth of earthworms without processing to compost and when supplemented with cellulose material such as rice straw was superior as a worm growth medium to cow dung.

Eisenia fetida produced cocoons 10 days after being transferred to the sludge cake feed, and hatchlings appeared at 30 days. Thereafter, on average, four progeny were produced from each adult worm every 5 days for 120 days, after which the rate declined.

The organic matter/ash ratio of the sludge cake feed decreased during the experiment from 4.5 to 2.1 at 120 days, a similar level to that observed in natural composts.

Chapter 29

The earthworm bait market in North America

A. D. TOMLIN

29.1 INTRODUCTION

There is a strong demand from about 50 million anglers in North America for bait worms that is met almost completely by the worm-harvesting industry which collects 'Canadian night crawlers' (*Lumbricus terrestris* Linnaeus) from golf courses and pastures. During 1980, at least 500 million Canadian night crawlers were exported from Ontario to the US. At an average wholesale price of $35 (US) per thousand worms, the total value of the 'crop' was $17.5 million which is obviously a large and attractive market. However, note that these are 'Canadian night crawlers' or 'dew worms.' Anglers prefer this species of worm as bait and, as far as scientists know, and as past experience has repeatedly shown, the dew worm cannot be cultured or grown cheaply enough to compete with the harvested price from golf courses and pastures.

The large earthworm bait market has continued to expand rapidly over the past few years because of the shortage of traditional bait such as shrimp and coarse fish. This has increased earthworm prices and has attracted earthworm growers who have attempted to place cultured worms in competition with 'harvested' worms. Unfortunately some companies have set up worm-growing franchise operations which prey upon a wide variety of people who have little understanding of how the earthworm bait market operates.

The two most widely cultured species in North America are the red worm (*Eisenia foetida* Savigny) and the African night crawler (*Eudrilus eugeniae* Kinberg) (Gates, 1970; Reynolds, 1977). Neither species is very desirable as bait.

This chapter describes some of the intricacies of the North American bait market and why the cultured worm industry has had such limited success.

29.2 THE DEW WORM (*L. TERRESTRIS*) INDUSTRY

The dew worm industry is geographically centred in the area between Toronto and Windsor in Ontario which lies between several of the Great Lakes. The soil from which dew worms are picked is generally a light clay loam in flat or gently rolling countryside well supplied by water from rainfall and the spring snow melt. Vegetation, exclusive of the large area of maize and wheat, is grassland and mixed deciduous forest.

Worms are picked during the night from about 22:00 h to 05:00 h, depending on time of sunset and sunrise. The best weather conditions for worm harvesting are mild overcast nights with high relative humidity (low-pressure cells), following a good rainfall. Clear cool moonlit nights with northerly breezes (high-pressure cells) are poor picking nights. The best harvesting periods usually occur between April and mid-July and mid-August to late October.

Pickers use miner's lamps for illumination and strap an open tin can to each ankle, one of which contains hardwood maple sawdust, used to help the picker grip the slippery mucus-producing worms. The other larger can acts as a receptacle for approximately 500 worms. As each group of 500 worms is picked, it is packed into a small (~ 1 litre) fine mesh bag, a number of which are left around the picking site. On a favourable night a good picker will obtain up to 10 000 worms in 7 hours. As day breaks the pickers bring their night's collection to an air-conditioned van where each bag of 500 worms is poured into a 30 cm square (4 cm deep) wooden container, a 'flat', along with each picker's dated code number and some sawdust to keep the worms in the container. The foreman, from experience, is able to ensure a count of about 500 by the weight of the flat and the size of the worms. The flats are stacked in the van, up to 400 flats a night, and transported to a refrigerated storeroom where the worms are transferred 500 at a time along with the picker code number into foam plastic containers (30 cm square) half-filled with moist black muck. These containers are stacked diagonally one upon another in the refrigerated storeroom and maintained at 2–5°C with the lights on to keep the worms in the containers. The diagonal stacking is required for ventilation and observing that the worms move into the soil. Worm quality is maintained by removing worms that do not move into the soil within one hour following transfer. If the soil is changed every 2–3 weeks worms can be stored for up to four months using these techniques in order to meet increased demand and reduced supply (higher prices) in the mid-summer months.

The worms seem sensitive to changes in relative humidity and the dry air of the refrigerated storage assists in reducing worm activity and escape from the open-topped containers. Opening the doors of the storage area on rainy days induces considerable unrest among the worms. Cases have occurred where storage doors left open accidentally for several hours on rainy days have resulted in several hundred thousand worms escaping from their containers.

The operators of the worm cooler usually sell to wholesale distributors in lots of at least 100 000 and often 500 000 worms. The distributors, usually American, service areas in the United States northeast and mid-west. They bring refrigerated trucks to the cooler and load a contracted number of worms which are then trans-shipped, duty-free, to US points where they are sold to retailers. Increasingly, American wholesale distributors are prepackaging worms in waxed cardboard containers (12 or 24 worms/container) which are sold from automatic vending machines. Otherwise, retailers package worms as needed using the foam plastic containers for stock storage. This is the time when the picker's code number is used to check the integrity of the original count. If the worm counts of containers are consistently below the accepted trade minimum of 475, the picker's code number and date slip is returned to the cooler operators who take appropriate action.

A medium-sized bait worm operation might operate as follows. A picking crew at any one site will have up to 20 pickers and the total harvest on an average night will be 60 000–100 000 worms; on a good night this total will approach 200 000 worms. Larger bait-worm companies operate several crews simultaneously. Larger, better-organized companies also try to avoid harvesting the same site year after year, since leaving a site unharvested for one or two years allows the worms to mature to larger sizes of 10–12 g. Besides reducing worm size, constant annual picking also reduces the number picked from a site, although apparently not enough to make picking uneconomic. Annual fees from several hundred to several thousand dollars are paid by the company to golf courses or farmers for 'harvesting rights.'

Each picker is paid from C$20 to 24 ($0.8 US = $1C) for each 1000 worms. A good picker can earn $240–300 on a good night, and a total of $10 000–12 000 per season.

A medium-sized bait company will process 30–50 million worms during the season in this manner and sell them for C$30 to 44 (average C$35). Prices for worms are high when supplies are short, and vice versa. Matching picking requirements and storage capacity with market demand requires considerable nimbleness on the part of the company, and a very good understanding of the market. The business is not regulated by government or industry to any great extent, with no marketing boards to stabilize supply and demand.

The Ontario worm supply market is further complicated by a number of free-lance pickers who have no cooler storage capacity. They must sell their worms, particularly on hot days within three or four hours of completing their picking. This is done by means of a daily informal auction in a west end Toronto restaurant where free-lance pickers meet with cooler operators. Millions of worms can change hands weekly at this auction as the market players attempt to cover their supply and demand positions in a highly volatile market. The numbers suggest that the bait supply companies can make a very good living from the business if they are well organized and keep in close touch with market requirements – in other words, follow good business practice. The Canadian government has attempted to determine the value of worm exports and their destination since 1978. Its figures, which are almost certainly under-stated, are given in Table 29.1 and represent more than C$50 million at the retail level (Table 29.2) in the US market alone.

Table 29.1 Canadian export markets for bait worms (*L. terrestris*).

Country	Value (1000s of C$)		
	1978	1979	1980
USA	10 700	12 400	12 900 (Est)
W. Germany	159	235	532
Switzerland	56	130	12
Netherlands	8	25	29
Total	10 923	12 790	13 473

Table 29.2 Retail value of Canadian bait worm exports to the USA (1980).

Canadian export value	No. of worms shipped at C$35/1000 worms	Value at retail level if price C$1.75/doz.
C$12 900 000	370 000 000	C$ 54 000 000

29.3 THE CULTURED WORM INDUSTRY

The two species of earthworms commonly cultured in North America, the African night crawler and the red worm (= tiger worm = brandling = manure worm = yellow tail), have been offered to the bait industry, but for a variety of reasons have been found wanting. Both species are much smaller than the dew worm; the African rarely weighs more than 3 g and the red worm rarely more than 1.5 g, and most anglers perceive this as a disadvantage. There is also the anglers' perception that neither of these species stay on the fish hook as well or are as active as the dew worm once on the hook. Both cultured species have been widely available in the US for several years, but neither species has made inroads into the market share of dew worms.

Both species can be reared at high density as long as soil temperatures are in the range of 22–28°C. These worms do not require refrigeration, an advantage to the angler, and in fact will not tolerate cooling or frost. However, their requirement for warm temperatures means that they respire at a higher rate, requiring a constant supply of food and ventilation to sustain their higher respiration rate. This introduces difficulties in handling these worms in transport as they tend to leave containers and require extensive ventilation on long shipments. It also reduces their shelf life if they cannot be frequently re-fed which itself is an additional expense and inconvenience.

The worm culture industry has had other problems. In the US and, to a lesser extent in Canada, unscrupulous operators advertise worm culture as an easy road to riches. Operators talk of huge market requirements for worms for disposal of sewage sludge and municipal garbage. This promise has not been met and has no real likelihood of being practical on a large scale for many years, if ever (Prince *et al.*, 1981). The usual approach of the unscrupulous operators is to place an advertisement in newspapers suggesting a profitable business sideline, particularly for rural dwellers. A phone call answering the ad. brings a salesman who describes sales possibilities in glowing terms, and the ease with which large numbers of worms can be grown by a 'secret' technology known only to his company. The sales-person may also suggest that the worm species to be cultured has one or several of the following attributes: (a) a hybrid pedigree genetically different from any previously described worm; (b) the pedigree of the worm is top secret but that it was developed by a high-technology biological company (usually foreign); (c) the worm has been 'genetically engineered' to be easily reared using the company's technology; (d) the worm species in question has been designed as a bait worm and is better as bait than the dew worm; (e) the worm is a hybrid of the dew worm and one or more species of other worms. These statements are either misleading or untrue and in the case of (e) impossible.

The sales-person may play on the prospect's sense of 'ecological responsibility' by suggesting, for example, that these worms will improve soil fertility; in fact neither of these two species will survive the Canadian winter. The sales-person may further suggest that his company will buy back the produce – at a price not necessarily set at this time, but a value of 5c per worm ($50/1000) may be mentioned in passing. The sales-person may also imply that certain respected financial institutions have 'an interest' in the company (translation: a bank has loaned money to the company), and that the company is in good standing with a credit rating agency (translation: the company meets its own financial obligations). Neither of these facts has any bearing on how successful the subsidiary franchise business will be or on the disposition of worm productivity. The parent distributing company may be very profitable, because many people have invested their savings and bought worm production franchises.

A franchise may be offered at two levels. The more expensive one, which may cost up to $20 000 or more gives 'exclusive' rights to sell the worm growing 'technology' plus worms and hardware to potential growers in a certain region. In turn a regional franchiser would sell franchises to growers to produce worms which would be sold for final marketing. The cost of this grower franchise varies from about $700 to $3000. For this price, the franchise kit includes a 'technology manual' usually described as 'secret'; 500 earthworm cocoons which will hatch to about 1000 small worms, a couple of dozen plastic buckets in which to grow the worm, a starter kit of 'soil substrate', e.g. peat moss and manure, and a food mixture, usually of grain and vitamins. Exclusive of the 'technology manual' which is often inaccurate, wrong or next to useless, the value of the rest of the materials rarely exceeds $50. Incidentally, these worms are available from other sources for as little as $15/1000.

The sales-people may suggest sale of the worm castings as a profitable by-product. Wormcastings can be an excellent soil adjuvant depending on the food the worms were fed [a 'market-basket' average of four different, commercially packaged castings was 0.61, 0.08 and 0.16 % of N, P and K respectively (J. B. Robinson, pers. comm.)], but castings do not qualify as a field fertilizer because in Canada the minimum acceptable standard of N, P, K is 1, 1, 1 %. Castings are also very low in nitrogen, the expensive component of fertilizers. It is also economically impractical, considering the high cost of transportation, to ship a bulky low-value product around a large country such as Canada. Some companies dispose of castings profitably on a local or regional basis to nurseries or flower shops.

It also seems that worms are not nearly as easy to culture as the sales-people suggest. Agriculture extension specialists in southern Ontario have been barraged by phone calls and letters from disenchanted worm growers whose worms are sick or dying from a variety of ailments. The original enfranchising company is unable to provide the technical expertise to solve these problems.

Another problem worm growers have is worm size. Fishermen want large worms for bait and the average maximum weight of an African night crawler is about 3 g. To attain this size, a grower may have to keep them for up to five or six months longer than he anticipated which increases costs. Sometimes the worm distributing company which originally sold the franchise decides not to buy worms under 5 g each, a target which may be impossible to reach (a fact not unknown to the company!).

Even if larger worms can be grown economically, the number available can be important. A distributing company may require a minimum of perhaps 100 000 worms which is beyond production capacity. Even if a separate sale with a wholesaler handling dew worms can be made, a minimum of 100 000–500 000 worms over a period of two or three months may be required to make it worthwhile for the dew worm wholesaler. Because of the storage and shelf life difficulties encountered with the African night crawler, production of this number of worms in such a short time would be beyond the capacity of any but the largest worm culture operation.

The failure rate for worm culture operations is very high. Similar problems occurred several years ago in the US and, as far as I am aware, there have been very few winners, but many losers. The original enfranchising companies may do very well, mostly by selling starting kits to many other people. Unfortunately these people then get left with the worms for which there is no market at a price at which they can recover their original investment, let alone turn a profit.

29.4 GOVERNMENT INVOLVEMENT WITH LARGE-SCALE EARTHWORM CULTURE SCHEMES

The 'ecologically attractive' features of earthworms have tempted several government agencies in Canada to provide public money to underwrite earthworm growing schemes, variously purported to provide such things as inexpensive dog food, fertilizer, fish food and the solution to municipal garbage and sewage sludge disposal. Often even a superficial analysis of the proposals reveals gaping holes in the economic logic of the schemes. For example, red worms, probably the least expensive worm to culture, currently retail at $6 1000^{-1}, equivalent to $21 kg^{-1} live weight. Even at the current price of beef it would be less expensive to feed dogs tenderloin at about $7.20 lb^{-1} ($16 kg^{-1}). Taking water content into account, about 20 % less in dressed tenderloin than in live worms, redworms would still be an expensive form of protein even if their current price could be halved.

The fact that several hundred people in Ontario alone may have been defrauded of C$1000–10 000 each has drawn increasing interest on the part of police, legal and regulatory authorities, who are taking a close look at claims made by worm culture companies.

29.5 SUMMARY

There exists in North America a large well-organized bait worm industry supplied by dew worms picked from golf courses and pastures. The species harvested, *L. terrestris*, is preferred by anglers but cannot be cultured at prices competitive with the 'picked' price. The lucrative market generated by this bait species has attracted competition for cultured worms actively promoted by enfranchising companies. Cultured worms are perceived, by anglers, to be inferior to the dew worm. Claims for other markets (e.g. sludge reduction) are premature or exaggerated.

29.6 ACKNOWLEDGEMENTS

Lynch Bait Farms of Ilderton, Ontario, R & S Natural Projects Ltd. of Inverary, Ontario, and the Industry, Trade & Commerce Dept. of Canada provided technical and financial information on the bait worm industry in Ontario and North America.

29.7 REFERENCES

Gates, G. E. (1970) Miscellanea Megadrilogica VII. *Megadrilogica*, **1(2)**, 1–14.
Prince, A. R., Donovan, J. F. and Bates, J. E. (1981) Vermicomposting of municipal solid wastes and municipal wastewater sludges. In *Workshop on the Role of Earthworms in the Stabilization of Organic Residues* (M. Appelhof, compiler), Vol. 1, Proc. Beech Leaf Press, Kalamazoo, MI, pp. 207–219.
Reynolds, J. W. (1977) The earthworms (Lumbricidae and Sparganophilidae) of Ontario. *Life Sci. Misc. Pub., R. Ont. Mus.*, 141 pp.

Chapter 30

A simulation model of earthworm growth and population dynamics: application to organic waste conversion

M. J. MITCHELL

30.1 INTRODUCTION

Earthworms play an important role in processing organic material derived from natural (Edwards and Lofty, 1977; Satchell, 1967) and anthropogenic substrates such as sewage sludge (Neuhauser *et al.*, 1980b; Mitchell, 1979; Mitchell *et al.*, 1980). To determine how much organic matter can be utilized by earthworms, it would be useful if a predictive model were developed incorporating the effects of population dynamics, substrate quality and physical parameters on this conversion process. Most studies of population density and structure (e.g. Evans and Guild, 1948; van Rhee, 1967; Grant, 1956) lack the necessary detail for constructing mathematical descriptions of population dynamics and there is little quantitative information on the relationship of substrate quality and physical parameters to the growth rate of individual worms in natural populations.

However, recent work on the utilization of *Eisenia foetida* (Sav.) for processing anthropogenic wastes has provided detailed information on mortality, natality, ingestion and growth (Hartenstein *et al.*, 1979; Mitchell, 1979; Neuhauser *et al.*, 1980a; Watanabe and Tsukamoto, 1976).

The purpose of this chapter is to (1) synthesize published information on the role of *E. foetida* in waste conversion; (2) translate this information into mathematical formulations which describe population dynamics, growth and ingestion rate of this worm; (3) incorporate these formulations into a simulation model; (4) use the model to predict how much waste can be processed and how much earthworm biomass can be produced under specific conditions; and (5) establish what further information is needed to understand the role earthworms play in waste processing in natural and anthropogenic systems.

30.2 MODEL STRUCTURE

Mathematical models of animal populations generally require information on either the age or phenological structure of the population (Beddington, 1979; Usher, 1972). This is not available for earthworms which lack distinct phenological stages except cocoons, juveniles (non-clitellate) and adults (clitellate) and cannot be independently aged. Moreover, since the rates of growth and sexual maturation are dependent upon such factors as temperature (Michon, 1954) and substrate quality (Neuhauser *et al.*, 1980a), the chronological age of an earthworm would be a poor predictor of its reproductive potential.

For the present model, the population was divided into size classes on the assumption that worms of the same mass have the same population and energetic characteristics. The model does not account for phenotypic variation, the importance of which in earthworm populations has not been studied. The reproductive potential of earthworms within the same class may vary for example in relation to nutritional state but there are no data which would permit this to be formulated mathematically.

The model WORM.FOR is written in Fortran and has options for modifying specific parameters and the form of output. The user can obtain either tabular or graphical output at specified time intervals. In its present form most parameters, which can be varied, are designated in a subroutine (DATA) which can be compiled separately from other parts of the model. The complete coding is available from the author. An important feature of the model is the number of size classes chosen. By increasing the number of size classes, growth can be modelled more precisely; this is especially important at slow growth rates. Computer storage and running time increase with the number of size classes. After each iteration an algorithm redistributes individuals to an appropriate size class and if growth is not sufficient to permit a change in size class, they remain in their previous class. To simplify output, information is combined in ten size classes.

On the basis of calculated growth rates, the model gives the number of classes needed for a given level of precision under specific growth conditions. After the first iteration, the minimum number of classes required is determined by calculating how many classes are needed to detect growth of individuals at 90% of their maximum mass. This percentage can be altered within the program. Utilizing more size classes increases the precision of growth determination in the larger size classes in which the growth rate decreases. The effect of worm mass on growth rates is discussed in a later section. For the present study 500 size classes were used in all cases. This was more than adequate for the conditions simulated.

WORM.FOR is divided into four subroutines which calculate mortality, growth, reproduction and ingestion (Fig. 30.1). An iteration interval of one week was used.

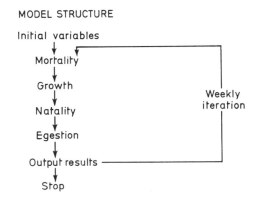

Fig. 30.1 Flow diagram of WORM.FOR simulation model.

30.2.1 Mortality and growth

Only laboratory measurements of mortality have been made for *E. foetida*. Hartenstein *et al.* (1979) and Tsukamoto and Watanabe (1977) found respectively that it ranged from 0.3 to 0.8 % week^{-1} and 0 to 1.0 % week^{-1}. It is difficult to extrapolate from these studies, which attempted to minimize mortality under optimal rearing conditions, to expected mortality in large-scale operations or field situations under different abiotic and biotic conditions. Mortality could also vary among size classes and although no data are available at present, differential mortality could easily be incorporated in the model. For comparative purposes a mortality rate of 1 % was used except as noted.

Earthworm growth is subject to such ambient conditions as moisture, temperature, substrate quality and availability, mortality agents such as toxins, and the size of the worm. The effects of temperature and organism size were incorporated in the model and substrate quality was accounted for by relating the maximum growth rate (K) to specified growth conditions.

The effect of temperature on growth was derived from data of Tsukamoto and Watanabe (1977) and incorporated into the temperature function of Shugart *et al.* (1974). This function assumes an exponential increase, expressed as a Q_{10}, in the rate of a process until a maximum is reached at an optimum temperature, followed by a decrease to an upper lethal temperature. The Q_{10} value was determined from growth of *E. foetida* at 10, 15, 20 and 25°C ($Q_{10} = 6.76 \pm 1.28$ (4); SE (*N*)). The optimum temperature was set at 25°C (Kaplan *et al.*, 1980) and upper lethal temperature at 35°C. The upper temperature limit for growth has not been established but is probably somewhat above 30°C.

The effect of size on growth rate was determined from Tsukamoto and Watanabe's (1977) data. The instantaneous growth rate was assumed (Brody, 1945) to fit:

$$\frac{dm}{dt} = K(X - m)$$

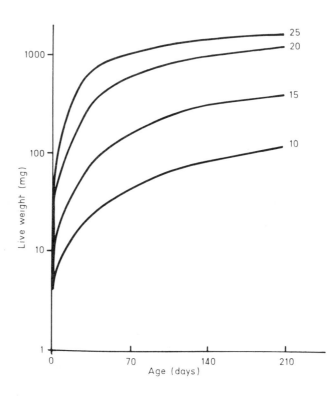

Fig. 30.2 Growth simulation of *E. foetida* at four temperatures using data from Tsukamoto and Watanabe (1977).

where m = mass of organism; K = maximum growth rate; X = maximum mass of organism. The growth rate decreased monotonically as the organism reached a maximum size, estimated at 2000 mg wet wt., a reasonable approximation for most conditions. The value for K was determined by linear regression using the following equation:

$$t = [\ln(1-m)/X] \times (1/-K) + C$$

where t = age in days and $K = 9.55 \times 10^{-3}$ mg mg^{-1} day^{-1} ($r = 0.994$, $n = 3$). The output of the growth portion of the model (Fig. 30.2) corresponds closely to the growth curves given by Tsukamoto and Watanabe (1977). This formulation was compared with data of Neuhauser *et al.* (1980a, b) in which *E. foetida* was grown at 25°C on horse manure, cow manure or activated sludge. There is a good fit to the model output for growth on manure (Fig. 30.3) which was the substrate used by Tsukamoto and Watanabe (1977). More rapid growth on activated sludge could be adjusted for by increasing the value of *XKMAX*.

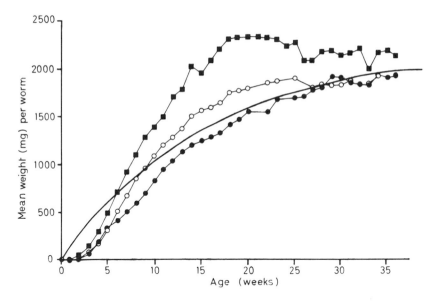

Fig. 30.3 Growth simulation of *E. foetida* (——) compared with laboratory growth (Neuhauser *et al.*, 1980a, b) on activated sludge (■), horse manure (●) and cow manure (○).

30.2.2 Reproduction

The production of new individuals is dependent on the number of cocoons produced, embryos per cocoon and survival of the embryos to hatching. Cocoon production was extrapolated from experiments by Hartenstein *et al.* (1979). At 25°C, cocoon production did not commence until the worms were five weeks old and weighed 329 mg. Production declined after nine weeks (618 mg) when the food was depleted. A maximum cocoon production of 5.5 cocoons/individual was assumed if the worm mass exceeded 618 mg. The following formulations describe cocoon production (Fig. 30.4):

$$COC = (SM \times 0.019) - 6.15$$
$$(r = 0.850, n = 5)$$
$$\text{If } SM \leqslant 329 \text{ then } COC = 0.0$$
$$\text{If } SM \geqslant 618 \text{ then } COC = 5.5$$

where SM = mass of *E. foetida* in mg and COC = cocoons individual^{-1} week^{-1}. Progeny cocoon^{-1} (PC) was taken from the equation of Hartenstein *et al.* (1979):

$$PC = (0.002\,86 \times SM) + 1.17$$

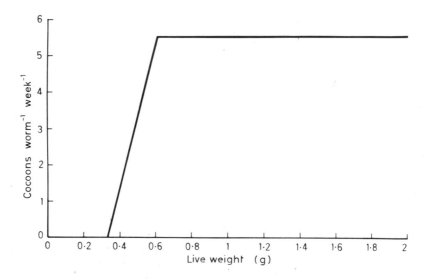

Fig. 30.4 Effect of *E. foetida* weight on cocoon production.

Since cocoon formation is temperature-dependent, the same formulation for describing the effect of temperature on growth was used. Optimal temperature was assumed to be 25°C; studies on reproduction suggest that maximum growth does not necessarily coincide with maximum reproduction. For the effect of temperature on embryo mortality, information from Tsukamoto and Watanabe (1977) was used:

$$FHATCH = (1.271 - 0.035) \times T$$
$$(r = 0.894, n = 4)$$

where T = temperature (°C) and $FHATCH$ = fraction of individuals which hatch.

The effect of temperature on the developmental rates of embryos was also determined from Tsukamoto and Watanabe's (1977) data:

$$PDAY^{-1} = (0.00278 \times T) - 0.0173$$
$$(r = 0.995, n = 4)$$

where $PDAY$ = velocity of development day^{-1}. The development of embryos was simulated with ten embryo classes, movement to a later class being temperature-dependent. After the tenth embryo class they entered the first worm size class corresponding to a size of 4 mg when there were 500 classes.

At present there is no quantitative information which could be utilized to relate such factors as nutrition and growth rate to reproductive rate although this would be desirable.

30.2.3 Ingestion

Relationships determining ingestion rate and assimilation efficiency formulated by Mitchell (1979) were partly incorporated in the present model. The effect of temperature on egestion was described by a modified Krogh curve (Mitchell, 1979, Equations 10 and 11). For earthworms feeding on sludge, moisture, if above a specific threshold, did not alter ingestion. There were marked differences in the ingestion rates on various substrates. In the WORM.FOR model the maximum ingestion rate ($XMAX$) expressed as a scalar can be modified. Hartenstein et al. (1981) found that neither the gut content expressed as a proportion of the body weight of E. foetida nor the gut transit time varied with worm size. It was therefore assumed that ingestion rate was linearly proportional to worm mass.

30.3 EFFECTS OF PARAMETER CHANGES ON MODEL OUTPUT

To determine how changing specific parameters would alter model output, various simulations were done. The standard conditions are given in Table 30.1. It was assumed that E. foetida was consuming a wet palatable sludge which would be optimal for growth. Four levels of

Table 30.1 Initial conditions used for the *E. foetida* simulation model.

Number of worms	100
Mass individual $^{-1}$	6 mg
Iterations	20 weeks
Survival	99 % week $^{-1}$
Reproduction	Maximum
Temperature	25°C
Classes harvested	None

mortality were tested with survival at 99, 90, 80 and 70% week $^{-1}$ (Table 30.2). Increasing mortality by 29% week $^{-1}$ decreased worm numbers, biomass and substrate consumed 135-, 151- and 86-fold respectively. An accurate assessment of earthworm mortality is clearly needed.

Temperature has a marked effect on the model output (Table 30.2). For example, a change of temperature from 25°C to 15°C decreased worm numbers, worm biomass and amount of substrate consumed over 10 000-fold. Even a small temperature change can have a major effect since it directly alters growth, ingestion and reproduction rate. Moreover there are various cumulative effects, for example, a decrease in growth rate slows the production of individuals capable of reproduction. Studies at temperatures near the optimum are needed to ascertain whether this would be found under experimental conditions and to establish optimal conditions for organic matter conversion.

Table 30.2 Effect of survival and temperature on output of WORM.FOR simulation.

Survival rate (% week $^{-1}$)	Number of worms	Biomass (kg wet wt.)	Substrate consumed (kg wet wt.)
99	1.01×10^6	303	94 800
90	2.59×10^5	75	2620
80	4.87×10^4	14	270
70	7.48×10^3	2	110

Temperature (°C)	Number of worms	Biomass (kg wet wt.)	Substrate consumed (kg wet wt.)
25	1.01×10^6	3.03×10^2	9.48×10^3
20	8.36×10^3	2.17	2.14×10^2
15	9.20×10	2.58×10^{-2}	9.00
5	8.20×10	1.84×10^{-5}	7.85×10^{-3}

It has been suggested (Hartenstein, 1981; Sabine, 1981) that since earthworms contain about 60% protein they may serve as a protein source. It is possible with the model to simulate harvesting various size classes to determine optimum yield under various harvesting regimes (Table 30.3). Harvesting individuals greater than 500 mg produces a larger yield than if those greater than 1000 mg are taken; in the latter case, however, the earthworm protein not harvested is about 20 times greater and constitutes a large unutilized resource. In some conditions, yield may be of secondary importance to waste.

Table 30.3 Effect of harvesting two size classes on output of WORM.FOR simulation.

Size class (mg)	Number of worms	Biomass remaining (kg wet wt.)	Biomass harvested (kg wet wt.)
> 1000	9.83×10^5	294	2
> 500	3.54×10^5	11	14

30.4 APPLICATION OF MODEL TO SEWAGE SLUDGE CONVERSION

Various workers have suggested that *E. foetida* could be used in sludge management (Mitchell *et al.*, 1980; Neuhauser *et al.*, 1980a, b) but there have been few large-scale field trials of the concept. The city of Lufkin, Texas (USA) initiated the use of earthworms in processing liquid sludge (Prince *et al.*, 1981) and predicted that 4500 kg of *E. foetida* would consume 900 kg dry sludge daily, the whole output of this plant which serves a population of 10 000 to 15 000. This assumption was tested with the WORM.FOR model. Assuming that the worms weigh 500 mg each, the initial population would be 9×10^6 worms. Reproduction was decreased to 10% of the maximum and survival to 90% week^{-1} to allow for suboptimal conditions. However, the optimal temperature of 25°C was used for the simulation over ten weeks. After this period an average of 1900 kg dry sludge would be consumed daily and earthworm biomass would increase to 18 500 kg. Since this quantity of sludge was not available, the worms would not grow at their optimal rate and produce this large increase in biomass. The earthworm population will change with time and the amount of conversion will change concomitantly.

30.5 POTENTIAL MODEL UTILIZATION

The four major divisions of the model can be altered independently and additional components such as respiration could easily be incorporated.

The model would be improved if such factors as assimilation and respiration rates and specific food items and environmental conditions could be employed in determining growth rate. Such information would allow the formulation of more general functions which could be directly linked to ingestion and egestion. In the present form, growth rates have to be determined empirically for different substrates.

The model does not consider the effects of intraspecific competition for resources under limiting conditions, and information on the relationship of density and/or food limitation in determining growth, reproduction and mortality is needed, as is information on field mortality and natality rates. Some worm growers get optimum hatching at 18°C and optimum growth at 25°C and more study on temperature optima is desirable. The model could be refined if further data were available on the size structure of field populations and the effect of food limitations on reproduction and growth.

This model could be applied to natural earthworm populations in soils if the relevant energetic and population information were obtained for other species. Better predictions could then be made of the amount of organic matter processed by earthworms and its effect in limiting population density. This would allow earthworms to be incorporated into general models of energy flow.

30.6 ACKNOWLEDGEMENTS

The comments of R. Burgess, R. Hartenstein, N. Ringler and J. Satchell were helpful in modifying this manuscript.

30.7 REFERENCES

Beddington, J. R. (1979) Harvesting and population dynamics. In *Population Dynamics* (eds. R. M. Anderson, R. D. Turner and L. R. Taylor), Br. Ecol. Soc. Symp., Blackwell Scientific Publishers, Oxford, pp. 307–320.

Brody, S. (1945) *Bioenergetics and Growth*. Reinhold Publishing Corp., 1023 pp.

Edwards, C. A. and Lofty, J. R. (1977) *Biology of Earthworms*. 2nd edn, Chapman & Hall, London. 333 pp.

Evans, A. C. and Guild, W. J. Mc. L. (1948) Studies on the relationships between earthworms and soil fertility. V. Field populations. *Ann. Appl. Biol.*, **35**, 485–493.

Grant, W. C. (1956) An ecological study of the peregrine earthworm, *Pheretima hupeiensis* in the eastern United States. *Ecology*, **37**, 648–658.

Hartenstein, F., Hartenstein, E. and Hartenstein, R. (1981) Gut load and transit time in the earthworm *Eisenia foetida*. *Pedobiologia*, **22**, 5–20.

Hartenstein, R. (1981) Production of earthworms as a potentially economical source of protein. *Biotechnol. Bioeng.*, **23**, 1797–1811.

Hartenstein, R., Neuhauser, E. F. and Kaplan, D. L. (1979) Reproductive potential of the earthworm *Eisenia foetida*. *Oecologia (Berl.)*, **43**, 329–340.

Kaplan, D. L., Hartenstein, R., Neuhauser, R. and Malecki, M. R. (1980) Physico-chemical requirements in the environment of the earthworm *Eisenia foetida*. *Soil Biol. Biochem.*, **12**, 347–352.

Michon, J. (1954) Influence de l'isolement à partir de la maturite sexuelle sur la biologie des Lumbricidae. *C. R. Hebd. Séances Acad. Sci., Ser. D*, **238**, 2457–2458.

Mitchell, M. J. (1979) Functional relationships of macroinvertebrates in heterotrophic systems with emphasis on sewage sludge decomposition. *Ecology*, **60**, 1270–1283.

Mitchell, M. J., Hornor, S. G. and Abrams, B. I. (1980) Decomposition of sewage sludge in drying beds and the potential role of the earthworm, *Eisenia foetida*. *J. Environ. Qual.*, **9**, 373–378.

Neuhauser, E. F., Hartenstein, R. and Kaplan, D. L. (1980a) Growth of the earthworm *Eisenia foetida* in relation to population density and food rationing. *Oikos*, **35**, 93–98.

Neuhauser, E. F., Hartenstein, R. and Kaplan, D. L. (1980b) Second progress report on potential use of earthworms in sludge management. *Proc. Natl. Conf. Sludge Composting.* Information Transfer Inc., Silverspring, MD, pp. 175–183.

Prince, A. B., Donovan, J. F. and Bates, J. E. (1981) Vermicomposting of municipal solid wastes and municipal wastewater sludges. In *Workshop on the Role of Earthworms in the Stabilization of Organic Residues* (M. Appelhof, compiler), Vol. 1, Proc. Beech Leaf Press, Kalamazoo, MI, pp. 207–219.

Sabine, J. R. (1981) Vermiculture as an option for resource recovery in the intensive animal industries. In *Workshop on the Role of Earthworms in the Stabilization of Organic Residues* (M. Appelhof, compiler), Vol. 1, Proc. Beech Leaf Press, Kalamazoo, MI, pp. 241–252.

Satchell, J. E. (1967) Lumbricidae. In *Soil Biology* (eds. A. Burges and F. Raw), Academic Press, London and New York, pp. 259–322.

Shugart, H. H., Goldstein, R. A., O'Neill, R. V. and Mankin, J. B. (1974) TEEM: A terrestrial ecosystem energy model for forests. *Oecol. Plant.*, **9**, 231–264.

Tsukamoto, J. and Watanabe, H. (1977) Influence of temperature on hatching and growth of *Eisenia foetida* (Oligochaeta, Lumbricidae). *Pedobiologia*, **17**, 338–342.

Usher, M. B. (1972) Developments in the Leslie matrix model. In *Mathematical Models in Ecology* (ed. J. N. R. Jeffers), Blackwell Scientific Publications, Oxford, pp. 29–60.

van Rhee, J. A. (1967) Development of earthworm populations in orchard soils. In *Progress in Soil Biology* (eds. O. Graff and J. E. Satchell), North Holland Publishing Co., Amsterdam, pp. 360–371.

Watanabe, H. and Tsukamoto, J. (1976) Seasonal change in size class and age structure of the lumbricid *Eisenia foetida* population in a field compost and its practical application as the decomposer of organic waste. *Rev. Ecol. Biol. Sol.*, **13**, 141–146.

Chapter 31

Earthworm microbiology

J. E. SATCHELL

31.1 INTRODUCTION

In the 1950s and 60s, in the heyday of ecosystem energetics, soil zoologists, who had done much to instigate this cult, were keenly disappointed to discover that the direct contribution of the mesofauna to total soil metabolism was negligible. A new goal for soil zoology was perceived – to demonstrate that invertebrate activity promotes microbial metabolism – but, like other babies, this proved easy to conceive but hard to deliver. Early work suffered from the limitations of dilution plate counting, selective media and general remoteness from field conditions where there are no pure cultures, temperatures fluctuate and energy sources are limited. Keenly interesting work was done with Collembola which consume fungi at one end and produce neat pellets at the other but earthworms, which in culture aerate the medium with burrows, mix organic and inorganic, living and non-living elements indiscriminately and smear their milieu with mucus, urine and faeces, are distressing subjects for microbiology. There are nevertheless numerous publications on earthworm microflora interactions, last critically reviewed in 1967 (this author) and 1972 (Edwards and Lofty). This chapter assesses present knowledge in this field as it applies to soil biota and to the microflora of other substrates, particularly the organic materials potentially available for vermiculture.

31.2 CHANGES IN MICROBIAL POPULATIONS IN THE EARTHWORM GUT AND FAECES

It is generally agreed that the earthworm gut contains essentially the same kinds of organisms as are present in the soil in which the worms are living. Bassalik (1913) isolated more than 50 species of bacteria from the alimentary canal of *Lumbricus terrestris* and found none which was not

351

present in the soil from which the worms came. Parle (1963a) examined the gut of three species of earthworm and similarly found no micro-organisms which are not common in soil or plant remains. It seems therefore that, apart from some parasitic protozoans (Dixon, 1975), earthworms possess no indigenous gut microflora.

A proportion of at least the vegetative cells of micro-organisms must be digested during transit through the earthworm intestine but the only demonstrations of their complete elimination have been of non-autochthonous forms introduced to the ingested soil by inoculation (Day, 1950; Brüsewitz, 1959). Kozlovskaya and Zhdannikova (1963) recorded a reduction in the numbers of fluorescing bacteria in the intestine of *Octolasion lacteum* compared with soil but, in uninoculated soils, in general, numbers of yeasts and fungi are little changed in passage through the earthworm gut, and bacteria and actinomycetes increase exponentially from fore-gut to hind-gut (see e.g. Parle, 1959).

Earthworm faeces emerge as a saturated paste, poorly aerated but rich in ammonia and partially digested organic matter. The ensuing changes in microbial numbers were first recorded by Stöckli (1928). He found that the total cell count doubled in the first week after the casts were formed and remained at about this level for a further three weeks. Parle (1963b) found no consistent changes in numbers of actinomycetes or bacteria in ageing casts but yeasts increased and fungi, present almost entirely as spores in the gut, started to germinate in the casts and hyphae were most abundant in casts 15 days old. As the casts age, an increasing proportion of the population forms resting stages, and oxygen uptake by the casts declines. Ruschmann (1953) attributed this decline in respiratory exchange to an increase in actinomycete populations antagonistic to aerobic bacteria.

These observations on the gut and cast microflora of earthworms of oligotrophic soils were extended by Kozlovskaya (1969) to species occupying podzols, including *O. lacteum* and *Eisenia nordenskioldi*. The microflora of the faeces of the latter species showed little difference from that of podzolic soil but substantial increases in the faeces of specimens from peat soils.

31.3 EFFECTS OF EARTHWORM ACTIVITY ON SOIL MICRO-ORGANISMS

The authors cited above and many others working on similar lines (see Edwards and Lofty, 1972) demonstrated that microbial numbers are generally higher in casts than in the surrounding soil and may remain so for some weeks. Some workers, e.g. Jeanson-Luusinang (1963), found no difference in microbial populations of worm excreta and uningested soil but their results are in some cases attributable to the experimental conditions in which glucose or some similarly easily degradable substrate was added to the soil. This does not account for all the discrepancies between the results of workers who found consistent increases in microbial populations in casts compared with uningested soil and those, e.g. Day (1950), who found no consistent differences. In a series of tests, Kulinska (1961) found that the total number of micro-organisms in the alimentary canal and in the fresh excrement of earthworms varied within wide limits from a decrease of 90 % to an increase of 570 %. She concluded that the immediate influence of earthworms on the soil microflora depends upon the physiological condition of individuals, among others, upon whether digestion of micro-organisms in the gut prevails over reproductive increases or the converse. Comparisons of casts and uningested soil tend, however, to confirm the view that the high microbial population of earthworm faeces is largely, though not wholly, attributable to selective feeding by the worms on plant residues and their associated microflora. An interesting converse of this is that, for certain groups, the soil may provide more favourable conditions than earthworm casts. Kamal and Singh (1970), in a comparison of earthworm casts and surrounding soil in a teak forest in India, found many more species of phycomycete fungi and higher numbers in the soil than in the casts although the casts were somewhat richer in total organic matter. They thought this probably reflected lower concentrations of sugars in the casts.

A notable contribution from Loquet et al. (1977) extended this early work to demonstrate differences in bacterial populations not only between casts and soil but between burrow walls and the intervening soil. Casts compared with the upper 6 cm horizon contained more cellulolytic aerobes, hemicellulolytic, amylolytic and nitrifying bacteria and less denitrifying bacteria. To a mean depth of 30 cm, burrow walls compared with intervening soil contained more *Azotobacter*, more N-fixing aerobes and anaerobes, denitrifiers, proteolytic and ammonifying bacteria. The effect of the earthworm burrows was to extend the sphere of activity of these groups deeper into the soil.

In a study of the changes in N status of casts produced by *Allolobophora longa* as they age under field conditions, Parle (1963b) observed a steady decline in the content of NH_4^+–N and an increase in NO_3^-–N. More recently, Syers et al. (1979), while confirming these findings have stressed the inefficiency of N assimilation from litter by *L. rubellus* and conclude that earthworm casting involves a 'pass-through' of most of the organic N ingested with some increase in mineralization but probably less than that which would occur in litter *in situ*.

It is well established that worm casts are richer in inorganic phosphorus compounds extractable in water than the surface soil ingested (Graff, 1970), and Sharpley and Syers (1976) found that exchangeable P, measured isotopically, was three times greater in casts than in the underlying soil. Mansell *et al.* (1981) extended this work to show that plant litter also contained more available P after ingestion by earthworms. These authors attributed the increase in available P to physical breakdown of the plant material and trituration of the mineral fraction for which the presence in casts of an increased proportion of fine particles provides support. Sharpley and Syers (1976) estimated the phosphatase activity in casts of *Allolobophora caliginosa* and found that *p*-nitrophenol release increased to a maximum after 18 hours incubation, a result not obtained with uningested soil of similar origin. This was followed by an increase in inorganic P released by mineralization of organic P. Studies in the field (Sharpley and Syers, 1977) showed a seasonal decrease in inorganic P, an increase in organic P and a subsequent increase in inorganic P, reflecting the effects of temperature changes on phosphatase activity. The indications from this New Zealand work were that the differences in phosphatase activity between wormcasts and soil reflect the effect of the earthworms on microbial activity although it had yet to be confirmed that the phosphatases observed were microbial and not the alkaline phosphatase of the gut.

31.4 EFFECTS OF EARTHWORM ACTIVITY ON THE MICROFLORA OF ORGANIC MATERIALS

The influence of earthworms on microbial activity on organic substrates are relevant both to the disposal of anthropogenic wastes by vermiculture and the decomposition of litter, particularly of resistant lignified material, in the field. The question of whether phosphatase production induced by earthworm activity renders organically bound P more labile is relevant to the stabilization of organic wastes in forms which could be used in horticulture no less than to wormcasts in pastures. It cannot readily be resolved with faecal wastes, human or animal, since the high concentration of gut phosphatase obscures any additional phosphatase which might be induced by microbial activity. Preliminary work has, however, been carried out recently by the author and colleagues on cellulose waste from a paper mill, a relatively phosphatase-free medium.

Four species were studied: *Eisenia fetida*, *Dendrobaena veneta*, *Lumbricus rubellus* and *A. caliginosa*. Each was tested separately in small polythene tubs containing sterilized paper waste, pH 6.8, with an approximate water content of 85%. Phytin (calcium inositol hexaphosphate) (1 g) was mixed with each 40 g of paper waste before the worms were introduced. The medium can be regarded as simulating the worms' natural diet insofar as paper waste is essentially cellulose and phytin is the principal form in which the organic P fraction of plant litter reaches the soil.

The cultures were left for 3–4 weeks until the worms had worked through the medium, and samples of fresh faecal material and of the control medium were then collected for phosphatase assay. They were incubated in their fresh condition at 20°C, the mean temperature at which the cultures had been maintained, in BDH universal buffer adjusted to the pH of the paper waste. Phosphatase was determined by a modified version of the disodium phenyl phosphate method (Hoffman, 1968) using dilutions of BDH standard phenol for the standards. Phosphatase activity was estimated as μg phenol liberated in three hours g^{-1} dry wt. of substrate and the following mean values were obtained: controls (two replicate assays) 50; *E. fetida* (five replicates) 631; *D. veneta* (five replicates) 756. The results showed a clear cut increase in phosphatase activity in the worm faeces from both species over that in the uningested material.

The experiment was then repeated as before and extended to include *L. rubellus* and *A. caliginosa*. In this test phosphatase activity was assayed with buffer which provided a pH range of 2–10. Duplicate assays were made at intervals of one pH unit. The results (Fig. 31.1) showed (a) a clear increase in phosphatase activity in the faeces of all four species over the uningested medium, (b) a double peak of phosphatase activity in the faeces of all species, at about pH 3–5 and about 9–10, and (c) some activity in the uneaten control substrates, particularly in the *E. fetida* controls at about pH 3–5. In all the controls, fungal conidia were observed on the surface of the paper waste.

We provisionally interpret these results to suggest that the peaks at pH 3–5 indicate phosphatase produced by microbial activity. We cannot exclude the possibility that the peaks at pH 9–10 also indicate increased microbial activity but, in view of the many literature references to alkaline phosphatases in earthworms, it seems more likely that these peaks indicate earthworm phosphatase.

The points graphed in Fig. 31.1 are all means of two replicates between which the variation, averaged over all eight sets of observations, was $\pm 15.5\%$ of the means. We therefore interpret the minor variations in the phosphatase levels in the uneaten substrate provisionally as experimental error. Overall, this work suggests that phosphatase activity in worm faeces may be substantially increased directly by the worms' own enzymes and indirectly by stimulation of the microflora.

In studies on the application of vermiculture to the stabilization of sewage sludge, Hornor and Mitchell (1981) recorded changes in microbial numbers and activity in containers of sludge with and without

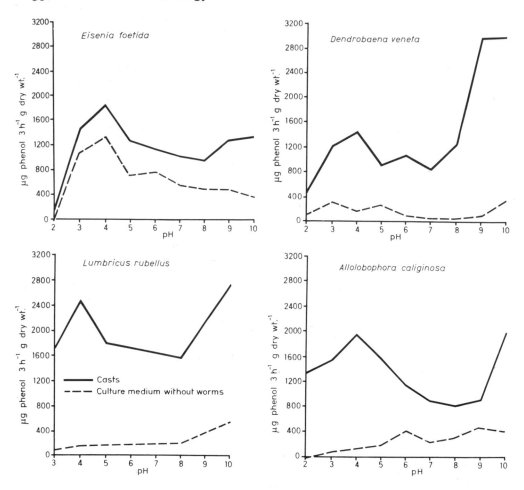

Fig. 31.1 Phosphatase activity in the casts of four earthworm species cultured in paper waste and phytin. ———, casts; – – – –, culture medium without worms.

introduced *E. fetida*. After 2–4 weeks, the faeces showed no difference from the uneaten substrate in populations of total anaerobes, nitrate reducers, sulphate reducers and methanogens but the rates of CO_2 and O_2 flux were higher in the faeces, and the production of CH_4 and volatile S was lower. As the counts were made on enriched culture media they did not differentiate active from inactive organisms but the authors conclude from their gas flux measurements that, under appropriate conditions, *E. fetida* feeding stimulates aerobic decomposition. Similar results were obtained by Mitchell *et al.* (1980) who found no differences in numbers of anaerobic bacteria or aerobic bacteria in sludge with and without *E. fetida* but recorded a small but significant increase in aerobes after 14 days of earthworm activity.

Comparable results have been obtained from vineyard residues by Anstett (1951) who incubated grape husks at 25–30°C in pots, some of which he inoculated with *E. fetida*. After 5 months, the microbial population was 4–5 times greater in the inoculated pots. Decomposition of the husks was measured by loss on ignition which was 75 % in the inoculated pots and 86 % in the controls at the end of the experiments. The additional decomposition in the presence of *E. fetida* appeared to result from the increased population of micro-organisms since the losses on ignition were the same from pots which contained worms throughout the experiment and from pots from which they were removed after six weeks.

The effects of earthworms on the fungal population of composted municipal waste and of cattle dung, recorded by Domsch and Banse (1972), show some interesting differences from their effects on soil fungi. The authors found that the fungal floras of an arable soil and the two organic materials differed markedly with six species isolated from the dung, 11 from the municipal compost and 25 from the soil. Only one of the species from the dung was also isolated from the compost and the soil, and only four from the compost were isolated from the soil. Casts produced by *L. terrestris* fed on mixtures of soil and either the compost or the dung contained only the soil fungi. None of the fungi specific to the compost and the dung were recovered from them. The reasons for this selective destruction are far from clear.

Recent studies on the effects of wood decomposition in forest ecosystems include a report by Ausmus (1977) on bacterial, actinomycete and fungal densities in sawdust substrate incubated for three weeks with *Octolasium* and *Bimastos* spp. with and without soil. Bacteria and actinomycetes increased and fungal populations were reduced in the presence of either worm, and nitrogen fixation was positively correlated with these changes. The effect of the earthworms and other invertebrates in shifting the competitive advantage temporarily from fungi to bacteria and allowing nitrogen fixation and more rapid C catabolism to occur is to turn the wood substrate from a nutrient sink to a nutrient source and is seen as the essential rate regulator in wood decomposition.

Eisenia lucens is a corticolous earthworm found in decomposing logs and tree stumps in central Europe. The bacterial gut flora of Hungarian specimens collected from rotting beech logs, examined by Marialigeti (1979), comprised 473 strains of which 73 % belonged to the genus *Vibrio*. The predominant strains are facultative anaerobes of a single species which dominates the intestinal flora and can produce enormous cell masses in the gut milieu. None of the strains were capable of lignin cleavage which is consistent with the findings of Neuhauser *et al.* (1978) that neither *E. fetida* nor its faecal bacteria is able to depolymerize lignin and degrade its aromatic constituents.

31.5 THE ROLE OF MICRO-ORGANISMS IN EARTHWORM NUTRITION

Digestion of micro-organisms ingested by earthworms was reported as early as 1914 when von Aichberger recorded the absence from earthworm intestines of diatoms, desmids, blue–green algae, rhizopods or live yeasts. Nekrasova *et al.* (1976) have subsequently recorded selective feeding by *L. rubellus* and *O. lacteum* on a range of algal species and digestion of a proportion of those ingested. Selective feeding by *L. terrestris* has also been demonstrated on bacteria by Wright (1972) and on soil fungi by Cooke and Luxton (1980) (see Chapter 32).

It is difficult to estimate for a general saprovore the extent to which its nutrition is dependent on digestion of the microbial population ingested with decomposing plant residues. From considerations of the amounts of N required for tissue production and the amounts excreted by some woodland populations of *L. terrestris*, Satchell (1963) argued that sufficient N could only be ingested if the worms were feeding not on plant litter alone but preferentially on associated microbial protein. Subsequent research (Lakhani and Satchell, 1970) showed that tissue production had been overestimated but as this is small relative to excreted N, the argument remains valid.

The demonstration by Hartenstein and others in many publications that *E. fetida* can sustain growth on activated sewage sludge shows that bacteria can provide an adequate basis for earthworm nutrition, sewage sludge being composed almost entirely of live and dead bacterial cells. Microbial feeding by earthworms can be even more readily observed in percolating sewage filters. Solbé (1971) fed *Dendrobaena subrubicunda* and *Eiseniella tetraedra* on bacterial and fungal film from a percolating filter which supported a combined population of $> 65\,000\,\mathrm{m}^{-2}$ of these worms and calculated that they could ingest the entire filter film once every 41 days.

From examination of the crop and gizzard contents of six common lumbricid species, Piearce (1978) concluded that algae and, to a lesser extent fungi, are possibly important in the nutrition of all six species and testaceans may be important in the nutrition of *Dendrobaena mammalis*.

Most earthworms have a gut flora of astomatous ciliates (Beers, 1938; Rees, 1960, 1961, 1962) and protozoa seem to be a necessary dietary requirement of earthworms. Protozoal protoplasm is highly assimilable

(Brandl and Fernando, 1975) and non-encysted protozoa may provide a more accessible source of nutrients than the detritus ingested with them. The role of protozoa in earthworm nutrition is discussed in Chapter 33.

31.6 MICRO-ORGANISMS PATHOGENIC TO EARTHWORMS

Most earthworms are infected with gregarine parasites. In a sample of 2645 worms of 27 species, Duhlinska (1977) found all species infected, with an infection rate for different species ranging from 70 to 96 %. Some 26 species of gregarines were reported. None of them is known to cause mortality and the effects on their hosts are not on the whole very great.

The coelomic fluid of earthworms has been known to have anti-bacterial properties since last century (Keng, 1895) and to exert this effect both within the coelom and when discharged through the dosal pores on to the body surface. Valembois *et al.* (1982) have shown that the growth of some soil bacteria is inhibited by culturing them in the presence of *E. fetida* coelomic fluid. The component responsible for the anti-bacterial activity is a polymorphic lipoprotein synthesized by the chloragogen cells. Not all bacteria are equally susceptible to it and Lassegues *et al.* (1981) found that of 22 strains of bacteria injected into the coelom of *E. fetida*, five caused the death of the worms.

In experimental conditions, *Bacillus thuringiensis* has been found (Smirnoff and Heimpel, 1961) to penetrate the gut of earthworms and enter the body cavity causing fatal septicaemia but preparations of *B. thuringiensis* as a biological pesticide for lepidopterous larvae have been found to have no effect on earthworms when applied at normal field rates (Benz and Altwegg, 1975). Otherwise the only bacterial pathogen known to attack earthworms is *Enterobacter aerogenes*, the aetiology of which is described in Chapter 34.

31.7 DISPERSAL OF PATHOGENS BY EARTHWORMS

A number of publications report transmission, or suspected transmission, by earthworms of a variety of plant, animal and human pathogens. Although few of these have been substantiated or are significant in field conditions, the subject is particularly important in establishing public confidence in the utilization of organic wastes by vermiculture.

The plant pathogen *Verticillium dahliae* has been isolated from casts of *L. terrestris* fed on ground peppermint (*M. piperita*) stems from a crop infected with *Verticillium* wilt. Colonies were obtained by laboratory incubation of casts (Melouk and Horner, 1976) but there is no evidence that earthworm activity plays any role, positive or negative, in the spread of the disease. The same applies to dwarf bunt, *Tilletia controversa*, the spores of which are unaffected by passage through the intestines of earthworms (Hoffman and Purdy, 1964). Spores of *Pithomyces chartarum*, a fungal saprophyte associated with facial eczema in grazing livestock, have been isolated by Keogh and Christensen (1976) from casts of *L. rubellus* but these authors conclude that development of the fungus is restricted by the earthworm which consumes the surface litter on which the spores develop. Similarly in apple orchards, earthworms may reduce the incidence of apple scab, *Ventura inequalis*, by burying the dead leaves on which the ascospores develop (Hirst *et al.*, 1955).

The transmission of the protozoan *Histomonas*, causing blackhead disease in chickens and turkeys, by several earthworm species is well documented, e.g. Lund *et al.* (1966). The only known gregarine infection reported in a vertebrate, a *Monocystis*-like species isolated from the liver of American woodcock, *Philohela minor*, is surmised (McGhee and Pursglove, 1977) to be transmitted by earthworms which are a major constituent of the bird's diet. The ubiquitous protozoan *Toxoplasma gondii* is common in many wild birds and in the faeces of cats which prey on them. Markus (1974) found that the viability of oocysts of *T. gondii* is not affected by passage through the gut of *L. terrestris* and surmised that a worm–bird–cat food chain might affect the transmission of the pathogen to man. *Aeromonas hydrophila*, a bacterial pathogen infecting frogs, snakes and fish, has been isolated from the coelom of *L. terrestris* (Marks and Cooper, 1977) but here too it has not been demonstrated that the organism in the worm is of the same strain as in vertebrates.

The effect of earthworms on human faecal micro-organisms is of particular importance in developing vermiculture as a method of sewage disposal. In the 1950s, two species of Enterobacteriaceae, *Serratia marcessens* and *Escherichia coli*, inoculated into soil were found (Day, 1950; Brüsewitz, 1959) to be killed when ingested by *L. terrestris*. Populations of the enteric pathogen *Salmonella enteriditis* ser. *typhyimurium* were cultured by Brown and Mitchell (1981) in a sterile commercial worm-rearing medium with and without worms. After four days, the population density was reduced 46 times in the cultures with worms to a level 2000-fold less than in the controls. A second experiment demonstrated that the reduction in the presence of worms could be ascribed partly to competition from the endemic microflora of the worm gut. When mixed bacteria from *E. fetida* faeces were introduced to cultures of *S. enteriditis*, the numbers of *Salmonella* were reduced by 4–5 orders of magnitude. Activated sludge does not generally support growth of human enterics but anaerobically digested sludge does (Taber, 1976). The effect of earthworm activity in sludge which is sufficiently aerobic for their survival is to aerate it and increase the rate at which it dries, thereby favouring aerobic bacteria. As noted above, Mitchell *et al.* (1980) found aerobic bacteria to be more than twice as numerous in sludge 14 days after introduction of *E. fetida* as in sludge without worms. As most human enteric pathogens are anaerobes, sludge conditioning by earthworms seems likely to be beneficial from the public health standpoint. Specific data are nevertheless required on other enteric bacteria, pathogenic viruses and parasites.

31.8 REFERENCES

Anstett, M. A. (1951) On the microbiological activation of humification phenomena. *C. R. Acad. Agric. Fr.*, **37**, 262–264.

Ausmus, B. S. (1977) Regulation of wood decomposition rates by arthropod and annelid populations. In *Soil Organisms as Components of Ecosystems* (eds. U. Lohm and T. Persson), *Ecol. Bull. (Stockholm)*, **25**, 180–192.

Bassalik, K. (1913) On silicate decomposition by soil bacteria. *Z. GärPhysiol.*, **2**, 1–32.

Benz, G. and Altwegg, A. (1975) Safety of *Bacillus thuringiensis* for earthworms. *J. Invertebr. Pathol.*, **26**, 125–126.

Beers, C. D. (1938) *Hysterocineta eiseniae* n. sp., an endoparasitic ciliate from the earthworm *Eisenia lonnbergi. Arch. Protistenk.*, **48**, 516–525.

Brandl, Z. and Fernando, C. H. (1975) Food consumption and utilization in two freshwater cyclopoid copepods (*Mesocyclops edax* and *Cyclops vicinus*). *Int. Rev. Gesamten, Hydrobiol. Syst. Beih.*, **60**, 471–494.

Brown, B. A. and Mitchell, M. J. (1981) Role of the earthworm, *Eisenia foetida*, in affecting survival of *Salmonella enteritidis* ser. *typhimurium. Pedobiologia*, **21**, 434–438.

Brüzewitz, G. (1959) Studies on the influence of earthworms on numbers of species and role of micro-organisms in soils. *Arch. Mikrobiol.*, **33**, 52–82. [In German.]

Cooke, A. and Luxton, M. (1980) Effect of microbes on food selection by *Lumbricus terrestris. Rev. Ecol. Biol. Sol.*, **17**, 365–370.

Day, G. M. (1950) Influence of earthworms on soil micro-organisms. *Soil Sci.*, **69**, 175–184.

Dixon, R. F. (1975) The astomatous ciliates of British earthworms. *J. Biol. Educ.*, **9**, 29–39.

Domsch, K. H. and Banse, H-J. (1972) Mycological studies on earthworm casts. *Soil Biol. Biochem.*, **4**, 31–38. [In German.]

Duhlinska, D. (1977) On the distribution of gregarines in lumbricid earthworms from Bulgaria. *Acta Zool. Bulg.*, **7**, 49–59.

Edwards, C. A. and Lofty, J. R. (1972) *Biology of Earthworms.* Chapman and Hall Ltd., London, 283 pp.

Graff, O. (1970) Phosphorus content of earthworm casts. *LandbForsch-Völkenrode*, **20**, 33–36. [In German.]

Hirst, J., Storey, M., Ward, W. C. and Wilcox, H. G. (1955) The origin of apple scab epidemics in the Wisbech area in 1953 and 1954. *Plant Pathol.*, **4**, 91.

Hoffman, G. (1968) Eine photometrische method zur bestimmung der phosphatase-aktivitat in Boden. *Z. PflErnähr., Dung., Bodenk.*, **118**, 161–172.

Hoffman, J. A. and Purdy, L. H. (1964) Germination of dwarf bunt (*Tilletia controversa*) teliospores after ingestion by earthworms. *Phytopathology*, **54**, 878–879.

Hornor, S. G. and Mitchell, M. J. (1981) Effect of the earthworm, *Eisenia foetida* (Oligochaeta), on fluxes of volatile carbon and sulfur compounds from sewage sludge. *Soil Biol. Biochem.*, **13**, 367–372.

Jeanson-Luusinang, C. (1963) Experimental study of the action of *Lumbricus herculeus* (Savigny) (Oligochaeta, Lumbricidae) on the microflora of an artificial soil. In *Soil Organisms* (eds. J. Doeksen and J. van der Drift), North Holland Publishing Co., Amsterdam, pp. 266–270. [In French.]

Kamal and Singh, C. S. (1970) Studies on soil fungi from teak forests of Gorakhpur. VIII-A comparative account of fungi of earthworm casts, termitarium and surrounding soil from a teak stand. *Ann. Inst. Pasteur, Paris*, **119**, 249–259.

Keng, L. B. (1895) On the coelomic fluid of *Lumbricus terrestris* in reference to a protective mechanism. *Philos. Trans. R. Soc. London Ser. B*, 186.

Keogh, R. C. and Christensen, M. J. (1976) Influence of passage through *Lumbricus rubellus* Hoffmeister earthworms on viability of *Pithomyces chartarum* (Berk. and Curt.) M. B. Ellis spores. *N. Z. J. Agric. Res.*, **19**, 255–256.

Kozlovskaya, L. S. (1969) Effect of earthworm casts on the activation of microbial processes in peat soils. *Pedobiologia*, **9**, 158–164.

Kozlovskaya, L. S. and Zhdannikova, E. N. (1963) The interaction of earthworms and soil microflora. In *Bogged Forests and Bogs of Siberia*, *Akademii Nauk S.S.S.R.*, 183–217.

Kulinska, D. (1961) The effect of earthworms on the soil microflora. *Acta Microbiol. Pol.*, **10**, 339–346.

Lakhani, K. H. and Satchell, J. E. (1970) Production by *Lumbricus terrestris* (L.). *J. Anim. Ecol.*, **39**, 473–492.

Lassegues, M., Roch, P., Valembois, P. and Davant, N. (1981) Action de quelques souches bactériennes telluriques sur le lombricien *Eisenia fetida andrei*. *C. R. Hebd. Séances Acad. Sci. Ser. D*, **292**, 731–734.

Loquet, M., Bhatnagar, T., Bouché, M. B. and Rouelle, J. (1977) Essai d'estimation de l'influence écologique des lombriciens sur les micro-organismes. *Pedobiologia*, **17**, 400–417.

Lund, E. E., Wehr, E. E. and Ellis, D. J. (1966) Earthworm transmission of *Heterakis* and *Histomonas* to chickens and turkeys. *J. Parasitol.*, **52**, 899–902.

McGhee, R. B. and Pursglove, S. R. (1977) Gregarine infection of the American woodcock, *Philohela minor*. *J. Parasitol.*, **63**, 160.

Mansell, G. P., Syers, J. K. and Gregg, P. E. H. (1981) Plant availability of phosphorous in dead herbage ingested by surface-casting earthworms. *Soil Biol. Biochem.*, **13**, 163–167.

Marialigeti, K. (1979) On the community structure of the gut-microbiota of *Eisenia lucens* (Annelida, Oligochaeta). *Pedobiologia*, **19**, 213–220.

Marks, D. H. and Cooper, E. L. (1977) *Aeromonas hydrophila* in the coelomic cavity of the earthworms *Lumbricus terrestris* and *Eisenia foetida*. *J. Invertebr. Pathol.*, **29**, 382–383.

Markus, M. B. (1974) Earthworms and coccidian oocysts. *Ann. Trop. Med. Parasitol.*, **68**, 247–248.

Melouk, K. A. and Horner, C. E. (1976) Recovery of *Verticillium dahliae* pathogenic to mints from castings of earthworms. *Proc. Am. Phytopathol. Soc.*, **3**, 265.

Mitchell, M. J., Hornor, S. G. and Abrams, B. I. (1980) Decomposition in sewage sludge drying beds and the potential role of the earthworm *Eisenia foetida*. *J. Environ. Qual.*, **9**, 373–378.

Nekrasova, K. A., Kozlovskaya, L. S., Domračeva, L. I. and Ština, E. A. (1976) The influence of invertebrates on the development of algae. *Pedobiologia*, **16**, 286–297.

Neuhauser, E. F., Hartenstein, R. and Connors, W. J. (1978) Soil invertebrates and the degradation of vanillin, cinnamic acid and lignins. *Soil Biol. Biochem.*, **10**, 431–435.

Parle, J. N. (1959) The effect of earthworms on soil micro-organisms. *Rothamsted Exp. Stn. Report 1958*, 70.

Parle, J. N. (1963a) Micro-organisms in the intestines of earthworms. *J. Gen. Microbiol.*, **31**, 1–11.

Parle, J. N. (1963b) A microbiological study of earthworm casts. *J. Gen. Microbiol.*, **31**, 13–23.

Piearce, T. G. (1978) Gut contents of some lumbricid earthworms. *Pedobiologia*, **18**, 153–157.

Rees, B. (1960) *Albertia vermicularis* (Rotifera) parasitic in the earthworm *Allolobophora caliginosa*. *Parasitology*, **50**, 61–66.

Rees, B. (1961) Three British species of the genus *Metaradiophyra* including a new species *M. gardneri*. *Parasitology*, **51**, 523–532.

Rees, B. (1962) *Mysterocineta davidis* sp. nov., an intestinal parasite of the earthworm *Allolobophora caliginosa*. *Parasitology*, **52**, 17–21.

Ruschmann, G. (1953), Antibioses and symbioses of soil organisms and their significance in soil fertility. Earthworm symbioses and antibioses. *Z. Acker-u.PflBau.*, **96**, 201–218.

Satchell, J. E. (1963) Nitrogen turnover by a woodland population of *Lumbricus terrestris*. In *Soil Organisms* (eds. J. Doeksen and J. van der Drift) North Holland Publishing Co., Amsterdam, pp. 60–66.

Satchell, J. E. (1967) Lumbricidae. In *Soil Biology* (eds. A. Burges and F. Raw), Academic Press, London and New York, pp. 259–322.

Sharpley, A. N. and Syers, J. K. (1976) Potential role of earthworm casts for the phosphorous enrichment of run-off waters. *Soil Biol. Biochem.*, **8**, 341–346.

Sharpley, A. N. and Syers, J. K. (1977) Seasonal variation in casting activity and in amounts and release to solution of phosphorus forms in earthworm casts. *Soil Biol. Biochem.*, **9**, 227–231.

Smirnoff, W. A. and Heimpel, A. M. (1961) Notes on the pathogenicity of *Bacillus thuringiensis* var. *thuringiensis* Bulinu for the earthworm *Lumbricus terrestris* Linnaeus. *J. Insect. Pathol.*, **3**, 403–408.

Solbé, J. F. de L. G. (1971) Aspects of the biology of the lumbricids *Eiseniella tetraedra* (Savigny) and *Dendrobaena rubida* (Savigny) f. *subrubicunda* (Eisen) in a percolating filter. *J. Appl. Ecol.*, **8**, 845–867.

Stöckli, A. (1928) Studien über den Einfluss der Regenwurmer auf die Beschaffenheit des Bodens. *Landw. Jb. Schweiz*, **42**, 121 pp.

Syers, J. K., Sharpley, A. N. and Keeney, D. R. (1979) Cycling of nitrogen by surface-casting earthworms in a pasture ecosystem. *Soil Biol. Biochem.*, **11**, 181–185.

Taber, W. A. (1976), Wastewater microbiology. *Annu. Rev. Microbiol.*, **30**, 263–277.

Valembois, P., Roch, P., Lassegues, M. and Davant, N. (1982) Bacteriostatic activity of a chloragogen cell secretion. *Pedobiologia*, **23**.

von Aichberger, R. (1914) Studies on the nutrition of earthworms. *Kleinwelt*, **6**, 53–88.

Wright, M. A. (1972) Factors governing ingestion by the earthworm *Lumbricus terrestris* with special reference to apple leaves. *Ann. Appl. Biol.*, **70**, 175–188.

Chapter 32

The effects of fungi on food selection by *Lumbricus terrestris* L.

A. COOKE

32.1 INTRODUCTION

As early as 1881, Darwin observed that earthworms could readily distinguish between leaves of various plant species. A detailed study of earthworm food preferences by Satchell and Lowe (1967) established that *Lumbricus terrestris* L. had a strong preference for leaves of elm (*Ulmus* spp.), ash (*Fraxinus excelsior*) and sycamore (*Acer pseudoplatanus*), whereas leaves of beech (*Fagus sylvatica*) and oak (*Quercus* spp.) were disliked. A positive preference was also shown for weathered litter as opposed to unweathered. They also investigated certain chemical and physical factors of the leaves which might have influenced feeding preference and found that the content of nitrogen, soluble carbohydrates and polyphenols could be broadly correlated with leaf palatability.

A microbial effect on food selection by earthworms was first demonstrated by Wright (1972) who noted that the preference for apple leaf discs shown by *L. terrestris* was considerably increased when the discs were inoculated with the bacterium *Pseudomonas aeruginosa*. Cooke and Luxton (1980) also investigated the effects micro-organisms may have on food selection by *L. terrestris*. Cellulose filter-paper discs were inoculated with one of two fungi, *Mucor hiemalis* and a species of *Penicillium*, or the bacterium *Pseudomonas fluorescens*. Preferential selection of the discs with fungi over uninoculated controls was observed and the conclusion drawn that microbial contamination of a potential food source may make it more attractive to earthworms.

To investigate further the effects fungi may have on food selection by earthworms, the following experiments were designed.

32.2 MATERIALS AND METHODS

Ten plastic bowls, each 30 cm ϕ, were filled to a depth of 10 cm with finely sifted garden soil. Into each were placed ten specimens of *Lumbricus terrestris*, each 1.5–2.0 g fresh wt. The soil was then watered and the containers were covered with polythene film to reduce evaporation. They were kept at approximately 15 °C for one week to allow the earthworms to acclimatize.

Eight species of fungi were used in the experiments: *Alternaria solani*, *Chaetomium globosum*, *Cladosporium cladosporioides*, *Fusarium oxysporum*, *Mucor hiemalis*, *Penicillium digitatum*, *Poronia piliformis* and *Trichoderma viride*. These were grown from pure culture on to filter-paper discs by the method described by Cooke and Luxton (1980).

Fifteen discs, all inoculated with the same species of fungus were placed at random on to the surface of each earthworm container, together with 15 control discs. After 24 h, any missing discs were presumed to have been selected by the earthworms and dragged into their burrows. This procedure was repeated twice for each species of fungus.

Samples of the inoculated discs together with control discs were analysed for moisture content and various chemical characteristics. Twenty sample discs of each treatment were weighed fresh, oven-dried at 105 °C, and then reweighed. For each treatment, fresh and dry weights of fungus per disc were determined together with moisture content. For the determination of Ca, N and C, approximately 100 discs per treatment were freeze-dried and then milled to a fine powder. Samples of this powder were used for the following analyses.

Ca was determined by the flame emission procedure as described by Allen *et al.* (1974). First, approximately 0.25 g of material was weighed into a 50 ml Erlenmeyer flask with 5 ml of 71 % 'Analar' nitric acid and the flask was covered with a watch glass. The flasks were then heated on a hotplate to 170 °C and maintained at this temperature until nitrogen dioxide fumes ceased to be evolved. The watch glasses were removed and any adhering drops washed into the flasks with distilled water. The temperature was increased to approximately 190 °C and the acid boiled off until an almost dry residue was left. Sample solutions were then prepared for flame emission analysis.

Total N was determined by the micro-Kjeldahl method following the procedure described by the Ministry of Agriculture, Fisheries and Food (1973). Organic carbon was estimated by the rapid titration method of Tinsley (1950) as modified by Allen *et al.* (1974).

Total water-soluble polyphenols were assayed by the Folin–Ciocalteau method (Lowry *et al.*, 1951) using catechol as a standard. Pure cultures of each fungus were grown in 2 % malt extract broth at 25 °C. Subsequently,

Table 32.1 Selection by *L. terrestris* of microbially contaminated cellulose paper discs. (Mean % number of discs taken, $n = 10$.)

Fungus inoculated	Test	Fungal discs taken (%)	Control discs taken (%)	Difference (%)	Significance of difference P
Fusarium oxysporum	1	58.7	23.3	+ 35.4	0.01
	2	50.7	8.0	+ 42.7	0.001
Mucor hiemalis	1	74.0	50.7	+ 23.3	0.05
	2	66.7	32.7	+ 34.0	0.001
Alternaria solani	1	37.3	24.7	+ 12.6	N.S.
	2	50.0	21.3	+ 28.7	0.01
Cladosporium cladosporioides	1	64.7	51.3	+ 13.4	N.S.
	2	47.3	34.0	+ 13.3	N.S.
Poronia piliformis	1	31.3	18.0	+ 13.1	N.S.
	2	39.3	27.3	+ 12.0	N.S.
Trichoderma viride	1	20.0	10.7	+ 9.3	0.1
	2	24.7	20.0	+ 4.7	N.S.
Chaetomium globosum	1	51.3	36.7	+ 14.6	N.S.
	2	21.3	20.0	+ 1.3	N.S.
Penicillium digitatum	1	40.0	36.0	+ 4.0	N.S.
	2	27.3	34.7	− 7.4	N.S.
Control	1	35.5*	28.7†	+ 6.6	N.S.
	2	38.0*	38.0†	0.0	N.S.

* Discs maintained on malt extract agar.
† Discs moistened with distilled water.

Table 32.2 Physical and chemical analyses of fungally contaminated discs.

Fungus inoculated	Means (n = 20)				Means (n = 3)		
	Fresh wt. fungus (mg disc⁻¹)	Dry wt. fungus (mg disc⁻¹)	Moisture (%)	Total N (%)	Organic C (%)	Total Ca*	Polyphenol content†
Fusarium oxysporum	73.8	8.3	71.31	0.80	45.36	397.85	1750.5
Mucor hiemalis	98.1	9.8	76.53	0.76	46.68	444.19	1302.5
Alternaria solani	56.3	6.6	71.76	0.75	45.72	388.77	1537.5
Cladosporium cladosporioides	67.5	15.0	68.21	1.10	45.72	384.74	2047.5
Poromia piliformis	60.1	7.1	72.01	0.87	45.72	414.00	857.5
Trichoderma viride	32.8	2.7	69.40	0.99	46.68	275.28	1640.0
Chaetomium globosum	42.2	12.9	62.91	1.33	45.24	338.94	2162.5
Penicillium digitatum	40.7	11.6	67.38	0.73	44.76	291.66	1597.5
Control‡	0.0	0.0	61.45	0.21	47.04	307.55	0.0
Autoclaved discs	0.0	0.0	0.0	0.01	45.54	119.58	0.0

* $\mu g\,g^{-1}$ dry wt. of sample.
† $\mu g\,g^{-1}$ equivalents of catechol g^{-1} dry wt. of pure fungus.
‡ Malt extract agar.

each fungus was washed three times in distilled water and filtered through Whatman GF/B filters under suction. The fungi were then freeze-dried and milled to a fine powder. Samples (0.1 g) of the powdered fungal cultures were macerated in 5 ml of distilled water in a micro-blender for approximately 5 min. The resulting suspension was initially filtered through Whatman No. 42 filter papers and then through a 0.45 μm micropore filter to produce a clear liquid which was then used in the polyphenol assay. Control levels were determined using freeze-dried filter-paper discs maintained sterile on malt extract agar. Blank determinations were carried out using distilled water in place of the fungal extracts.

32.3 RESULTS

Variance ratio and 'Students' t tests were applied to the data, with each selection test being analysed separately (Table 32.1). Only six of the tests were significantly different at $P < 0.1$. However, in all but one test the earthworms showed a positive preference for inoculated discs over the controls. The variances were large but a highly significant preference $P < 0.001$ was found for *Fusarium oxysporum* and *Mucor hiemalis*. Other fungi selected at a highly significant level were *Alternaria solani* and *Trichoderma viride*, but in each case this was only in one of the two repeated tests. Inoculated discs were preferred in three of the four remaining species although the differences were not statistically significant. The only fungus not appearing to increase selection was *Penicillium digitatum*.

The results of the chemical and physical analyses are shown in Table 32.2.

Association between selection and the measured variables was analysed with the correlation matrix present in Table 32.3. It indicates that fresh

Table 32.3 Correlation matrix (R) of selection and inoculated disc variables.

	Selection	Fresh wt.	Moist.	Dry wt.	N	C	Poly-phenols
Fresh wt.	0.6976						
Moisture	0.6212	0.8447					
Dry wt.	0.1070	0.5734	0.1376				
N	0.0434	0.3741	0.0825	0.6802			
C	0.0167	−0.2101	0.0455	−0.6906	−0.4681		
Polyphenols	0.1922	0.4597	0.1619	0.7527	0.8638	−0.6063	
Ca	0.7080	0.8334	0.6965	0.3382	0.1015	−0.0226	0.0615

weights of fungus, % moisture and calcium content were all significantly associated with selection. Linear regression analysis confirmed these positive correlations with palatability (Figs 32.1–32.3). However, the correlation matrix also revealed that a number of the independent variables were intercorrelated, and a stepwise regression analysis was carried out to account for the relative importance of the measured variables in selection (Table 32.4). Some 86% of the variance in selection could be accounted for by all parameters measured, with fresh weight, dry weight and moisture content together accounting for 75%. Introduction of the calcium parameter does not materially increase the coefficient of determination and the seemingly significant variables (fresh weight, % moisture and calcium content) together accommodate only 54% of the variance.

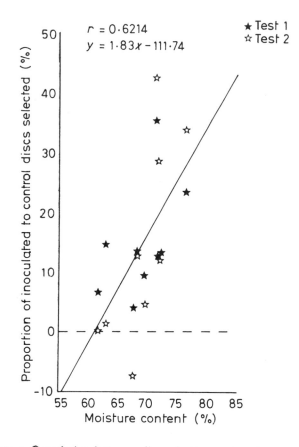

Fig. 32.1 Correlation between disc selection and moisture content.

Table 32.4 Stepwise regression analysis of selection on inoculated disc variables.

No. of variables entering analysis	New variable entering	Coeff. of determination	Coeff. of multiple regression
1	Ca	0.5013	0.7080
2	Fresh wt	0.5391	0.7342
3	Dry wt	0.6300	0.7938
4	moisture	0.7500	0.8660
5	C	0.8165	0.9036
6	Polyphenols	0.8304	0.9112
7	N	0.8636	0.9293

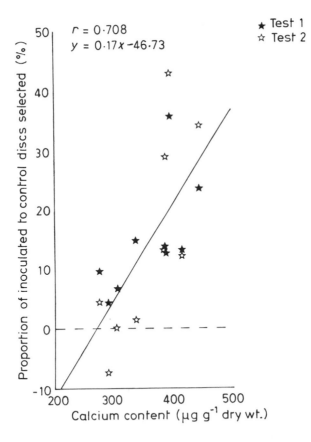

Fig. 32.2 Correlation between disc selection and calcium content.

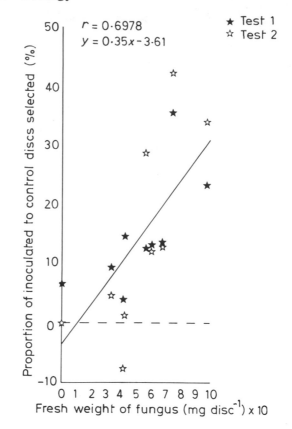

Fig. 32.3 Correlation between disc selection and fresh weight of fungus.

32.4 DISCUSSION

This study supports the finding of Cooke and Luxton (1980) that earthworms prefer fungally contaminated cellulose and reveals that palatability may be influenced by the nature of the fungal species. The preferences are not influenced by levels of polyphenols, nitrogen or carbon in the fungi but quantity of fungus and concentrations of Ca and moisture are all strongly correlated with selection. However, these last three variables are all intercorrelated and it is impossible to isolate any one as the prime causative agent in selection. Nevertheless, stepwise regression analysis discloses that fungus weight and moisture content together accommodate 75 % of the variance with addition of calcium not significantly altering the coefficient of determination. Thus, earthworms may find most attractive those potential foods bearing the greater growth of fungus either because this provides the richest food resource or because the fungal growth enables the food to retain higher moisture levels. Satchell and Lowe (1967) found that water-holding capacity played little part in determining earthworm preferences for tree litters, and other studies (Dash *et al.*, 1979; Neuhauser *et al.*, 1980) have observed that soil microbes may be a food source for some earthworms.

The particular characteristics of the inoculated discs which made them palatable to the earthworms were not unequivocally established in these experiments and a combination of factors is probably involved. Earthworms are likely to be drawn to moist foods as a result of their need to maintain body water. Moisture content may also indicate the state of decomposition of organic matter, leading the worm directly to a valuable microbial food substrate. The strong correlation between calcium and palatability may be fortuitous or may indicate an unmeasured pH effect. Another reason for selection, which is probable but difficult to measure, is that worms can taste the difference between fungi.

32.5 ACKNOWLEDGEMENTS

I thank Dr A. J. S. Whalley and Mr G. A. Brodie of Liverpool Polytechnic, and especially Dr M. S. Luxton for his help in preparing the manuscript.

32.6 REFERENCES

Allen, S. E., Grimshaw, H. M., Parkinson, J. A. and Quarmby, C. (1974) *Chemical Analysis of Ecological Materials*, Blackwell, Oxford. 565 pp.

Cooke, A. and Luxton, M. (1980) Effect of microbes on food selection by *Lumbricus terrestris. Rev. Ecol. Biol. Sol*, **17**, 365–370.

Dash, M. C., Mishra, P. C. and Behera, N. (1979) Fungal feeding by a tropical earthworm. *Trop. Ecol.*, **20**, 9–12.

Lowry, O. H., Rosebrough, N. J., Farr, A. L. and Randal, R. J. (1951) Protein measurement with the Folin phenol reagent. *J. Biol. Chem.*, **193**, 265–275.

Ministry of Agriculture, Fisheries and Food (1973) Technical Bulletin 27. *The Analysis of Agricultural Materials*. H.M.S.O., London.

Neuhauser, E. F., Kaplan, D. L., Malecki, M. R. and Hartenstein, R. (1980) Materials supporting weight gain by the earthworm *Eisenia foetida* in waste conversion systems. *Agric. Wastes*, **2**, 43–60.

Satchell, J. E. and Lowe, D. G. (1967) Selection of leaf litter by *Lumbricus terrestris*. In *Progress in Soil Biology* (eds. O. Graff and J. E. Satchell), North-Holland Publishing Co., Amsterdam, pp. 102–118.

Tinsley, J. (1950) Determination of organic carbon in soils by dichromate mixtures. *Trans. 4th Int. Congr. Soil Sci.*, **1**, 161.

Wright, M. A. (1972) Factors governing ingestion by the earthworm *Lumbricus terrestris* with special reference to apple leaves. *Ann. Appl. Biol.*, **70**, 175–188.

Chapter 33

Introduction of amoebae and *Rhizobium japonicum* into the gut of *Eisenia fetida* (Sav.) and *Lumbricus terrestris* L.

J. ROUELLE

33.1 INTRODUCTION

Earthworms consume large quantities of soil and in so doing disperse micro-organisms that can survive passage through the gut. Several pathogens have been shown to be eliminated in the gut while others, including *Bacillus anthracis* (Pasteur, 1880), *Mycobacterium tuberculosis* (Lortet and Despeignes, 1892a, b), *Fusarium* (Taylor, 1917; Rathburn, 1918), *Plasmodiophora brassicae* (Gleisberg, 1922) and *Vibrio* (Marialigeti, 1979), have been thought to be spread by worms. Amongst non-pathogens, *Serratia marcescens* (Day, 1950) and some algae (Atlavinyte and Pociene, 1973; Nekrasova *et al.*, 1976), are destroyed in the gut, and nitrogen-fixers have been shown to be lysed by a mycoplasma-like gut organism (Lepidi and Nuti, 1975). Hutchinson and Kamel (1956) showed that fungi were spread rapidly by worms in experimental containers and this has been confirmed for phycomycetes by Thornton (1970). In natural conditions, however, fungi, especially phycomycetes, were found by Kamal and Singh (1970) to be less abundant in casts than in neighbouring soil.

These studies imply that micro-organisms may play a significant role in earthworm nutrition and this has been demonstrated for protozoa by Miles (1963) and Duhlinska (1979) who showed that *Eisenia fetida* failed to reach maturity in protozoa-free soil. More recently Neuhauser *et al.* (1980) established that *E. fetida* grows better in the presence of live or dead protozoa (*Tetrahymena, Euglena*) even though the medium was rich in other micro-organisms. Most earthworms have a gut fauna of astomatous ciliates (de Puytorac, 1954) and Piearce and Phillips (1980) have demonstrated the lysis of a common freshwater ciliate, *Colpidium campylum*, in mid-gut fluid of *Lumbricus terrestris*.

The disappearance in the gut of *Escherichia coli* was attributed by Brüsewitz (1959) largely to antagonistic effects of actinomycetes but Khambata and Bhat (1957) detected no actinomycete antagonism in the worm-gut microflora, and Routien and Finlay (1952) commented that 'earthworm guts were disappointingly devoid of interesting antagonists'.

We have attempted to clarify these conflicting observations by feeding *Eisenia fetida* and *Lumbricus terrestris* with common soil amoebae and with *Rhizobium* since bacteria of this genus are known to be eaten by nematodes (Overgaard-Nielsen, 1949; Cayrol *et al.*, 1977).

33.2 MATERIALS AND METHODS

Three strains of amoebae were employed, two with a weak tendency to encyst under the conditions of the experiments (*Saccamoeba stagnicola* Page strain Ir Fld and *Thecamoeba* sp. strain F4) and one with a strong tendency to encyst with slow excystment (*Acanthamoeba triangularis* Pussard and Pons, strain SH621). Morphological peculiarities allowed these amoebae to be located even if other species were present. The three strains were maintained in monoxenic culture with *Klebsiella aerogenes* (Pasteur Institute 6086). The strain of *Rhizobium* used, which is specific to soya, was *R. japonicum* G33, (1000 μg ml^{-1}; streptomycin-resistant), provided by M. Obaton (INRA, Montpellier). Culture media were those of Vincent (1970).

Earthworms were initially sterilized by passage, over 2 days, through three lots of soil that had been sterilized by autoclaving (2 × 1 h at 115°C, interval 18 h) and then treated with antibiotics (actidion 0.1 % and streptomycin 0.2 %). They were then passed at least twice through autoclaved soil. A soil extract medium (Pochon and Tardieux, 1962) or water agar inoculated with *Klebsiella aerogenes* was used to verify the absence of undesirable micro-organisms in a sample of the soil in which earthworms had recently been kept.

Micro-organisms were introduced into the earthworm gut in three ways: (1) by injection of a dense suspension of micro-organisms in sterile water, or a suspension dilution (0.1–0.2 ml), by means of a thin catheter inserted through the earthworm's mouth (Bouché, 1966). In most cases the earthworms were previously anaesthetized by immersion in water containing acetone chloroform. Some injections were made into earthworms that had been killed by dipping in hot water or prolonged immersion in acetone chloroform. In a few cases, injections were made into living, active earthworms: (2) by national ingestion of little bits of paper (made when teleprinter cards are perforated) on which micro-organisms had been grown; these bits were easily recognizable in the gut contents: (3) by keeping earthworms in sterilized soil sprayed with microbial suspension.

Surviving micro-organisms were sought in faeces (indicating complete transit along the gut) and, more especially, in different sections of gut (incomplete transit). The sections distinguished were: fore-gut, generally up to the gizzard, mid-gut and hind-gut.

Surface sterilization of earthworms was accomplished by means of a technique similar to that used by Pellenard (1973) for crustaceans. Living earthworms were dipped into a crystal violet and 90 % alcohol solution. Death occurred in 15–30 seconds but earthworms were left somewhat longer to obtain good external sterilization. The dye allows one to check during dissection whether alcohol has penetrated to the gut. The same treatment was applied to earthworms killed before injection.

Faeces, gut contents or sections of gut were cultured on appropriate agar media to test for surviving micro-organisms. An agar medium covered with *K. aerogenes* was used for amoebae, and a specific agar medium (Vincent, 1970) containing streptomycin was used for the strains of *R. japonicum* resistant to streptomycin.

Juveniles of *E. fetida*, but not *L. terrestris*, were obtained from cocoons surface-sterilized by immersion for 20 min in 2.5 % Javel solution in sterile water with 5 % Teepol. After several rinses in sterile water, cocoons were put in individual dishes on soil extract agar medium with different combinations of antibiotics (actidion, streptomycin, rifamycin, neomycin). Asepsis of juveniles was checked by putting them on antibiotic-free soil extract medium. Juveniles were then put in individual Petri dishes with autoclaved soil, renewed every week. At the time of renewal, the soil was enriched with casein, oatmeal, egg yolk, *K. aerogenes* or *K. aerogenes* and amoebae. There were three juveniles per treatment, and the cultures were maintained for 4 months. Asepsis was checked at the end of the experiment or occasionally during the experiment.

It seems probable that in theses experiments juveniles could subsist on micro-organisms killed by autoclaving and on the products of previous microbial activity. Both are undoubtedly altered by the heat treatment. After autoclaving, it is necessary to wait several days before introducing juveniles into the soil; where this has not been done lysed juveniles have often been seen on the soil surface.

33.3 RESULTS AND DISCUSSION

Lumbricus terrestris and *Eisenia fetida* were placed in autoclaved soil after injection of a suspension of the three amoebae; searches of faeces after three days and ten days revealed only *Acanthamoeba triangularis*. *Rhizobium japonicum* was sought in the top layer of fresh faeces of *E. fetida* and *L. terrestris* placed on moist filter paper in a Petri dish. In all cases bacteria were recorded.

Culture of gut contents or whole sections of gut (from about 20 specimens of each species for mid- and hind-gut determinations and ten specimens for fore-gut determinations) revealed no trophozoites, i.e. active amoebae, of *Thecamoeba* or *Saccamoeba*, but did demonstrate the presence of *Acanthamoeba*. The latter was injected in the encysted form which is known to be very resistant. Its presence along the gut proves the transit of material along the gut in these experiments. Dissections were performed 3 and 16 hours after injection. Destruction of trophozoites was therefore early and swift. The fact that 16 hours after injection, cysts were still found in the mid-gut and sometimes in the fore-gut shows that, at least for some of these small particles, transit was very slow under these experimental conditions.

Examination of the mid-gut and hind-gut of three earthworms of each species freshly removed from soil, which therefore supported a complex microflora and microfauna, gave the same results. In contrast, living specimens of all three amoebae were found in each section of the guts of 20 *E. fetida* and 20 *L. terrestris* killed before injection and dissected as described above.

Perhaps trophozoites are mechanically destroyed in the living earthworm, or mixing of gut contents is necessary to ensure effective contact with digestive secretion. In the work of Piearce and Phillips (1980), conditions encountered by the ciliate *Colpidium campylum* in alimentary fluid removed from *L. terrestris* appear comparable with those met by trophozoites in the dead earthworms in the present study. However, the ciliates were destroyed whereas some trophozoites survived. Perhaps contact between protozoans and digestive fluid was better in the former case.

Fifteen sterilized earthworms of each species which had been injected with *Rhizobium japonicum* suspensions of dilutions 3×10^5, 3×10^4 and 3×10^3 cells ml^{-1} (for the latter dilution about 300 bacterial cells were injected) were dissected 4–20 hours after injection; 20 hours after injection, *R. japonicum* was isolated from the hind-gut, mid-gut and often the fore-gut, showing that transit for some *R. japonicum*, as for cysts, can be slow. Further investigation is needed, however, since the technique of dissecting from hind- to fore-gut makes contamination possible, though improbable.

Ingested pieces of paper were generally found in the gizzard, rarely in the pharynx, which is often devoid of solid contents (Piearce, 1978). In about 20 earthworms of each species neither *Saccamoeba* nor *Thecamoeba* were isolated from the mid-gut or fore-gut. However, in about 25 % of cases trophozoites were found in the hind-gut; this may indicate penetration of micro-organisms through the anus where earthworms remain for 1–4 days on agar plates with the pieces of paper and a large density of micro-organisms. When about 20 *E. fetida* and *L. terrestris* were exposed to *R. japonicum* in the above way the bacterium was isolated from all parts of the gut.

Since earthworms do not destroy *Rhizobium* during its transit along the gut it may be that these bacteria are not sought as a direct source of food, although the earthworms could be attracted to polysaccharides secreted by bacteria or sugar solution included in the inoculum. Morren, quoted in Darwin (1882), pointed out that earthworms ingest sugar.

Each of the three amoebae was isolated from all parts of the gut in three specimens of *E. fetida* and *L. terrestris* which had ingested infected soil. This suggests that soil may protect some protozoans from destruction, as indicated earlier.

In media containing crude egg yolk or cultures of amoebae (amoebae + *K. aerogenes*), sterilized *E. fetida* became nearly adult (about 0.3 g) after 4 months, while other diets allowed only a little growth (0.08 g). At the end of this experiment most Petri dishes were accidentally polluted, principally by *Penicillium*.

The presence of *L. terrestris* in pots containing various kinds of soil led to the spread of nodules on the root system of soya. In pots without *L. terrestris*, nodules were more numerous but restricted to the upper part of the root system. Reduction in nodule abundance might result from general dilution of micro-organisms in the soil or increased antagonism of autochthonal microflora and microfauna. Spreading of nodules might reflect dispersal of *Rhizobium* by earthworms.

The two species of earthworm used, *E. fetida* and *L. terrestris*, occupy very different habitats, respectively compost and manure, and the mineral soils of grasslands, forest and cultivated land. Our work shows that trophozoites of amoebae can be used as food by these two species despite their different ecological requirements. If the ingested amoebae are encysted or protected by soil they may be spread by earthworms as too may *Rhizobium*. Further work remains to be done on the nutritional role of astomatous ciliates and perhaps other resident gut micro-organisms.

33.4 SUMMARY

A procedure is described for studying the consequences to amoebae (*Thecamoeba* sp., *Saccamoeba lignicola* and *Acanthamoeba triangularis*) and the bacterium *Rhizobium japonicum* of being introduced into the alimentary canal of earthworms which have had their microflora and microfauna removed. Two ecologically different earthworm species were used: *Eisenia fetida* Sav. and *Lumbricus terrestris* L. Non-encysted amoebae (trophozoites) were destroyed except when soil was present to protect them, but the bacteria survived. Amoebae seem to be a necessary dietary requirement of earthworms.

33.5 REFERENCES

Atlavinyte, O. and Pociene, C. (1973) The effect of earthworms and their activity on the amount of algae in the soil. *Pedobiologia*, **13**, 445–455.

Bouché, M. B. (1966) Sur un nouveau procédé d'obtention de la vacuité artificielle du tube digestif des Lombricides. *Rev. Ecol. Biol. Sol*, **3**, 479–482.

Brusewitz, G. (1959) Untersuchungen über den Einfluss des Regenwurms auf Zahl, Art und Leistungen von Mikroorganismen in Boden. *Arch. Mikrobiol.*, **33**, 52–82.

Cayrol, J. C., Couderc, C. and Evrard, I. (1977) Études des relations entre les nématodes libres du sol et les bactéries des nodosités des légumineuses. *Rev. Zool. Agric. Pathol. Veg.*, **76**, 77–89.

Darwin, C. R. (1882) (French translation). Rôle des vers de terre dans la formation de la terre végétale. Reinwal, Paris. 257 pp.

Day, G. M. (1950) Influence of earthworms on soil microorganisms. *Soil Sci.*, **69**, 175–184.

de Puytorac, P. (1954) Contribution a l'étude cytologique et taxonomique des infusoires astomes. *Ann. Sci. Nat.*, **b16**, 85–270.

Duhlinska, D. D. (1979) Cultivation of some Lumbricidae uninfected with monocystid gregarines. *Acta Zool. Bulg.*, **12**, 78–80.

Gleisberg, W. (1922) Das Rätsel der Hernieverbreitung. *NachrBl. dt. Pflchutzdienst, Berl.*, **2**, 89–90.

Hutchinson, S. A. and Kamel, M. (1956) The effects of earthworms on the dispersal of soil fungi. *Soil Sci.*, **7**, 213–218.

Kamal and Singh, C. S. (1970) Studies on soil fungi from teak forests of Grakhpur. VIII. A comparative account of fungi of earthworm casts, termitarium and surrounding soil from a teak stand. *Ann. Inst. Pasteur, Paris*, **119**, 249–259.

Khambata, S. R. and Bhat, J. V. (1957) A contribution to the study of the intestinal microflora of Indian earthworms. *Arch. Mikrobiol.*, **28**, 69–80.

Lepidi, A. A. and Nuti, M. P. (1974) Aspetti particolari della fissazione microbiologica dell'azoto nel territorio del Gran Sasso. In *Omaggio al Gran Sasso* (ed. C. A. I. L'Aquila), Arti Grafiche Tamari, Bologna, pp. 10–20.

Lortet, L. and Despeignes, V. (1892a) Les vers de terre et les bacilles de la tuberculose. *C. R. Hebd. Séances Acad. Sci., Ser. D.* 1er semestre, **114**, 186–187.

Lortet, L. and Despeignes, V. (1892b) Vers de terre et tuberculose. *C. R. Hebd. Séances Acad. Sci., Ser. D* 2e semestre, **115**, 66–67.

Marialigeti, K. (1979) On the community-structure of the gut microbiota of *Eisenia lucens* (*Annelida, Oligochaeta*). *Pedobiologia*, **19**, 213–220.

Miles, H. B. (1963) Soil protozoa and earthworm nutrition. *Soil Sci.*, **95**, 407–409.

Nekrasova, K. A., Kozlovskaya, L. S., Domraceva, L. I. and Stina, E. A. (1976)

The influence of invertebrates on the development of algae. *Pedobiologia*, **16**, 286–297.

Neuhauser, E. F., Kaplan, D. L., Malecki, M. R. and Hartenstein, R. (1980) Materials supporting weight gain by the earthworm *Eisenia foetida* in waste conversion systems. *Agric. Wastes*, **2**, 43–60.

Overgaard-Nielsen, C. (1949) Studies on the soil microfauna. II. The soil inhabiting nematodes. *Natura Jutlandica*, **2**, 131 pp.

Pasteur, L. (with the collaboration of C. Chamberland and P. Roux) (1880) Sur l'étiologie du charbon. *C. R. Hebd. Séances Acad. Sci. Ser. D* 2e semestre, xci, 86–94.

Pellenard, P. (1973) *Les matiéres organiques de l'argile souterraine et l'alimentation de Niphargus virei (crustacé amphipode hypogé).* Thèse de spécialité de l'Université de Lyon.

Piearce, T. G. (1978) Gut contents of some lumbricid earthworms. *Pedobiologia*, **18**, 153–157.

Piearce, T. G. and Phillips, M. J. (1980) The fate of ciliates in the earthworm gut: an *in vitro* study. *Microb. Ecol.*, **5**, 313–319.

Pochon, J. and Tardieux, P. (1962) *Techniques d'analyse en microbiologie du sol.* Editions de la Tourelle, St. Mandé, France, 108 pp.

Rathburn, A. E. (1918) The fungus flora of pine seed beds, Part I. Fungus flora of the soil. *Phytopathology*, **8**, 469–483.

Routien, J. B. and Finlay, A. C. (1952) Problems in the search for micro-organisms producing antibiotics. *Bact. Rev.*, **16**, 51–67.

Taylor, M. W. (1917) Preliminary report on the vertical distribution of *Fusarium* in soil. *Phytopathology*, **7**, 374–378.

Thornton, M. L. (1970) Transport of soil-dwelling aquatic phycomycetes by earthworms. *Trans. Br. Mycol. Soc.*, **55**, 391–397.

Vincent, J. M. (1970) *A manual for the practical study of the root-nodule bacteria.* IBP Handbook no. 15, Blackwell Scientific Publications, Oxford and Edinburgh, 164 pp.

Chapter 34

Enterobacter aerogenes infection of *Hoplochaetella suctoria*

B. R. RAO, I. KARUNA SAGAR and J. V. BHAT

34.1 INTRODUCTION

Knowledge of the microbial pathogens of earthworms is very fragmentary. Though many internal and external parasites have been reported, information about the relationships between earthworms and their parasites is scanty. Gates (1972) gave a good account of different bacteria, ciliates, sporozoans, cestoda, nematoda and insects encountered in different species of Burmese earthworms without indicating their pathogenicity. Though *Spirochaeta* and *Clostridium botulinum* have been reported to be associated with earthworms, little is known of their effects (Edwards and Lofty, 1972). To our knowledge, the only clear-cut

Fig. 34.1 Two diseased *H. suctoria.*

383

description of a bacterial pathogen of earthworms available today is that of Smirnoff and Heimpel (1961) who found that *Bacillus thuringiensis* penetrates the gut and enters the body cavity causing fatal septicaemia.

The present chapter describes studies on a bacterial disease which occurs in *Hoplochaetella suctoria* towards the end of the rainy season (August–October) resulting in considerable mortality. The disease was also noted in *Hoplochaetella kempi* and *H. affinis*. It generally commences as pin-hole dots in the clitellar region and then appears as irregular patches. The infection from the clitellar region spreads to other parts of the body, especially towards the posterior. The morbid worms become sluggish and lose weight (Fig. 34.1).

34.2 MATERIAL AND METHODS

Sample earthworms were collected from a cashew garden, an area of kitchen drainage, grassland and land fertilized with cow dung and decaying leaves, during the first week of each month of the rainy season for three consecutive years from August 1977 to October 1979. Three samples were taken every time in each plot from areas thought to contain abundant earthworms. The samples were dug up in 100 cm × 100 cm sections to a depth of 40 cm and the earthworms were collected by handsorting. The diseased and healthy *H. suctoria* were separated after close examination. During the first year of study, the worms were grouped into three age categories.

To ascertain the contagiousness of the disease, one diseased worm was introduced into a sterile pot containing six healthy worms. In another series of experiments, six diseased worms were maintained in a sterile pot for four days; they were then removed and six healthy worms were introduced in their place.

The diseased worms were dissected to note any morphological changes. The affected area of skin was examined histologically after staining with haematoxylin and eosin as well as with periodic acid Schiff stain.

For bacteriological study, the diseased worms were washed thoroughly with tap water and subsequently with sterile distilled water. Using a sterile scalpel, skin scrapings from diseased portions were inoculated into the following culture media: (a) nutrient agar containing peptone (BDH), 1 %; meat extract (Oxoid), 0.5 %; sodium chloride (BDH), 0.5 %; agar (Difco), 2 %, and of pH 7.4; (b) soil extract agar of pH 7.2 containing soil extract (prepared as described by Rangaswami, 1966), 100 ml; glucose (BDH), 1 g; K_2HPO_4 (BDH), 0.5 g; agar (Difco), 15 g; tap water, 900 ml; (c) soil extract broth; (d) glucose broth containing peptone (BDH), 1 %; meat extract (Oxoid), 0.5 %; sodium chloride (BDH), 0.5 %; glucose

(BDH), 1 % and pH 7.4; (e) Sabouraud's glucose agar containing peptone (BDH), 1 %; glucose (BDH), 4 %; agar (Difco), 2 % and pH 5.6.

The scrapings were streaked on nutrient agar, soil extract agar and Sabouraud's glucose agar and inoculated into soil extract broth and glucose broth. All cultures were incubated at room temperature (27–28 °C). For isolation of bacteria, plates were observed after 24 hours and 96 hours and for isolation of fungi after incubation for 4–10 days. Organisms growing in the glucose broth were subcultured into nutrient agar and soil extract. Isolated colonies were further purified on the same media and maintained on slopes for identification following Bergey's Manual of Determinative Bacteriology (Buchnan and Gibbons, 1974).

Fungi growing on Sabouraud's glucose agar were stained with Lactophenol Blue to observe morphological characters, and Riddell's slide culture technique (Cruickshank, 1970) was used to observe mycelial growth and sporulation at various stages of growth.

Microbial isolates obtained by all these methods were tested for their ability to produce the disease individually or in association with others in healthy worms. Healthy worms collected from natural habitats were maintained in sterile pot culture for about 10 days before use to ensure that they were not already infected. Six worms were used in each experiment. They were washed first in tap water and three times in sterile distilled water. The microbial culture to be tested was then smeared over the body surface, particularly the clitellar region, with a sterile swab. The worms were then put into sterile culture pots watered on alternate days with a dilute suspension of the test micro-organism in distilled water. The control worms were smeared with sterile distilled water. The cultures were maintained for a month.

34.3 RESULTS

Though the infection is common in clitellate worms, aclitellates as well as juveniles are not free from disease (Table 34.1). The disease starts appearing in August, especially in clitellate worms, and spreads to other healthy worms. High mortality may be the reason for the decrease in total number of worms recorded in October (Table 34.1).

When a diseased worm was transferred to a culture pot containing six healthy worms (Pot A, Table 34.2), the latter became infected in the course of a week. Even when healthy worms were reared in pots inhabited earlier by diseased worms (Pot B, Table 34.2) all of them became infected.

Dissected diseased earthworms did not reveal any change in internal organs. Histological sectioning of diseased portions showed patchy necrosis of the epidermal region.

Microbiological studies revealed the presence of *Pseudomonas*,

Table 34.1 Incidence of disease among different age groups of *Hoplochaetella suctoria* in different environments. (All figures are means).

Place	Month	No. of worms m^{-2}						Total no. of worms
		Healthy			Diseased			
		J*	A	C	J	A	C	
Cashew garden	August	08	16	14	01	09	11	59
	September	00	07	11	00	14	23	55
	October	00	02	05	00	08	18	33
Kitchen drainage region	August	12	16	10	00	17	28	83
	September	09	13	08	02	11	30	73
	October	00	01	06	00	07	19	33
Grassland	August	24	19	08	04	05	06	66
	September	07	11	25	02	09	10	64
	October	02	06	13	03	08	12	44
Cow dung with decaying leaves	August	18	21	09	05	07	32	92
	September	04	13	12	00	14	21	64
	October	00	10	15	00	06	25	56

* J, Juvenile; A, aclitellate; C, clitellate.

Table 34.2 Transference of disease to healthy *Hoplochaetella suctoria*.

No. of days	Pot A*		Pot B†	
	No. of healthy worms	No. of diseased worms	No. of healthy worms	No. of diseased worms
0	6	1	6	0
2	6	1	6	0
4	4	3	3	3
6	1	6	0	6
8	0	7	0	6

* One diseased worm transferred to the sterile pot containing six healthy worms.
† Six healthy worms transferred to pot inhabited earlier by diseased worms.

Micrococcus, Alcaligenes, Cephalosporium, Enterobacter and yeasts. Table 34.3 shows that only *Enterobacter* was able to produce the disease in healthy worms. Signs of infection appeared on the third day in 30 % of the worms as black spots in the clitellar region. These then spread to other

Table 34.3 Ability of different microbial isolates to reproduce disease in healthy *Hoplochaetella suctoria* culture pots. (H, healthy; D, diseased.)

No. of days	Cephalosporium		Yeast		Pseudomonas		Micrococcus		Alcaligenes		E. aerogenes		Control	
	H	D	H	D	H	D	H	D	H	D	H	D	H	D
1	6	—	6	—	6	—	6	—	6	—	6	—	6	—
2	6	—	6	—	6	—	6	—	6	—	6	—	6	—
3	6	—	6	—	6	—	6	—	4	—	4	2	6	—
4	6	—	6	—	5	—	6	—	4	—	2	4	6	—
5	6	—	6	—	4	—	6	—	3	—	—	6	6	—
6	6	—	6	—	4	—	6	—	2	—	—	6	6	—
7	6	—	6	—	3	—	6	—	1	—	—	4	6	—
8	6	—	6	—	2	—	6	—	0	—	—	0	6	—
9	6	—	6	—	2	—	6	—	0	—	—	0	6	—
10	6	—	6	—	0	—	6	—	0	—	—	0	6	—

Table 34·4 Ability of combinations of microbial isolates to reproduce disease in healthy *Hoplochaetella suctoria* in culture pots.

No. of days	C & Y		C & P		C & M		C & A		C & E		Y & P		Y & M		Y & A		Y & E		P & M		P & A		P & E		M & A		M & E		A & E	
	H	D	H	D	H	D	H	D	H	D	H	D	H	D	H	D	H	D	H	D	H	D	H	D	H	D	H	D	H	D
1	6	—	6	—	6	—	6	—	6	—	6	—	6	—	6	—	6	—	6	—	6	—	6	—	6	—	6	—	6	—
2	6	—	6	—	6	—	6	—	6	—	6	—	6	—	6	—	6	—	6	—	6	—	6	—	6	—	6	—	6	—
3	6	—	6	—	6	—	6	—	5	1	6	—	6	—	6	—	6	—	6	—	6	—	6	—	6	—	6	—	5	1
4	6	—	6	—	6	—	6	—	4	2	6	—	6	—	6	—	4	2	6	—	5	—	2	4	6	—	3	3	2	3
5	6	—	6	—	6	—	4	—	2	4	6	—	6	—	6	—	3	3	5	—	4	—	0	4	5	—	3	3	0	3
6	6	—	4	—	6	—	4	—	—	6	6	—	6	—	6	—	1	5	5	—	4	—	—	2	3	—	1	5	—	3
7	6	—	4	—	6	—	3	—	—	4	4	—	6	—	4	—	—	6	4	—	3	—	—	2	3	—	—	4	—	3
8	6	—	3	—	6	—	1	—	—	1	3	—	6	—	2	—	—	4	4	—	0	—	—	2	1	—	—	2	—	1
9	6	—	3	—	6	—	0	—	0	—	3	—	6	—	1	—	0	—	3	—	0	—	0	—	1	—	0	—	0	—
10	6	—	2	—	6	—	0	—	0	—	3	—	6	—	0	—	0	—	2	—	0	—	0	—	0	—	0	—	0	—

Key: *Cephalosporium*; Y, Yeast; P, *Pseudomonas*; M, *Micrococcus*; A, *Alcaligenes*; E, *E. aerogenes*; H, healthy; D, diseased.

parts of the body. By the sixth day, the disease had spread to almost all parts of the body; the worms became inactive and death ensued. Almost all the worms died, notably in the reinfection experiment with *Enterobacter* in which in three batches all of the six worms died within 10 days. The course of the disease was similar in all the three series of experiments (Table 34.3). The worms in the control culture pot remained healthy. *Pseudomonas* and *Alcaligenes* did not produce any noticeable infection but caused death in 8–10 days. *Micrococcus,* yeast and *Cephalosporium,* on the other hand, did not harm the earthworms in any way. In fact, the earthworms in cultures inoculated with yeast increased noticeably in girth.

When the microbial isolates were tried in various combinations for their ability to reproduce the disease in healthy worms, it was observed that only those pairs containing *Enterobacter* could bring this about (Table 34.4). Combinations containing either *Pseudomonas* or *Alcaligenes*

Table 34.5 Identifying characters of *Enterobacter aerogenes*

Test	Reaction
Gram stain	Gram-negative bacilli
Motility	Present
Oxidase	Negative
Nitrate reductase	Present
Catalase	Present
Indol	Not produced
Voges Proskauer	Positive
Methyl Red test	Negative
Fermentations	
Arabinose	Acid and gas
Galactose	Acid and gas
Glucose	Acid and gas
Mannose	Acid and gas
Xylose	Acid and gas
Lactose	Acid and gas (late)
Maltose	Acid and gas
Sucrose	Acid and gas
Dulcitol	Not fermented
Mannitol	Acid and gas
Sorbitol	Acid and gas
Salicin	Acid and gas
Rhamnose	Acid and gas
Citrate utilization	Positive
Lysine decarboxylase	Present
Arginine dehydrolase	Absent
DNAase	Negative

caused considerable mortality without producing symptoms of the disease.

To prove the fourth postulate of Koch, experimentally infected worms were cultured and *Enterobacter* could be reisolated in pure culture from all of them. This culture was further studied for its biochemical characters, and, on the bases of the results obtained (Table 34.5), the culture has been conclusively identified as *Enterobacter aerogenes*.

34.4 DISCUSSION

The results unequivocally establish that *Enterobacter aerogenes* is the causative agent of the disease of earthworms noted in *Hoplochaetella suctoria*, Koch's postulates being met completely. This is the first report, to our knowledge, of *Enterobacter* infection in earthworms. The pathogenicity of this bacterium to man is known. *E. agglomerans* has been implicated in wound infections, bacteraemia, urinary tract infection, meningitis, brain abscess, septicaemia and osteomyelitis (Pien *et al.*, 1972), but its pathogenicity for invertebrates is not known. This study showed the incubation period of the disease to be 3 days commencing, as in nature, with the appearance of black dots in the clitellar region. The disease characteristically spreads rapidly over the entire body surface and the death of the worms is perhaps attributable to respiratory failure.

Worms in pots infected with *E. aerogenes* along with other bacteria, fungi and yeasts also developed disease. This shows that the pathogenecity of *E. aerogenes* is not affected by the presence of other micro-organisms tested.

The observation that earthworms gain weight in yeast cultures indicates that, as in other species, yeast serves as a nutritious diet for worms. The marked toxicity of *Pseudomonas* and *Alcaligenes* to earthworms is surprising since they are normal constituents of the soil microflora. The explanation may be that the low counts of this organism occurring in the soil may be innocuous but at the concentration smeared on the body surface of the worms, they may be pathogenic. In mixed culture experiments with *Pseudomonas* and *E. aerogenes* and with *Alcaligenes* and *E. aerogenes*, it was not clear which caused the worms to die. While some of the worms developed characteristic symptoms of *E. aerogenes* disease, others died without showing any external disease symptoms.

Figure 34.2 shows that the disease occurs less frequently in grassland compared with other habitats. The reason for this is not known. The disease appears to be more prevalent in kitchen drainage and in cow dung with decaying leaves, attributable perhaps to the occurrence of *E. aerogenes* in large numbers from human and animal sources.

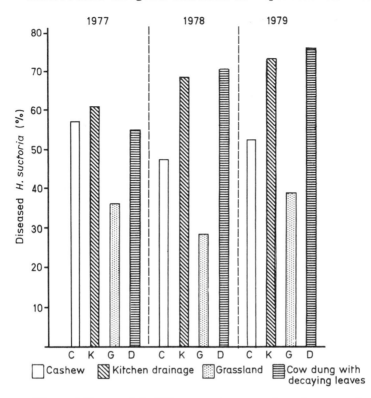

Fig. 34.2 Diseased *H. suctoria* in different environments from August to October in Moodabidri.

34.5 REFERENCES

Buchnan, R. E. and Gibbons, N. E. (1974) *Bergey's Manual of Determinative Bacteriology*, 8th edn, Williams & Wilkins, Baltimore. 325 pp.

Cruickshank, R. (1970) *Medical Microbiology*, 11th edn, English Language Book Society, 514 pp.

Edwards, C. A. and Lofty, J. R. (1972) *Biology of Earthworms*, Chapman & Hall, London, 283 pp.

Gates, G. E. (1972) Burmese earthworms. An introduction to the systematics and biology of megadrile oligochaetes with special reference to southeast Asia. *Trans. Am. Philos. Soc.*, **62**, 326 pp.

Pien, F. D., Martin, W. J., Hermans, P. E. and Washington, A. J., II (1972) Clinical and bacteriologic observations on the proposed species *Enterobacter agglomerans* (the *Herbicola-Lathyri* bacteria). *Mayo Clin., Proc.*, **47**, 739–745.

Rangaswami, G. (1966) *Agricultural Microbiology*. Asiatic Publishing House, Bombay. 381 pp.

Smirnoff, W. A. and Heimpel, A. M. (1961) Notes on the pathogenicity of *Bacillus thuringiensis* var. *thuringiensis* Bulinu for the earthworm *Lumbricus terrestris*, Linnaeus. *J. Insect Pathol.*, **3**, 403–408.

Chapter 35

Predation on earthworms by terrestrial vertebrates

D. W. MACDONALD

35.1 INTRODUCTION

The earthworm *Lumbricus terrestris* is widespread and abundant. Its biomass may reach 1000 kg ha^{-1}, and > 20 worms m^{-2} may surface simultaneously at night. Earthworms are thus potentially available to subterranean predators, to those that dig them up and those which catch them on the surface. Earthworm tissue has a high protein content and is rich in essential amino acids (Sabine, Chapter 24). It contains a considerable amount of fat variously estimated at 1.5% (Lawrence and Millar, 1945), 4.6% (French *et al.*, 1957) and 17.3% (Durchon and Lafon, 1951) and this is reflected in a high energy content of about 22.24 kJ g^{-1} dry wt. (Bolton and Phillipson, 1976). This abundant food resource is utilized by diverse predators and for some of them is seasonally the principal food. Earthworms feature in the diets of hundreds of species of terrestrial vertebrates but the adaptations of these predators to securing earthworms as prey have been studied in very few of them. This chapter reviews (a) a selection of studies of the importance of earthworms in the diets of sympatric species, (b) the few studies that have investigated predation on *Lumbricus*, and (c) the relationships between the ecology of the earthworm and the behaviour and social organization of these predators.

35.2 LUMBRICIDS AS A FOOD SOURCE

35.2.1 Amphibia and reptilia

The majority of amphibians and reptiles feed above ground and are only likely to prey upon earthworms that have surfaced at night. Earthworms

393

are generally of secondary importance in their diet, e.g. Lescure (1966) found that they comprised 0.5 % of the gut contents of a sample of 50 toads, *Bufo bufo*, in contrast to beetles and ants which together amounted to 77 %. Smith (1951) supports the view that worms occur at low frequency in the diets of toads and frogs, *Rana temporaria*, but suggests that newts, *Triturus* spp., are voracious feeders on worms. Worms eaten by toads are stuffed into the mouth using the forelimbs, and soil is cleaned from the prey by squeezing it between two fingers. Bas Lopez (1979) lists oligochaetes amongst the major prey of *Salamandra salamandra*.

Butler's garter snake, *Thamnophis butleri*, feeds almost exclusively on earthworms (Catling and Freedman, 1980). In contrast the sympatric Eastern garter snake, *T. sirtalis*, takes a mixed diet of earthworms and anuran amphibia while the third member of the guild, the brown snake, *Storeria dekayi*, feeds primarily on slugs and only takes earthworms occasionally. In a study of three other sympatric Garter snakes on Vancouver Island, Gregory (1978) found that *Thamnophis ordinoides* specialized on earthworms and slugs, whilst *T. sirtalis* fed on amphibians and earthworms and *T. elegans* ate slugs and small mammals. Of these three species, respectively 43.5, 30.8 and 3.4 % of full stomachs contained some earthworms.

Steward (1971) reports that, of European snakes, the viperine snake, *Natrix maura*, the grass snake, *Natrix natrix*, and the blind snake, *Typhlops vermicularis*, eat earthworms. In Kansas, USA, the worm snake, *Carphophis vermis*, eats lumbricids of the genus *Allolobophora* (Clark, 1971) and in Ontario earthworms identified from the stomachs of *T. butleri* included *A. chlorotica*, *Aporrectodea tuberculata* and *Lumbricus terrestris* (Reynolds, 1977a). Logier (1958) records four or five other American snakes which prey extensively on earthworms. In South Africa, microchaetid worms which may extend several metres in length are attacked by the night adder, *Causus rhombeatus* (Ljungström and Reinecke, 1969).

35.2.2 Birds

In 1913, Collinge found an earthworm in the stomach of a tawny owl, *Strix aluco*, and later Southern (1969) and Hirons (1976) recorded worms as frequent prey of the owls in Wytham Woods near Oxford. Delmee *et al.* (1979) found only occasional *L. terrestris* in the prey of tawny owls in Belgium. These owls catch earthworms on the surface at night. They appear to locate the worms aurally whilst perched on fenceposts as the worms crawl above ground. After craning the neck whilst pinpointing the worm, the owl bounds forward, with wings open, in two or three long hops before pecking up the worm. Foraging in this way, tawny owls catch

0.4 worms min^{-1}, rather less than some mammalian predators on earthworms under comparable conditions (Macdonald, 1976a).

Yalden and Warburton (1979) found that 5.5 % of the diet of kestrels, *Falco tinnunculus*, comprised earthworms in an area of northern England where 73 % of the diet was vertebrates (mainly *Microtus* spp.). Shrubb (1980) found a seasonal maximum of invertebrate prey of kestrels of 7 % (by wt.) of which earthworms were a component and suggested that they could be important seasonally in arable habitats.

Earthworms have also been recorded in the food of rooks, *Corvus frugilegus*, where they may predominate in the food taken to nestlings (Coleman, 1971), and crows, *Corvus corone*, where they are an intermediary host for the nematode parasite, *Capillaria oesophagealis*. Eleven out of 18 jay (*Garrulus glandarius*) stomachs contained worms (Collinge, 1913) which are also recorded in the diet of wryneck, *Jynx torquilla* Koral (1976), song thrush, *Turdus musicus*, blackbirds, *T. merula*, and starlings, *Sternus vulgaris* (Harlin, 1977; see Binder, 1971 and Lesinch and Melbarde, 1974, for nematode parasites). Harlin noted that blackbird and starling fed on a similar range of prey throughout the year. In spring, both feed mainly on animal food and, in summer, especially in the case of the starling, they take more fleshy fruit. Overall, lumbricids were estimated as making up 2.4 % and 0.3 % by volume of blackbird and starling diets. Of waders, lapwings, *Vanellus vanellus* (Collinge, 1913), and oystercatchers, *Haematopus ostralegus*, both take earthworms, and lapwings foraging on earthworms have been the subject of an important behavioural study (Barnard and Stephens, 1981, see below). Heppleston (1971) found that the frequency of occurrence of earthworms in the stomachs of 57 adult oystercatchers varied from 5.1 to 20.0 to 76.9 % depending on whether they were shot in the locality of mussel beds, mud flats or fields. Similarly, Sibly and McCleery (in press) found that radio-tracked Herring gulls, *Larus argentatus*, foraged for worms around dawn and at that time the gulls were found in pastures on 87 % of occasions, mown grass on 7 %, ploughed fields on 4 %, and long grass on 2 %. These figures are compatible with the known habitat preferences of *L. terrestris*, which the gulls were seen to catch, apparently having detected them by hearing, on or near the surface. Later in the day, the gulls' use of pasture diminished to only 41 % – at times when the circadian rhythm of *Lumbricus* would have driven them further below the surface. McCleery and Sibly saw gulls catch worms in three ways: most often they walked rapidly across the pasture, with head held high and gaze fixed on the ground before grabbing at a worm. Secondly the gulls followed tractors tilling soil and thirdly they 'paddled' with their feet on the ground, apparently attracting worms to the surface. Study of these radio-tracked gulls revealed that specific worming expeditions were undertaken when the gulls left the colony shortly before dawn and returned on average 1.5 hours later. The birds returning from such expeditions were weighed automatically when they stood on a balance built into their nest (Sibly and McCleery, 1980) and this revealed that they caught worms at a rate of $33\,\mathrm{g\,h^{-1}}$.

The hazard of gulls on airfields is determined largely by the availability of earthworms. In a study at Christchurch Airport, New Zealand, Moeed (1976) recorded seasonal changes in the gizzard contents of birds shot while feeding on the airfield. Black-backed gulls, *Larus dominicanus*, fed exclusively on earthworms. Black-billed gulls, *L. bulleri*, took a more varied diet but in a nearby breeding colony, 43 % of the invertebrates consumed by young gulls were earthworms (Dawson, 1958).

Earthworms constitute an important food item for various species of migratory birds including mistle thrushes, *Turdus viscivorous*, fieldfare, *T. pilaris* (Collinge, 1913), American robin, *T. migratorius*, starling, *Sturnus vulgaris*, Wilson's snipe, *Capella gallinago*, and American wood-cook, *Philohela minor*, and by 1928 the US Bureau of Biological Survey had already identified fragments of worms in the stomach contents of 45 species of birds (Walton). Literature on the diet of the American Woodcock spans many decades (Sheldon, 1967) but which species of earthworms are eaten has been clarified only recently (Reynolds, 1977b). In the northern part of its breeding range, 12 species of worms are present. Of these, *Lumbricus terrestris* is generally deep in its burrow beyond the probing depth of the woodcock during the daylight hours when it feeds and *Dendrodrilus rubidus* which lives mainly under the bark of decaying logs is also generally unavailable to soil-probing woodcock. Five other species may be present in the soil within its probe depth but they are the least abundant and the smallest. Reynolds suggests that these are rarely eaten because the energy cost of probing for them is likely to exceed the return. Of the worms found in 50 woodcock stomachs, 49 % were *Dendrobaena octaedra*, 42 % *Aporrectodea tuberculata* and all but one of the remainder were *Lumbricus rubellus*. *D. octaedra* is a small species and *A. tuberculata* must be considered the most significant contributor to the woodcock energy budget. In the southern part of the woodcock breeding range a different assemblage of earthworm species is present and their contribution to woodcock diet has yet to be elucidated. In a study in Central Maine, Reynolds (1977c) demonstrated that woodcock tended to use second-growth hardwood cover in preference to coniferous or mixed stands and that the extent to which different types of habitat were used was correlated with the biomass of the earthworm population. He concluded that the vegetation of the most used habitats produced the kinds of litter most palatable to earthworms and that management of secondary woodland to maximize earthworm biomass would increase its carrying capacity for woodcock.

In Iceland, Bengtson *et al.* (1978) also demonstrated selection of earthworm prey by Golden plover, *Pluvialis apricaria*. The earthworm fauna of a hayfield, as estimated by formalin sampling, comprised 37% *A. caliginosa*, 29% *L. rubellus* and 33% *L. terrestris*. In plover stomach contents, *L. terrestris* comprised only 11%, *L. rubellus* 28% and *A. caliginosa* 61%. As for the American woodcock, *L. terrestris* would be mainly below the probe depth of plovers feeding diurnally. The excess of *A. caliginosa* over *L. rubellus* in the prey may have been attributable to cryptic coloration in the latter and its absence in the former. In an experiment with caged rooks, *Corvus frugilegus*, Satchell (1967) demonstrated that the green morph had advantage over the pink morph of the polymorphic lumbricid *Allolobophora chlorotica*, showing that at least the rooks could distinguish between different degrees of pigmentation. In one year, Bengtson *et al.* (1976, 1978) also found the smallest and largest size classes of *L. rubellus* to be under-represented in the stomachs, a result comparable with observations on *Larus ridibundus* discussed in Section 35.3.

When phloem necrosis and Dutch elm disease threatened American elms, *Ulmus americana*, at Urbana, Illinois, in 1949, DDT was sprayed in attempts to control suspected insect vectors. Subsequently, dying robins, *T. migratorius*, were found containing about 60 μg of DDT g^{-1}. Analyses of earthworms showed that a 50 g robin could accumulate this quantity of DDT if it consumed 11 *L. terrestris* or about 60 of the smaller species of earthworms examined (Barker, 1958). Since robins had been reported to take 10–12 earthworms in as many minutes, the poisoned birds appeared to have ingested the DDT from this source. Later work, notably by Dimond *et al.* (1970), demonstrated a clear increase in DDT residues in American robins over those in earthworms in other habitats. The robins they analysed had brain levels of DDT residues well below concentrations considered lethal but they concluded that certain birds are at risk in habitats exposed to light applications of DDT over many years. Concentration factors of 3–4 times of heptachlor residues ingested in earthworms by American woodcocks have been recorded by Stickel *et al.* (1965) and Jefferies and Davis (1968) have demonstrated that dieldrin concentrations may reach lethal concentrations in song thrushes. Migratory birds are particularly at risk from pesticides ingested in earthworms and released from storage fat during long migratory flights.

35.2.3 Mammals

Amongst the mammals, most records of predation upon earthworms occur in the Order Insectivora, and especially the Soricidae. Hamilton (1930, 1941) examined the stomach contents of four species of North

American shrews and found at least some worms were eaten by each species. Of 244 specimens of the short-tailed shrew, *Blarina brevicauda*, earthworms were never the major component of the gut contents and overall made up only 7.2 % by volume of the diet. Similarly, 4.3 % of the diet of *Sorex cinereus* and 3.4 % of that of *S. fumeus* was of earthworms, and only one stomach of 13 examined of *S. palustris* contained any sign of worm remains. Two of these species of shrew were studied by Whitaker and Schmeltz (1973) in an area of St. Louis, USA, where they were sympatric and there the water shrew, *S. palustris*, was found to eat rather more worms (Table 35.1).

Table 35.1 The proportion of the diet, measured from stomach contents of two sympatric species of shrew, that was composed of earthworms.

	S. palustris		S. cinereus	
	Volume (%)	Frequency (%)	Volume (%)	Frequency (%)
Earthworms	21.2	30.7	7.2	11.1
Gryllidae	7.7	7.7	30.0	33.3
Caterpillars	1.2	7.7	28.8	44.4

Whitaker and Mumford (1972) also studied the diets of four species of shrews in Indiana, each of which took at least some earthworms and which again seemed to be partitioned on the basis of diet (Table 35.2). *Blarina brevicauda* primarily inhabits woodland, wheras *S. cinereus*, *S. longirostris* and *Cryptotis parva* are found mainly in old pastures. In another habitat in New York, Whitaker and Ferraro (1963) found that earthworms were present in the diet of *Blarina brevicauda* at 54.3 % by frequency and 31.4 % by volume.

Of the Old World Soricidae, Okhotina (1969) examined the stomach contents of 25 *S. mirabilis* trapped in woodland over 22 years. This species consumes as much as 28.7 g of invertebrate food each day (213 % of its body weight), and of this 82.5 % is earthworms. Amongst the shrews, *S. mirabilis* is said to be second only to *S. unguiculatus* as a burrower, which may account for the importance of worms in its diet. Lumbricids were also the most common type of prey in the stomachs of 10 water shrews, *Neomys fodiens*, found by Churchfield (1979).

In Britain, Rudge (1968) studied the diet of *S. araneus*, the common shrew. In the laboratory these shrews invariably ate earthworms when presented with them. Rudge compared the diets of common shrews from three localities (Table 35.3) and found both seasonal and geographical

Table 35.2 Comparison of the representation in the diet of four species of shrew of earthworms and of other prey, expressed as percentage volume of stomach contents and as percentage frequency of occurrence in stomachs (after Whitaker and Mumford, 1972).

	Blarina n = 125		Cryptotis n = 109		S. cinereus n = 50		S. longirostris n = 7	
	Volume (%)	Frequency (%)	Volume (%)	Frequency (%)	Volume (%)	Frequency (%)	Volume (%)	Frequency (%)
Earthworms	35.7	48.1	11.2	15.6	3.3	6.0	1.4	14.3
Slugs and snails	8.5	14.4	3.3	3.7	10.9	14.0	14.3	14.3
Lepidopterous larvae	8.2	16.8	17.9	29.4	17.2	28.0	19.3	42.9
Gryllidae	6.2	8.8	3.2	4.6	8.1	10.0	—	—
Chilopoda	4.5	8.0	3.6	7.3	1.2	4.0	11.4	14.3
Spiders	0.5	2.4	6.8	11.0	9.0	14.0	20.0	42.9
Vegetation	1.2	8.8	0.5	1.8	0.2	2.0	14.3	14.3

variation in the diet which included up to 55 % occurrence of earthworms, some of which were 10 cm long.

Table 35.3 The importance of earthworms in the diet of shrews, *Sorex araneus*, in three regions, expressed as percentage frequency of occurrence in stomachs and as percentage volume of the total diet (after Rudge, 1968).

	Scotland $n = 66$	Oxfordshire $n = 51$	Devon $n = 327$
Occurrence (%)	29	55	40
Diet (%)	10	10	12

Other members of the Insectivora also take worms, some specializing on them. Morris (1963) reports that European hedgehogs, *Erinaceous europaeus*, prey upon worms, catching them on the surface at night and freeing them from their burrows by a gentle rocking motion. Reeve (1982) also observed hedgehogs hunting for worms whilst he radio-tracked them. Moles, *Talpa europaea*, are the most specialized insectivoran worm eaters and eat large quantities of worms (Mellanby, 1966; Raw, 1966). Each mole consumes 18–36 kg of prey y^{-1} of which at least half is worms. This has made them especially susceptible to poisoning from vermicides (Shilova *et al.*, 1971; Ivanter, 1969). In Poland, Skoczen (1965) found a rather constant (87.7–100%) frequency of occurrence of worms in the stomachs of moles irrespective of season, although there was a slight peak in spring and summer. In contrast, Larkin (in Southern and Corbett, 1977) found that worms comprised 90–100% of mole diet in winter and 50% in summer. Funmilayo (1979) studied the diet of moles in pastures in SE Scotland where he also sampled the abundance of prey in the soil fauna. He found that moles, irrespective of age or sex, ate almost exclusively earthworms and insect larvae. The worms were most important in January and were taken in direct proportion to their availability.

Considerable interest has focused on the moles' habit of storing mutilated worms in caches within their fortresses. The worms are immobilized by bites to their anterior segments. Barrett-Hamilton (1910) wondered whether the putative caches might be worms which had fallen into the moles' tunnels but Evans (1948), Raw (1966), Funmilayo (1979) and others have confirmed that both *T. europea* and *S. araneus* store worms near their nests, and that water shrews, *N. fodiens*, hoard them under stones. Skoczen (1970) found a mole's cache containing 470 worms (total weight 820 g). Clearly the storage of these worms depends on incapacitating them, and some species of vermivores, such as foxes and badgers, apparently do not have a technique to immobilize worms and so never cache them (Macdonald, 1976b).

North American species such as the Hairy-tailed mole, *Parascalops breweri*, are also heavily dependent on worms (Eadie, 1939; Hamilton, 1941), as is the Star-nosed mole, *Condylura cristata*, to a lesser extent (Hamilton, 1931). Arlton (1936) described how *Scalopus aquaticus*, the American mole, squeezed worms through its front digits as it ate them, so removing soil from both the exterior and interior of the worm. Rodents also eat earthworms, although generally infrequently. Watts (1968) found occasional traces of worm in the stomachs of wood mice, *Apodemus sylvaticus*, and bank voles, *Clethrionomys glareolus*, as did Hamilton (1941) for *Peromyscus leucopus* and *P. maniculatus*, and Obrtel and Holisová (1980) and Obrtel (1973) for the yellow-necked mouse, *Apodemus flavicollis*, in summer. Holisová and Obrtel (1979) also studied the diet of *Clethrionomys glareolus*, and predation on worms by *A. flavicollis*, *A. sylvaticus* (Obrtel and Holisová, 1979) and *C. glareolus* can be compared from their various publications (Table 35.4).

Table 35.4 Frequency of earthworm remains found in stomachs of three species of small rodent in Europe.

	Frequency (%)			
	Spring	Summer	Autumn	Winter
C. glareolus	6.0	0.0	2.1	no data
A. flavicollis	2.9	3.1	1.1	
A. sylvaticus	8.8	5.2	5.1	

Holisovà (1967) found 30% of 318 stomachs of *Apodemus agrarius* had remains of Lumbricidae (see also Obrtel, 1973, for further notes on *A. flavicollis*). Circumstantial evidence suggests that the grasshopper mouse, *Onychomys torridus*, feeds on earthworms (Horner *et al.*, 1965) and Landry (1970) cites many references to rodents preying on earthworms, including *Dremomys everetti* (Liat and Heyneman, 1968), *Tamias striatus* (Yerger, 1955), *Citellus grammerus* (Bradley, 1929), *Mus musculus*, (Whitaker, 1966), *Apodemus microps* (Holisovà *et al.*, 1962), *Atherurus africanus* (Rosevear, 1969) and *Cryptomys* spp. (Shortridge, 1934).

Funmilayo (1981) has studied the incidence of earthworms in the stomachs of some 500 mammals of 17 species captured in the south of Nigeria. In the stomachs of nine of these species there were at least traces of earthworms (Table 35.5), but only one rodent and one insectivore took worms in any numbers: 69.8% of 73 Rufous-bellied rats, *Laphuromys sikapusi*, and 41.9% of 62 musk shrews, *Crocidura manni*, had eaten

Table 35.5 The incidence of remains of earthworms in the stomachs of each of 14 species of mammal sampled by Funmilayo (1981) in southern Nigeria.

Mammal		No. of stomachs examined	No. of stomachs containing earthworms	Percentage of stomachs containing earthworms
Musk shrew	Crocidura manni	62	26	41.9
Pigmy mouse	Mus musculoides	20	—	—
Dalton's mouse	Myomys daltoni	2	—	—
Spotted grass mouse	Lemniscomys striatus	36	2	5.6
Striped mouse	Hybomys trivirgatus	16	3	18.8
Ship rat	Rattus rattus	29	1	3.4
Multimammate rat	Mastomys natalensis	110	9	8.2
Soft-furred rat	Praomys tullbergi	9	—	—
Target rat	Stochomys longicaudatus	1	—	—
Shining Thicket rat	Thamnomys rutilans	2	—	—
Shaggy rat	Dasymys incomtus	29	1	3.4
Nile harsh-furred rat	Arvicanthis niloticus	12	—	—
Long-footed rat	Malacomys longipes	1	—	—
Rufous-bellied rat	Lophuromys sikapusi	73	51	69.9
Brush-furred rat	Uranomys foxi	56	8	12.5
Giant rat	Cricetomys gambianus	24	—	—
Kemp's gerbil	Tatera kempi	8	—	—
	All species	490	101	20.6

earthworms. In the same study, Funmilayo examined stomachs of 11 species of bird and found that only two of them ate worms. These were the Glossy-backed drongo, *Dicrurus adsimilis*, and the Kurrichane thrush, *Turdus libonyanus*. Funmilayo concluded these vertebrate predators were unlikely to have much effect on numbers of earthworms.

Amongst the Carnivora, the Fanalouka, *Fossa fossa*, reputedly specializes on earthworms and has a correspondingly reduced dentition amongst comparable viverrids such as, for example, the Fossa, *Cryptoprocta ferox*. Of Mustelidae, the European badger, *Meles meles*, eats worms frequently (Chichikin, 1968; Skoog, 1970; Neal, 1977; Kruuk and Parish, 1981) and is behaviourally highly specialized as a predator upon *L. terrestris* (Kruuk, 1978a; Kruuk and Parish, 1981). Many of these specializations are shared by red foxes, *Vulpes vulpes* (Macdonald, 1980) and under similar circumstances both species have capture rates of *L. terrestris* of up to 10 worms per minute. The foraging behaviour of foxes and badgers hunting for earthworms has led Kruuk (1978a), Kruuk and Parish (1982) and Macdonald (1977, 1981) to argue that for these and other carnivores, the pattern of availability of earthworms and some other prey underlies the social organization of the predators. Furthermore, since badger population density seems to depend heavily on availability of earthworms, Kruuk *et al.* (1979) have suggested that badger numbers and movements might be influenced by grassland management which in turn affects availability of earthworms to badgers. Such manipulation might be relevant to the control of bovine tuberculosis (Cheeseman *et al.*, 1981).

The weasel, *Mustela cicognani*, is reported to feed its young on earthworms (Osgood, 1936). Earthworms are also seasonally important in the diet of raccoon, *Procyon lotor*, and opossum, *Didelphis marsupialis* (Dexter, 1951).

35.3 FORAGING FOR WORMS

The behaviour of blackbirds, *Turdus merula*, and song thrushes, *T. philomelos*, searching for earthworms on a lawn was investigated by Smith (1974a,b) in unparalleled detail. In the Botanical Gardens at Oxford he tape-recorded commentaries of the birds' foraging path, noting changes of direction, pauses, and failed and successful attempts to capture worms. Computer analyses of each commentary enabled him to reconstruct each bird's path and so to investigate changes in several parameters before and after a capture. Smith was primarily interested in the adaptive nature of searching behaviour and used his detailed commentaries to test the hypothesis that blackbirds follow a search tactic known as area-restricted searching (Tinbergen *et al.*, 1967; Croze, 1970). Area-restricted searching involves combing an area in greater detail having discovered prey there and is clearly adaptive when searching for prey which are patchy in distribution, as *L. terrestris* usually are. Smith showed that, following a capture, the thrushes changed their pattern of movements over the succeeding 12 turns so that they probably searched the area in the vicinity of the capture more thoroughly than they would otherwise have done. By giving the birds artificial prey, he compared their behaviour under known conditions of 'prey' availability. He found that when the thrushes discovered a patch of prey they began to make turns through a greater number of degrees, and were less inclined to alternate left and right turns, both of which changes would help to keep them within the patch of prey.

There are many similarities in the way in which red foxes and blackbirds search for earthworms: both move in a series of definable searches separated by pauses some of which are followed by changes in foraging path in a fairly well-defined direction. Since evidence for area-restricted searching had previously been studied only in birds and invertebrates (Banks, 1957; Murdie and Hassel, 1972) and since many prey of foxes are heterogeneous in their availability (Macdonald, 1981), I applied Smith's (1974a) methodology to foxes. The foxes were found and observed by night using a combination of radio-tracking, luminescent markers (beta lights) and night-vision equipment (Macdonald, 1978; Macdonald and Amlaner, 1980). A fox hunting for earthworms walks slowly ($0.4 \, \mathrm{m \, s^{-1}}$), constantly moving its ears and frequently pausing. When an earthworm is located, apparently from the scraping sound made by its chaetae, the fox pauses over it and then plunges its snout into the grass, grabs the worm with its incisors and swiftly raises its head. If the worm retains a hold on its burrow, the fox after a slight pause draws its head up in a slowly accelerating arc. Capture rates reached $10 \, \mathrm{min^{-1}}$, but were typically $2\text{--}5 \, \mathrm{min^{-1}}$. In some areas, earthworms caught in this way may comprise seasonally the bulk of the foxes' diet (Macdonald, 1977, 1980) and analysis of individual faeces shows that on given nights foxes feed almost exclusively upon *L. terrestris*. These faeces consist mainly of soil, derived from the earthworms' gut content, and when stained with picric acid, lignin pink or carbol fuchsin and examined under the microscope are seen to be heavily loaded with chitinous lumbricid chaetae and oesophageal collars.

The study of fox predation upon earthworms involved sampling to find where worms were most abundant, investigating the conditions under which (a) they were most available and (b) foxes foraged for them, and finally analysing the foxes' tactics while foraging for earthworms (Macdonald, 1980). Earthworm abundance was measured in each of ten fields by sampling with formalin extraction. Even neighbouring fields differed greatly in the abundance of worms which, in a July sample, varied from 0.5 to 14.5 *L. terrestris* $\mathrm{m^{-2}}$. This variation corresponded with the known effects of soil type and agricultural practice. In contrast, worm availability to the foxes, defined as the number of worms on the surface, was largely affected by microclimatic variables such as relative humidity, the number of hours which had elapsed since the last rainfall, and the temperature. Availability was studied by walking transects at midnight and, in the light of a red-screened torch, counting the *L. terrestris* on the surface. Figure 35.1 shows results for one such area. Such relationships were found in each sample area, although the slopes of relationships between worm availability and climatic factors varied according to the population density. These relationships were best described ($r = 0.73$, $P < 0.01$) by the expression:

$$\text{No. available worms } \mathrm{m^{-2}} = 10.31 - 0.3979(\mathrm{h}) - 1.045 \, (\text{delta } T)$$
$$+ \, 11.43 \, (\mathrm{RH}) - 0.7124 \, (\mathrm{mm})$$

where h = hours since rain, delta T = departure in degrees C from

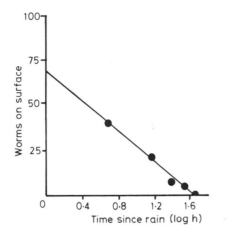

Fig. 35.1 Numbers of *L. terrestris* on the surface on 15 m² of a lawn in relation to time since rainfall.

10.5°C, RH = hours of over 95 % relative humidity, mm = rainfall (mm) during last shower.

Observations on the foraging behaviour of foxes showed that they were responsive to the abundance of available *L. terrestris*. Some 77.6 % of 121 different sightings of foxes foraging for earthworms during more than 92 nights were, out of the ten fields sampled, in the two which had the greatest worm abundance. Foxes were observed foraging for worms on 63 out of 76 rainy nights, defined as those on which it had rained during the previous 24 hours, but on only 29 out of 61 dry nights. Their capture rate on those nights when they hunted worms was influenced ($r = 0.39$, $P < 0.01$) by the same factors that largely determined worm availability: worms caught per min = 1.48 − 0.001 (h) + 0.009 (RH) − 0.003 (delta T) of the variables, the number of hours of more than 90 % relative humidity preceding the observation (RH) being the single most important determinant of capture rate.

Foxes foraging for earthworms varied in the way they moved; highly convoluted search paths were associated with a mean earthworm capture rate of 1.04 worms min⁻¹, whereas fewer worms were caught (0.63 worms min⁻¹) when they followed more meandering, on-going tracks. Figure 35.2(a) shows the track of a vixen which foraged in a small patch for about half an hour, catching 46 worms. There were no significant changes in direction or duration of movement or pauses before or after captures. Figure 35.2(b) shows that her captures were made in short sequences separated by pauses in which she listened for another worm or searched without success. By following convoluted tracks, a fox might forage within an area of no more than 25 × 25 m for 10–20 min. This constitutes searching in a restricted area but is not evidence of an adaptive search tactic since the limits of the foraging patch could be those of a restricted area of available prey. The proof of a search tactic lies in any change in the fox's behaviour which is not solely a direct response to each individual prey. To search for any such tactic I split each fox's track into a series of moves, direct distances and angles preceding and following each capture. A detailed comparison of these parameters at successive moves on either side of a capture revealed (Macdonald, 1980) that foxes, unlike Smith's blackbirds, were not altering their searching tactics on the basis of individual worm captures. Within a patch of worms on the surface, the foxes seemed to move back and forth in direct response to the sound of individual worms. However, to keep within patches of high earthworm availability they responded to clues as to the location of the entire patch.

Two clues contributed to the foxes' ability to find potential patches of high worm availability: (1) wind speed and direction and (2) the grazing pattern of horses. First, worm numbers on the surface were much reduced in windy conditions and foxes regularly foraged in the lee of copses and spinneys. Irrespective of their recent capture rate, foxes would move out of the wind and so restrict their search to the more rewarding leeward patches. Secondly, horses in the two fields with the highest worm population did not graze uniformly but avoided areas of the field where they defaecated. However, these dunged areas harboured significantly more worms. Where the horses dunged they did not graze and the grass grew so long that although the abundance of *Lumbricus* was high, even those on the surface were beyond the fox's reach. The farmer periodically cut this long grass and when the grass was short enough, the foxes methodically hunted the dunged areas. When the grass was longer, they hunted the edges of these grassy islands, a compromise between grass length and worm availability. These two factors prompted foxes to use area-restricted search tactics but there were some occasions when they foraged intensively for earthworms in small areas to the limits of which I could find no external cue.

Having found a patch, the fox faces the question of when to leave for another. The success rates of foxes foraging for earthworms changed from night to night. On nights when worms were scarce, foxes would persist within a patch at a success rate lower than that at which they would have left it on a night when worms were more abundant. Analysis of eight foraging tracks from the moment of the fox's arrival within a patch until its departure showed that the foxes invariably left a patch after their success rate fell below the average for that patch on that night (Fig. 35.3). Another analysis showed that foxes left a patch of earthworms and moved to another when their success in the first patch fell below the average over

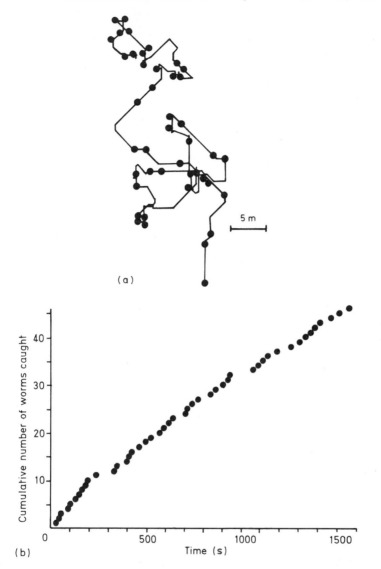

Fig. 35.2(a) Convoluted track of a vixen foraging in a small patch of pasture. Points (●) indicate worm captures. **(b)** Distribution of worm captures in time.

all patches that night, including travel time between them. These findings are broadly in accord with the theory of optimal foraging (Krebs, 1978).

Barnard and Stephens (1981) have studied the ways in which lapwings increase their feeding efficiency upon worms by flocking behaviour. By

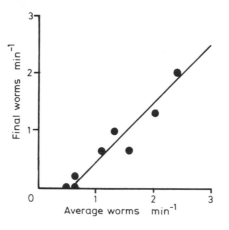

Fig. 35.3 Relationship between average and final capture rates within single patches.

recording detailed commentaries of the predators' behaviour, these authors established that lapwings in larger flocks tended to capture larger worms than did those in smaller flocks. They estimated earthworm length in lapwing bill length units and found that the size of captured worms increased rapidly with the increasing number of birds up to about 20. In attempting to calculate the profitability of foraging for earthworms of different sizes, Barnard and Stephens had to surmount the confusing fact that larger worms were more likely to break during withdrawal from the soil. They resolved this problem with an index which incorporated the calorific value of worms of different sizes, a factor calibrating the risk of breakage, and the proportion of the worm likely to be lost if it broke. They concluded that earthworms of 17–48 mm were more profitable than either longer or shorter ones. Lapwings in larger flocks were more efficient foragers partly because they selected for more worms of this size class. In large flocks they frequently adopt a crouching posture prior to pecking up a worm, seemingly enabling them to select the better size of worm. Crouching was more common in big flocks largely because big flocks were more common in areas of high worm availability where the lapwings could afford to be more selective. There may also be a reduced need in a big flock for continued vigilance by every bird for predators, allowing them more time for selective searching. When Black-headed gulls, *Larus ridibundus*, infiltrated the lapwing flock, foraging success fell. The gulls ate solely by pilfering worms from lapwings and were more likely to steal larger than smaller worms. Barnard and Stephens showed that when gulls were present, lapwings swallowed their worms more quickly. Lapwings crouch less frequently when gulls are in the flock, presumably because crouching informs the kleptoparasitic gulls of the likely capture of a large worm. Hence, lapwings eat smaller worms when Black-headed gulls are present.

The foraging behaviour of birds feeding on earthworms may have a significant impact on the dynamics of the ecosystem. Ashton's (1975) study of Australian *Eucalyptus regnans* forests provides an interesting example. The annual litter fall in some of these forests approaches 8 t ha^{-1} and although earthworms contribute to the incorporation of leaf material, lyrebirds, *Menura nova hollandiae*, foraging for earthworms and other invertebrates rake over the litter which thus becomes thoroughly mixed with the topsoil. Ashton estimated that the whole area in his study sites could be raked over once every two years, and concluded that the activities of this bird would assist in the rapid breakdown of litter and contributed materially to soil erosion on the steeper slopes.

The European badger, *Meles meles*, is a specialist upon *Lumbricus* species and forages for them in a way that closely parallels the behaviour of foxes (Kruuk, 1978a; Kruuk and Parish, 1981). There are fundamental similarities in the social organization of these two, often sympatric, species. Both occupy group territories which vary widely in size and live in social units which may involve several adult animals, each of which forages alone (Kruuk, 1978b; Macdonald, 1981). The spatio-temporally heterogeneous availability of the earthworm and other prey of both foxes and badgers has led to the parallel development of hypotheses to explain their social systems (Kruuk, 1978a; Kruuk and Parish, 1982; Macdonald, 1981, 1983).

Kruuk (1978a) found that adult badgers ate between 130 and 200 *L. terrestris* each night, and that, like foxes, they foraged for these worms within small patches on good nights and followed on-going paths when the weather was unsuitable for earthworm emergence. In spite of this specialization on *Lumbricus*, badgers took only about 5% of the estimated biomass of earthworms and in several study areas took roughly the same number of worms throughout the year (Kruuk and Parish, 1981). As weather conditions vary, worms become available in different patches. Kruuk hypothesized that a badger territory would require sufficient patches of potential worm availability to ensure that one would be fruitful on any given night. If earthworms are potentially available on both north and south faces of a hill, and if a chilling wind may blow from either the north or the south, then to ensure that its territory always encompasses a leeward slope irrespective of wind direction, a badger must defend both faces of the hill. However, worms are very abundant even in small patches of availability and, as estimated by Kruuk (1978a), one patch 20 × 20 m would feed more than 30 badgers for one night. The smallest economically defensible territory that would support a pair of badgers or similar predators would thus also contain enough resources to support additional individuals. Testing this hypothesis, Kruuk and Parish (1982) produced convincing evidence that the dispersion of potential patches of earthworm availability determines the size and configuration of badger territories, whereas the average abundance of available earthworms in those patches determines the group size within each territory.

35.5 REFERENCES

Arlton, A. V. (1936) An ecological study of the mole. *J. Mammal.*, **17**, 349–371.

Ashton, D. H. (1975) Studies of leaf litter in *Eucalyptus regnans* forests. *Austr. J. Bot.*, **23**, 413–433.

Banks, C. J. (1957) The behaviour of individual coccinellid larvae on plants. *Br. J. Anim. Behav.*, **5**, 12–24.

Barker, R. J. (1958) Notes on some ecological effects of DDT sprayed on elms. *J. Wildl. Mgmt.*, **22**, 269–274.

Barnard, C. J. and Stephens, H. (1981) Prey size selection by lapwings in lapwing/gull associations. *Behaviour*, **77**, 1–22.

Barrett-Hamilton, G. E. H. (1910) *A History of British Mammals*. Gurney and Jackson, London. 748 pp.

Bas Lopez, S., Rivera, J. S., Lorenzzo, A. de C. and Sandez Canals, B. L. (1979) Data on the feeding of *Salamandra salamandra* L., in Galicia, N.W. Spain. *Bol. Estac. Cent. Ecol.*, **8**, 73–78.

Bengtson, S. A., Nilsson, A., Nordström, S. and Rundgren, S. (1976) Effect of bird predation on lumbricid populations. *Oikos*, **27**, 9–12.

Bengtson, S. A., Rundgren, S., Nilsson, A. and Nordström, S. (1978) Selective predation on lumbricids by golden plover, *Pluvialis apricaria*. *Oikos*, **31**, 164–168.

Binder, N. (1971) Contributions to morphology, invasion dynamics and development of helminths of the blackbird (*Turdus merula* L.) *Zool. Beitr.*, **17**, 83–150.

Bolton, P. J. and Phillipson, J. (1976) Burrowing, feeding, egestion and energy budgets of *Allolobophora rosea* (Savigny) (Lumbricidae). *Oecologia (Berl.)*, **23**, 225–245.

Bradley, R. M. (1929) Habits and distribution of the rock squirrel in southern New Mexico. *J. Mammal.*, **10**, 168–169.

Catling, P. M. and Freedman, B. (1980) Food and feeding behaviour of sympatric snakes at Amherstburg, Ontario. *Can. Fld. Nat.*, **94**, 28–33.

Cheeseman, C. L., Jones, G. W., Gallagher, J. and Mallinson, P. J. (1981) The population structure, density and prevalence of tuberculosis (*Mycobacterium bovis*) in badgers (*Meles meles*) from four areas in south-west England. *J. Appl. Ecol.*, **18**, 795–804.

Chichikin, Y. N. (1968) The ecology and economic importance of the badger in the South of Kirgiziya. *Tr. Sary-chelekskogo Gos. Zapoved.*, **3**, 109–120.

Churchfield, J. S. (1979) A note on the diet of the European water shrew, *Neomys fodiens bicolor*. *J. Zool. (London)*, **188**, 294–296.

Clark, D. R. (1971) Ecological study of the worm snake, *Carphophis vermis*. (Kennicott). *Univ. Kans. Publs. Mus. Nat. Hist.*, **19**, 85–194.

Coleman, J. D. (1971) The distribution, numbers, and food of the rook, *Corvus frugilegus frugilegus* L. in Canterbury, New Zealand. *N.Z. J. Sci.*, **14**, 494–506.

Collinge, W. E. (1913) *The Food of some British Wild Birds: a Study in Economic Ornithology*. Dulau, London. 109 pp.

Croze, J. (1970) Searching image in carrion crows. *Z. Tierpsychol.*, 1–85.

Dawson, E. W. (1958) Food of young black-billed gulls *(Larus bulleri)* in a breeding colony, North Canterbury. *Notornis*, **8**, 1–7.

Delmee, E., Dachy, P. and Simon, P. (1979) Comparative study of a feeding regime of a forest population of tawny owls, *Strix aluco. Gerfaut*, **69**, 45–78.

Dexter, R. W. (1951) Earthworms in the winter diet of the opossum and racoon. *J. Mammal.*, **32**, 464.

Dimond, J. B., Belyea, G. Y., Kadunce, R. E., Getchell, A. S. and Blease, J. A. (1970) DDT residues in robins and earthworms associated with contaminated forest soils. *Can. Ent.*, **102**, 1122–1130.

Durchon, M. and Lafon, M. (1951) Quelques données biochimiques sur les annélides. *Ann. Sci. Nat., Zool.–11 Ser.*, **13**, 427–452.

Eadie, W. R. (1939) A contribution to the biology of *Parascalops breweri. J. Mammal.*, **20**, 150–173.

Evans, A. C. (1948) The identity of earthworms stored by moles. *Proc. Zool. Soc.*, **118**, 356–359.

French, C. E., Liscinsky, S. A. and Millar, D. R. (1957) Nutrient composition of earthworms. *J. Wildl. Mgmt.*, **21**, 16–19.

Funmilayo, O. (1979) Food consumption, preferences and storage in the mole. *Acta. Theriol.*, **24**, 379–389.

Funmilayo, O. (1981) Earthworms in the diet of small mammals and birds in Nigeria. Unpublished paper presented at the Darwin Centenary Symposium on Earthworm Ecology, Grange-over-Sands, Cumbria, UK.

Gregory, P. T. (1978) Feeding habits and diet overlap of three species of garter snakes *(Thamnophis)* on Vancouver Island. *Can. J. Zool.*, **56**, 1967–1974.

Hamilton, W. J. (1930) The food of the Soricidae. *J. Mammal.*, **11**, 26–39.

Hamilton, W. J. (1931) Habits of the star-nosed mole, *Condylura cristata. J. Mammal.*, **12**, 345–355.

Hamilton, W. J. (1941) The food of small forest mammals in Eastern United States. *J. Mammal.*, **22**, 250–263.

Harlin, J. (1977) Comparison of diets of the blackbird and the starling. *Zool. Listy*, **26**, 45–56.

Heppleston, P. B. (1971) The feeding ecology of oystercatchers, *(Haematopus ostralegus* L.) in winter in Northern Scotland. *J. Anim. Ecol.*, **40**, 651–672.

Hirons, G. (1976) A population study of the tawny owl *Strix aluco* and its main prey species in woodland. PhD thesis, Oxford University.

Holisová, V. (1967) The food of *Apodemus agrarius* (Pall). *Zool. Listy*, **16**, 1–14.

Holisová, V. and Obrtel, R. (1979) The food eaten by *Clethrionomys glareolus* in a spruce monoculture. *Fol. Zool.*, **28**, 219–230.

Holisová, V., Pekikan, J. and Zejda, J. (1962) Ecology and population dynamics in *Apodemus microps* (Krat. and Ros) (Mamm. Muridae). *Pr. Brn. Zakl. Čsl. Akad. Věd.*, **34**, 493–540.

Horner, B. E., Taylor, J. M. and Padykula, H. A. (1965) Food habits and gastric morphology of the grasshopper mouse. *J. Mammal.*, **45**, 513–535.

Ivanter, E. H. (1969) A contribution to the study of the mole *(Talpa europea europea* L.) in Karelia. *Uch. Zap. Petrozavodsk. Univ.*, **16**, 186–202.

Jefferies, D. J. and Davis, B. N. K. (1968) Dynamics of dieldrin in soil, earthworms, and song thrushes. *J. Wildl. Mgmt.*, **32**, 441–456.

Koral, N. F. (1976) Data on the ecology of the wryneck in the gardens of the Middle Dneiper. *Vestn. Zool.*, **4**, 87–90. [In Russian].

Krebs, J. R. (1978) Optimal foraging: decision rules for predators. In *Behavioural Ecology: an Evolutionary Approach* (eds. J. R. Krebs and N. B. Davies), Blackwell, Oxford, pp. 23–63.

Kruuk, H. (1978a) Foraging and spatial organisation of the European badger, *Meles meles* L. *Behav. Ecol. Sociobiol.*, **4**, 75–89.

Kruuk, H. (1978b) Spatial organisation and territorial behaviour of the European badger, *Meles meles* L. *J. Zool. (London)*, **184**, 1–19.

Kruuk, H. and Parish, T. (1981) Feeding specialisations of the European badger, *Meles meles* in Scotland. *J. Anim. Ecol.*, **50**, 773–788.

Kruuk, H. and Parish, T. (1982) Factors affecting population density, group size and territory size of the European badger, *Meles meles*. *J. Zool. (London)*, **196**, 31–39.

Kruuk, H., Parish, T., Brown, C. A. J. and Carrera, J. (1979) The use of pasture by the European badger (*Meles meles*). *J. Appl. Ecol.*, **16**, 453–459.

Landry, S. O. (1970) The Rodentia as omnivores. *Q. Rev. Biol.*, **45**, 351.

Lawrence, R. D. and Millar, H. R. (1945) Protein content of earthworms. *Nature (London)*, **155**, 517.

Lescure, J. (1966) The food of the common toad, *Bufo bufo* (Linnaeus 1758). *Vie Milieu*, **15**, 757–764.

Lesinch, K. P. and Melbarde, R. E. (1974) Development aspects of *Syngamus trachea*, as part of the earthworm development cycle. *Latv. P.S.R. Zinat. Akad. Vestis*, **11**, 34–37. [In Russian]

Liat, L. B. and Heyneman, D. (1968) A collection of small mammals from Tuaran and the southwest face of Mt. Kinabalu, Sabah. *Sarawak Mus. J.*, **16**, 257.

Ljungström, P. O. and Reinecke, A. J. (1969) IV. Studies on influences of earthworms upon the soil and the parasitological questions. Ecology and natural history of the microchaetid earthworms of South Africa. *Pedobiologia*, **9**, 152–157.

Logier, E. B. S. (1958) *The Snakes of Ontario*. University of Toronto Press.

Macdonald, D. W. (1976a) Nocturnal observations of tawny owls, *Strix aluco*, preying upon earthworms. *Ibis*, **118**, 579–580.

Macdonald, D. W. (1976b) Food caching by the red fox and other carnivores. *Z. Tierpsychol.*, **42**, 170–185.

Macdonald, D. W. (1977) *The behavioural ecology of the red fox, Vulpes vulpes: a study of social organisation and resource exploitation*. D.Phil. Thesis. Oxford University.

Macdonald, D. W. (1978) Radio tracking: some applications and limitations. In *Animal Marking: Recognition Marking of Animals in Research* (ed. B. Stonehouse), Macmillan, London, pp. 192–204.

Macdonald, D. W. (1980) The red fox, *Vulpes vulpes*, as a predator upon earthworms, *Lumbricus terrestris*. *Z. Tierpsychol.*, **52**, 171–200.

Macdonald, D. W. (1981) Resource dispersion and the social organisation of the red fox, *Vulpes vulpes*. In *Worldwide Furbearer Conference Proceedings* (eds. J. A. Chapman and D. Parsley), Maryland, USA, pp. 918–949.

Macdonald, D. W. (1983) The ecology of carnivore social behaviour. *Nature (London)*, **301**, 379–384.

Macdonald, D. W. and Amlaner, C. J. (1980) A practical guide to radio tracking. In *A Handbook on Biotelemetry and Radio Tracking* (eds. C. J. Amlaner and D. W. Macdonald), Pergamon Press, Oxford, pp. 143–159.

Mellanby, K. (1966) Mole activity in woodlands, fens and other habitats. *J. Zool. (London)*, **149**, 35–41.

Moeed, A. (1976) Birds and their food resources at Christchurch International Airport, New Zealand. *N.Z. J. Zool.*, **3**, 373–390.

Morris, P. (1963) *Biology of the Hedgehog*, Ph.D. thesis. London University.

Murdie, G. and Hassel, M. P. (1972) In *The Mathematical Theory of the Dynamics of Biological Populations* (eds. M. S. Bartlett and R. W. Hiorns), Academic Press, London and New York, pp. 87–107.

Neal, E. G. (1977) *Badgers*. Blandford Press, Poole.

Obrtel, R. (1973) Animal food of *Apodemus flavicollis* in a lowland forest. *Zool. Listy*, **22**, 15–30.

Obrtel, R. and Holisová, V. (1979) The food eaten by *Apodemus sylvaticus* in a spruce monoculture. *Fol. Zool.*, **28**, 299–310.

Obrtel, R. and Holisová, V. (1980) The food eaten by *Apodemus flavicollis* in a spruce monoculture. *Fol. Zool.*, **29**, 21–32.

Okhotina, M. (1969) Some data on the ecology of *Sorex (Ognevia) mirabilis* (Ognev, 1937). *Acta Theriol.*, **14**, 273–284.

Osgood, F. C. (1936) Earthworms as a supplementary food of weasels. *J. Mammal.*, **17**, 64.

Raw, F. (1966) The soil fauna as a food source for moles. *J. Zool. (London)*, **149**, 50–54.

Reeve, N. (1982) Radio tracking studies of the hedgehog. In *Radio-Tracking of Vertebrates* (eds. C. L. Cheeseman and R. Mitson), Proc. Symp. Zool. Soc. London, (in press).

Reynolds, J. W. (1977a) The earthworms (Lumbricidae and Sparganophilidae) of Ontario. *Royal Ontario Museum*, 141 pp.

Reynolds, J. W. (1977b) Earthworms utilized by the American woodcock. *Proc. Woodcock Symp.*, **6**, 161–169.

Reynolds, J. W. (1977c) Earthworm populations as related to woodcock habitat usage in Central Maine. *Proc. Woodcock Symp.*, **6**, 135–146.

Rosevear, D. R. (1969) The Rodents of West Africa. *Br. Mus. Publ.*, No. 677.

Rudge, M. R. (1968) The food of the common shrew, *Sorex araneus* L. (Insectivora: Soricidae) in Britain. *J. Anim. Ecol.*, **37**, 565–581.

Satchell, J. E. (1967) Colour dimorphism in *Allolobophora chlorotica* Sav. (Lumbricidae). *J. Anim. Ecol.*, **36**, 623–630.

Sheldon, W. G. (1967) *The Book of the American Woodcock*. University of Massachusetts Press, Amherst. 227 pp.

Shilova, S. A., Denisova, A. V., Dmitriev, G. A., Voronova, L. D. and Gardier, M. N. (1971) Effect of some insecticides upon the common mole. *Zool. Zh.*, **50**, 886–892.

Shortridge, G. C. (1934) *The Mammals of South West Africa*. Heinemann, London. 779 pp.

Shrubb, M. (1980) Farming influences on the food and hunting of kestrels. *Bird Study*, **27**, 109–115.

Sibly, R. and McCleery, R. H. (1980) Continuous observation of individual herring gulls during the incubation season using radio tags. In *A Handbook on Biotelemetry and Radio Tracking* (eds. C. J. Amlaner and D. W. Macdonald), Pergamon, Oxford, pp. 345–352.

Sibly, R. M. and McCleery, R. H. (in press) The distribution between feeding sites of herring gulls breeding at Walney Island, UK, *J. Anim. Ecol.*

Skoczen, S. (1965) Stomach contents of the mole, *Talpa europaea*, L. 1758, from southern Poland. *Acta Theriol.*, **11**, 551–575.

Skoczen, S. (1970) Food storage of some insectivorous mammals (Insectivora). *Przegl. Zool.*, **14**, 243–248.

Skoog, P. (1970) The food of the Swedish badger, *Meles meles* L. *Viltrevy (Stockholm)*, **7**, 1–120.

Smith, J. N. M. (1974a) The food searching behaviour of two European thrushes. I. Description and analysis of search paths. *Behaviour*, **48**, 276–302.

Smith, J. N. M. (1974b) The food searching behaviour of two European thrushes. II. The adaptiveness of search patterns. *Behaviour*, **49**, 1–61.

Smith, M. (1951) *The British Amphibians and Reptiles*. Collins, London. 318 pp.

Southern, H. N. (1969) Prey taken by tawny owls during the breeding season. *Ibis*, **111**, 293–299.

Southern, H. N. and Corbett, G. B. (eds.) (1977) *The Handbook of British Mammals*. 2nd edn, Blackwell Scientific Publ., Oxford. 520 pp.

Steward, J. W. (1971) *The Snakes of Europe*. Fairleigh Dickinson, Newton Abbot, 238 pp.

Stickel, W. H., Hayne, D. W. and Stickel, L. F. (1965) Effects of heptachlor-contaminated earthworms on woodcocks. *J. Wildl. Mgmt.*, **29**, 132–146.

Tinbergen, N., Impekovan, M. and Frank, D. (1967) An experiment on spacing out as a defence against predation. *Behaviour*, **28**, 307–321.

Walton, W. R. (1928) Earthworms as pests and otherwise. *USDA Farmers' Bull.* No. 1569, 14 pp.

Watts, C. H. S. (1968) The foods eaten by wood mice *(Apodemus sylvaticus)* and bank voles *(Clethrionomys glareolus)* in Wytham Woods, Berkshire. *J. Anim. Ecol.*, **37**, 25–41.

Whitaker, J. O. (1966) Food of *Mus musculus, Peromyscus maniculatus bairdi* and *Peromyscus leucopus* in Vigo County, Indiana. *J. Mammal.*, **47**, 473–486.

Whitaker, J. O. and Ferraro, M. G. (1963) Summer food of 220 short-tailed shrews from Ithaca, New York. *J. Mammal.*, **44**, 419–422.

Whitaker, J. O. and Mumford, R. E. (1972) Food and ectoparasites of Indiana shrews. *J. Mammal.*, **53**, 329–335.

Whitaker, J. O. and Schmeltz, L. L. (1973) Food and external parasites of *Sorex palustris* and food of *Sorex cinereus* from St. Louis County, Minnesota. *J. Mammal.*, **54**, 283–285.

Yalden, D. W. and Warburton, A. B. (1979) The diet of the kestrel in the Lake District. *Bird Study*, **26**, 163–170.

Yerger, R. W. (1955) Life history notes on the eastern chipmunk *Tamais striatus lysteri* (Richardson), in central New York. *Am. Midl. Nat.*, **53**, 312–323.

Chapter 36

Predation on earthworms by the Black-headed Gull (*Larus ridibundus* L.)

G. CUENDET

36.1 INTRODUCTION

This study, carried out in Switzerland, aimed to provide information for numerous farmers who, fearing that predation by Black-headed Gulls might significantly decrease earthworm populations in their cultivated fields, had sought permission to shoot the gulls. Earlier studies on the topic were only qualitative and were done during a period when earthworms seemed less important in the gulls' diet (e.g. Jirsik, 1945; Creutz, 1963; Schlegel, 1977).

36.2 THE BLACK-HEADED GULL IN THE CANTON OF VAUD

The Black-headed Gull is a gregarious bird which breeds, feeds and roosts in groups. It is able to exploit numerous environments and its diet is very varied. It is migratory and in Central Europe moves southwest after breeding.

Since the end of the last century, its wide ranging adaptability has allowed it to use a large number of new environments created or transformed by man, hence there has been considerable expansion in both its numbers and its range.

In the Canton of Vaud in the west part of Switzerland, a great extension of its foraging area over the countryside occurred during the 1960s and 1970s and it is now common over cultivated fields far from the lakes. Possible causes of this extension include increases in the size of cultivated fields and associated advances in agricultural mechanization. During August to November or December, when the soils begin to freeze, most Black-headed Gulls forage over the countryside, taking advantage of ploughing and other soil preparations. Figure 36.1 shows the foraging area during this period. Gulls are present almost daily in areas which are, characteristically, quite flat regions of intensive agriculture, with few hedges or small woods. Areas where presence is occasional are characterized by less intensive agriculture, higher altitude and more varied ground. By regularly watching Black-headed Gulls flying to the Lake of Geneva (Lac Leman) to roost, it was possible to obtain an estimate of the numbers foraging. Over a region around Echallens with about 2210 ha of ploughed fields and 1570 ha of permanent or non-permanent meadows, about 6000 Black-headed Gulls were foraging at the end of September. These observations made it possible to assess the impact of the gulls on earthworm populations.

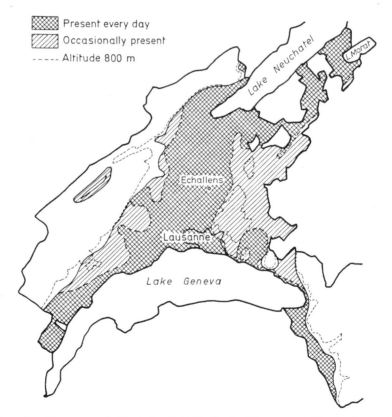

Fig. 36.1 Presence of Black-headed Gulls in Canton of Vaud, August to November (1976–1977).

36.3 EARTHWORM POPULATIONS IN CULTIVATED FIELDS

Earthworm populations in cultivated fields were studied by the sampling method of Bouché (1972), which uses formalin sprinkling followed by digging, handsorting and wet sieving. In the principal sites, samples were taken at three locations. Sites between them were sampled later during ploughing and harrowing. Earthworm biomass varied between 1000 and 2000 kg ha^{-1} (Fig. 36.2).

Several studies have assessed the damage done to earthworms by ploughing and harrowing, and it is clearly established that ploughing permanent meadows produces a decrease of earthworm biomass (Evans and Guild, 1948). The present study tried to assess the impact of ploughing and harrowing on earthworms brought to the surface and thus made available to Black-headed Gulls.

By following immediately behind the plough, 46 samples of 100 earthworms were collected, of which an average of $31 \pm 5\%$ were damaged. Five of these samples were kept in large pots, each containing 25 individuals. After one or two weeks, mortality was recorded (Table 36.1). Taking account both of the effect of the wheels of the tractor passing back in the furrow and a possible over-assessment of the mortality in the experimental conditions, it is reasonable to assess the mortality due to ploughing at $25 \pm 5\%$.

The additional effects of harrowing are more difficult to measure because fewer earthworms are brought to the surface; but the effects seem to be slight because, after ploughing, most of the surviving earthworms go deep enough to escape being damaged.

Sprinkling of liquid manure (slurry) over meadows was studied by Nebiker (1974). Liquid manure is usually rich in ammonia (NH_3) and

Table 36.1 Mortality after ploughing.

Date	Vegetation	Soil	Percentage damaged	Percentage dead after (n) days
24.09.76	Meadow	Loam	?	28 (7)
19.09.78	Meadow	Loam, gravel	40	27 (12)
05.10.78	Meadow	Loam, gravel	40	28 (13)
09.10.78	Meadow	Loam, gravel	41	19 (11)
24.10.78	Cereal stubble	Loam	27	18 (13)

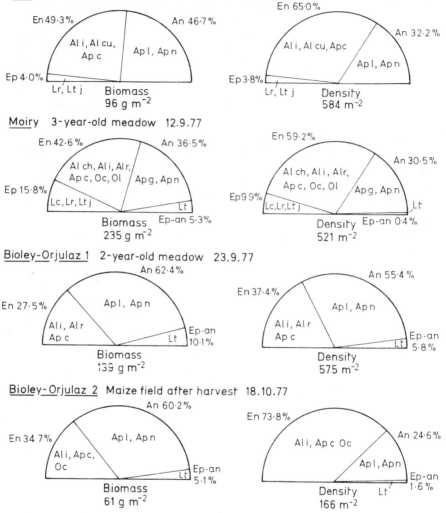

Fig. 36.2 Earthworm populations and biomass (wet wt.) in cultivated fields.
Key:

Ep, Epigeic species:

Lc,	*Lumbricus castaneus*	
Lr,	*Lumbricus rubellus*	
Lt j,	*Lumbricus terrestris* small juveniles	

En, Endogeic species:

Al ch,	*Allolobophora chlorotica*	
Al cu,	*Allolobophora cupulifera*	
Al i,	*Allolobophora icterica*	
Al r,	*Allolobophora rosea*	
Ap c,	*Aporrectodea caliginosa*	
O c,	*Octolasion cyaneum*	
O l,	*Octolasion lacteum*	

hydrogen sulphide (H_2S) produced by anaerobic fermentation, and sprinkling it over the soil causes some earthworms to come to the surface and die. From Nebiker's observations, it is possible to estimate this biomass at between 10 and 20 m^{-2}. Black-headed Gulls often feed on these worms.

36.4 PREDATION BY THE BLACK-HEADED GULL

Predation by the Black-headed Gull on earthworm populations was studied by observing feeding gulls through binoculars (36.4.1); analyses of stomach contents (36.4.2); observing gull behaviour during ploughing and harrowing (36.4.3); estimating earthworm biomasses brought to the surface by the farm machinery and made available to the gulls (36.4.4); assessment of the gulls' daily food requirements to permit an estimate of the biomass of worms needed to meet it (36.4.5).

36.4.1 Direct observations

Direct observations provided estimates of earthworm quantities eaten in different kinds of foraging area. It was possible to record the ingestion of all the large to middle-sized earthworms, and the weight of worms eaten was estimated from the observed rate of feeding and the mean weight of adult worms obtained by sampling the field population, assessed as 1.5 g. During liquid manure sprinkling, an average of 4.3 g of earthworms were eaten per minute per individual. Averages of 0.9 and 0.5 g min^{-1} individual^{-1} were recorded respectively on meadows and fields where people were not working. A gull, which was observed for 2 hours, foraging by low flying over fields, ate about 28 g h^{-1} so it appears that the biomass of earthworms eaten by gulls may be quite important.

36.4.2 Stomach content analyses

Ninety-five Black-headed Gulls were shot in the study area and stomach content analyses showed that they were eating mostly earthworms. Worms were present in all individuals, insects in 75 %, cereals in 31 % and other seeds in 46 % (Fig. 36.3). Expressed in terms of dry weight biomass, earthworms constituted the bulk of the intake, some 92 % of the total

An, Anecic species:	Ap g,	*Aporrectodea giardi*
	Ap l,	*Aporrectodea longa*
	Ap n,	*Aporrectodea nocturna*
Ep-an, Epi-anecic species,	L t,	*Lumbricus terrestris*

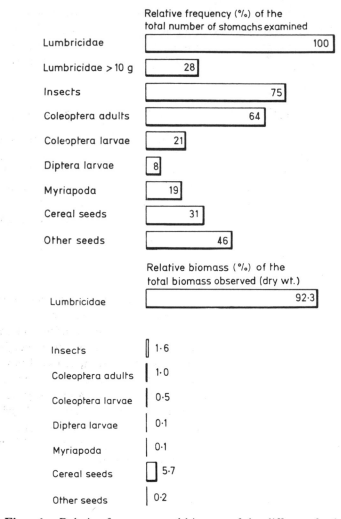

Fig. 36.3 Relative frequency and biomass of the different food items.

observed food. In fact the importance of earthworms is slightly under-estimated by this method, because they are more rapidly digested than insects (Cuendet, 1979). The maximal quantities of earthworms (about 40 g) were recorded in birds that had foraged behind a plough or during liquid manure sprinkling.

Although insects were present in 75 % of the stomach contents, they represented only 1.6 % of the observed food. They were mostly Coleoptera living or moving on or in the soil. Another important food was cereal seeds, about 6 % of the observed food, apparently picked up over stubble. In some stomachs, it proved possible to identify the earthworms to species or at least to ecological category. It was therefore possible to see that the importance of these different categories varies according to the area foraged. The χ^2 test applied to the proportions of the categories in the sites illustrated in Fig. 36.4 showed that they were significantly different.

1. Gulls foraging over meadows are able to eat only earthworms which are moving over the soil or are active very near the surface, such as *Lumbricus castaneus*, *L. rubellus* and immature *L. terrestris*. These are relatively unimportant in cultivated soils where anecics and endogeics are more abundant. The stomachs of gulls foraging during ploughing and harrowing contained earthworm species quantitatively representative of the populations of cultivated soils, but in those of gulls foraging over meadows, endogeic and anecic species were proportionately fewer than in the soil.

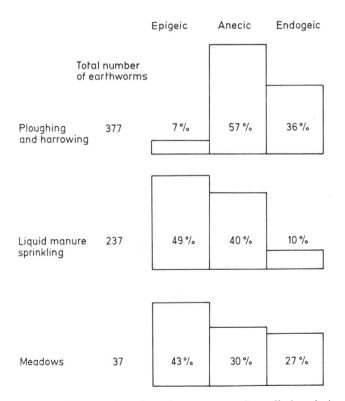

Fig. 36.4 Ecological categories of earthworms eaten by gulls in relation to the type of foraging area.

2. Liquid manure sprinkling particularly affects epigeic worms near the surface and anecics and epi-anecics whose vertical burrows allow the liquid to penetrate deep soil. The endogeic worms are less influenced by the toxic liquid and, being more lucifugous, come less readily to the soil surface.

36.4.3 Presence of Black-headed Gulls during cultivation

Farmers from two areas were asked if Black-headed Gulls were present during ploughing and harrowing. Two-thirds of the replies were positive. However, Black-headed Gulls followed the machines on average in only about one-third of the area of the fields being ploughed or harrowed (Table 36.2). There are two reasons for this. First, earthworm biomass available after ploughing and harrowing is much greater than the immediate needs of the gulls. Second, hedges or woods are present in some fields and appear to be disliked by the gulls.

Table 36.2 Predation during cultivation.

Region	Cultivation	Total area (ha)	Percentage of area with gulls present	Percentage of area with gulls behind the machines
Bioley-Orjulaz	Ploughing	83	66	32
Bioley-Orjulaz	Harrowing	96	64	30
Rennaz	Ploughing	53	74	35
Rennaz	Ploughing	77	66	29

36.4.4 Earthworm biomasses available during cultivation

Earthworms were sampled during cultivation in areas where the total biomass had been assessed a few days prior to agricultural operations. Earthworm biomass brought to the surface by ploughing formed 5–10% of the total in the soil (Fig. 36.5). Harrowing uncovered a further 0.2–2.5%. Thus, if gulls systematically follow all cultivation in a field, they are presented with less than 13% of the total earthworm biomass. A quarter of these earthworms would die anyway because of injury caused by the machinery.

36.4.5 Assessment of food requirement

Energy requirements of captive Black-headed Gulls (535 kJ day^{-1} individual^{-1}) and data from other studies (Jirsik, 1945; Creutz, 1963;

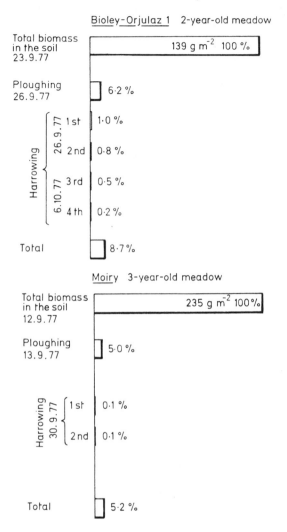

Bioley-Orjulaz 1   2-year-old meadow

Fig. 36.5 Earthworm biomasses available for gulls during cultivation.

Landry, 1978) indicate that, if Black-headed Gulls meet 90% of their
energy needs with earthworms, each gull will require 150–200 g of worms
per day. This makes it possible to estimate the total demand for
earthworms in a region where the number of foraging gulls is known.

In a region around Echallens, during the autumn, ploughing, harrowing and liquid manure sprinkling alone make available a quantity of
earthworms which must approximate 240 t. The estimated demand by the
Black-headed Gulls is 65–85 t. The supply must therefore exceed the

demand by a factor of 3–4, higher if account is taken of food collected other than behind the machines.

36.5 CONCLUSION

In the Canton of Vaud, earthworms are an important part ($\sim$ 90% by weight) of the diet of Black-headed Gulls, and modern cultivation makes them readily available to gulls in autumn and winter when other foods are scarce. The gulls eat only a small fraction, a quarter to a third, of the worms made available by cultivation which is itself only about 10% of the biomass. It is hence most unlikely that predation by Black-headed Gulls appreciably reduces earthworm biomass in the soil.

36.6 REFERENCES

Bouché, M. B. (1972) Lombriciens de France. Ecologie et systématique. I.N.R.A. *Ann. Zool. Ecol. Anim.*, no. spécial, **72–2**, 671 pp.

Cuendet, G. (1979) *Étude du comportement alimentaire de la Mouette rieuse (Larus ridibundus L.) et de son influence sur les peuplements lombriciens.* Thèse de doctorat. Conservation de la faune et Section protection de la nature et des sites du canton de Vaud. 111 pp.

Creutz, G. (1963) Ernährungsweise und Aktionsradius der Lachmöwe (*Larus ridibundus* L.). *Beitr. Vogelk.*, **9**, 3–58.

Evans, A. C. and Guild, W. J. McL. (1948) Studies on the relationship between earthworms and soil fertility. V. Field populations. *Ann. Appl. Biol.*, **35**, 485–493.

Jirsik, S. C. (1945) The importance of the Black Headed Gull (*Larus ridibundus* L.) in the economy of agriculture and fisheries. *Acta Soc. Sci. Nat. Moravosiles.*, 17, 28 pp. [In Czech.]

Landry, P. (1978) *Modélisation et comparaison de l'impact énergétique de deux populations francaises de Mouette rieuse (Larus ridibundus L.)* Thése de 3e cycle. 117 pp.

Nebiker, H. (1974) Neues Verfahren zur Aufbereitung von Flüssigdünger. *Schweiz. Landw. Mh.*, **52**, 57–87.

Schlegel, R. (1977) Zur Nahrung der Lachmöwe an Oberlausitzer Karpfenteichen. *Der Falke*, **24**, 198–206.

Chapter 37

Agastrodrilus Omodeo and Vaillaud, a genus of carnivorous earthworms from the Ivory Coast

P. LAVELLE

37.1 INTRODUCTION

Agastrodrilus has been described as a genus of the Benhaminae with calciferous glands in XV-XVI-XVII, rudimentary oesophageal gizzards, intestinal caeca, posterior displacement of the ovaries, clitellum attenuated through numerous segments, lack of penial setae and great development of setae *ab*. This genus appears very close to *Millsonia*, a very widespread genus in this region; the only differences are the clitellum extension (no more than seven segments in *Millsonia* against 23 in *A. opisthogynus* and 31 in *A. multivesiculatus*), the extreme reduction of intestinal gizzards and development of setae *ab*. Jamieson (1971) placed *Agastrodrilus* in his tribe Dichogastrini (in which *Millsonia* was unequivocally placed), in the subfamily Megascolecidae, but queried this placement owing to lack of data on the nephridial system.

Since then, two new species have been discovered that can be related to this genus: *A. dominicae* in central Ivory Coast (Lavelle, 1981) and a species from close to the Voltaic boundary here referred to as *A.* sp.

These species do not present all the characters of the genus as they have less rudimentary gizzards than the type species and the clitellum on only seven segments. The two additional species have been shown to be carnivorous, feeding, at least partially, on Eudrilidae of the species *Stuhlmannia porifera* Omodeo and Vaillaud.

An individual of *A. dominicae* was observed while ingesting a small eudrilid earthworm. The earthworms collected for a sampling programme had been placed in a Petri dish with some water before being weighed and put in formalin for identification. We did not observe the capture of the small worm, but only the end of its ingestion. This was very rapid as the anterior half, measuring several centimetres, disappeared in less than a minute.

Attempts to repeat this observation were unsuccessful but the result of another experiment tends to confirm that *A. dominicae* is really specialized in earthworm predation. Two individuals were placed in culture with 36 small eudrilid worms from the species *Chuniodrilus zielae* and *Stuhlmannia porifera*. After 24 days, only six of these worms were found again and some of them had part of the caudal region amputated. The two adult *A. dominicae* had increased their weight from respectively 0.30 and 0.50 g to 0.50 and 0.75 g. At the same time, the density of small eudrilids in a control culture had passed from 18 to 16 individuals.

The carnivorous behaviour of *A.* sp. has been confirmed also by the observation of a well-recognizable eudrilid worm in the fore-gut of one individual. Such remains will not be found commonly, however, as the digestion of the prey is very rapid leaving only fragments of chetae and the soil that was contained in the gut. It is interesting to note that a little eudrilid earthworm ingested accidentally by an individual of *Millsonia lamtoiana*, a litter-feeding species of the same savannas, was found intact in the posterior part of the gut (Kanyonyo, 1981); this could mean that the rapid digestion of small worms in the gut of *Agastrodrilus* is the result of a digestive adaptation.

A. opisthogynus and *A. multivesiculatus* have not yet been shown to be carnivorous but they present the same adaptive characters; some of them more developed than in the two more recently discovered species. Moreover, it has never been possible to rear them in laboratory cultures in which all the geophagous species from other genera exhibited normal growth and reproduction (Lavelle, 1978).

It thus appears that *Agastrodrilus* has adaptive characters that enabled it to become an earthworm predator. This seems now to allow validation of this genus, even if some of these characters (camphor-smelling mucus, setal enlargement, rudimentary gizzards and different behaviour in cultures) are not among those generally considered as decisive.

37.2 ADAPTATIONS TO SMALL EARTHWORM PREDATION

Examination of morphological, anatomical, physiological and ethological characters suggest that these earthworms are adapted for moving fast in the soil and for searching for prey, utilizing a highly nutritious and smooth food and avoiding cannibalism (Table 37.1).

These worms move fast in the soil with the aid of strong *ab* setae,

Table 37.1 Evolution of some characters from *Millsonia* to the different species of *Agastrodrilus*.

	Millsonia spp.	*Agastrodrilus* sp.	*A. dominicae*	*A. opisthogynus*	*A. multivesiculatus*
1. Setae *ab*	Normal	Strong	Strong	Strong	Strong
2. Clitellum extension (no. of segments)	up to 7	7	7	21	32
3. Camphor-smelling mucus	Absent	Absent	Present in young	Young & adults	Young & adults
4. Thick funnel-shaped septa	Generally between 6 & 10	7	9	8	14
5. Gizzards	Generally two strong ones	Reduced	Reduced	Absent	Absent
6. Intestinal caeca	2 to 24 according to size	15	19	19	24
7. Typhlosole	Present	Absent	Absent	Absent	Absent

especially well developed in the first thirty and the last five segments. They also have a very geodynamic shape with a sharp anterior part unlike the typical geophagous species whose anterior part is swollen because the gizzard and clitellum are voluminous. *Agastrodrilus*, on the contrary, has rudimentary gizzards and the clitellum is always very long and correspondingly slender even when it occupies only seven segments as in *A. dominicae* and *A.* sp. Moreover, this clitellum is only apparent in the reproductive period and soon disappears after cocoon laying, thus minimizing interference with locomotion.

The funnel-shaped septal musculature also allows fast movements in the soil. It is strongly developed as the gizzards are very small and, as in the case of *A. multivesiculatus*, the reproductive system is displaced posteriorly, allowing the number of funnel-shaped septa to increase from eight to fourteen. In culture, these worms do not burrow by ingestion as do all the geophagous ones; they move very fast in the soil without ingesting it. This is compatible with predatory behaviour.

Other characters show adaptations for feeding on smooth and energy-rich foods: the reduction of the gizzards, which begins in *A.* sp and is almost total in *A. opisthogynus* and *A. multivesiculatus*, and the absence throughout the genus of a typhlosole. In contrast, the number of intestinal caecae appears relatively greater in *Agastrodrilus* than in *Millsonia*. Sims (1965) showed that the number of intestinal caecae seems to be correspondingly greater, the larger the species: *M. hemina*, the smallest *Millsonia*, has two pairs of caecae whereas the biggest ones, *M. lamtoiana* and *M. caecifera* which reach a live weight of 20 to 30 g, have respectively 19 and 24 pairs, the same number as *Agastrodrilus* species which weigh from 0.60 to 3 g fresh weight.

A further noteworthy character is the camphorous smell of the mucus. This is found in both adult and young worms of *A. opisthogynus* and *A. multivesiculatus*, only in the young of *A. dominicae*, and is lacking in *A.* sp. This is tentatively interpreted as an olfactory signal to avoid cannibalism.

Thus it seems that *Agastrodrilus* has separated from *Millsonia* by progressive adaptation to small earthworm predation. Four stages in this adaptation may be discerned in the known species although it is not suggested that they form a direct lineage. The first step is represented by *A.* sp. which presents strong *ab* setae, reduced gizzards and no typhlosole; the second by *A. dominicae* whose young have a camphor-smelling mucus; the third by *A. opisthogynus* with a total reduction of gizzards and enlargement of the clitellum to 23 segments; and the last by *A. multivesiculatus* which has acquired intercalary segments and with this acquisition has increased the number of strong funnel-shaped septa and has extended the clitellum.

37.3 CONCLUSION

It is concluded that the genus *Agastrodrilus* has been derived from *Millsonia* by progressive adaptation to small earthworm predation. Different steps in this adaptation can be seen in the four known species.

In cultures these species cannot grow when fed on soil even if it is relatively rich in organic matter; we need to know more now about their real diet. It is exceptional to see a recognizable worm in the gut of these species as ingested worms seem to be very quickly digested. Remains of setae can sometimes be identified with soil from the gut of the prey.

As it is difficult to recognize remains of ingested prey these earthworms have long been considered as oligohumic geophages (Lavelle, 1978). It is also possible, if not probable, that in their forward movement they ingest soil which is not always smooth enough to allow displacement only by force.

37.4 ACKNOWLEDGEMENTS

This work was supported by CNRS (LA no. 258 and ATP 'Ecosystemès') and Ecole Normale Supérieure de Paris. I am most grateful to Dr J. E. Satchell and Dr B. G. M. Jamieson for valuable criticism and revision of the text.

37.5 REFERENCES

Jamieson, B. G. M. (1971) A review of the megascolecoid earthworm genera (Oligochaeta) of Autralia. I: Reclassification and checklist of the megascolecoid genera of the world. *Proc. R. Soc. Wd.*, **82**, 75–86.

Kanyonyo, J. (1981) Étude préliminaire du régime alimentaire du ver de terre detritivore *Millsonia lamtoiana* (Acanthodrilidae – Oligochétes) dans la savane de Lamto (Côte d'Ivoire). DEA, Paris, Multigraphié, 46 pp.

Lavelle, P. (1978) *Les vers de terre de la savane de Lamto (Côte d'Ivoire): peuplements, populations et fonctions dans l'ecosysteme.* Thèse Doctorat, Paris VI. *Publ. Lab. Zool. ENS*, **12**, 301 pp.

Lavelle, P. (1981) Un ver de terre carnivore des savanes de la moyenne Cote d'Ivoire: *Agastrodrilus dominicae* nov. sp. (Oligochètes-Megascolecidae). *Rev. Écol. Biol. Sol*, **18**, 253–258.

Sims, R. W. (1965) Acanthodrilidae and Eudrilidae (Oligochaeta) from Ghana. *Bull. Br. Mus. Nat. Hist.*, **D12**, 285–311.

Chapter 38

The establishment of earthworm communities

M. B. BOUCHÉ

38.1 INTRODUCTION

The communities, roles and morphology of present-day earthworms are the fruits of a long period of evolution: present abilities or inabilities to perform certain functions or to saturate available niches are the expression of a genome inherited from their past. This chapter attempts to trace the origins of present earthworm communities from geological times.

38.2 THE DATA BASE

38.2.1 Fossils

Annelids consist mostly of soft tissues and usually decay quickly, leaving no fossils. They have chitinous setae capable of preservation but these are not peculiar to annelids and are difficult to interpret. With one recorded exception (Schwert, 1979), earthworm cocoons too are not fossilized. Traces of burrows and other marks of annelid activity survive but their origins are incapable of proof. Palaeomoders and particularly palaeomulls indicate the evolution of whole ecosystems similar to present-day types in which earthworms play a major role – nevertheless it is possible to imagine other animals playing the same role in the past.

38.2.2 Morphology

Morphology, based on a genetic code inherited from the past, provides evidence of adaptation to function in present-day environments. It provides a basis for classifications which, though arbitrary at first, may later be used to indicate phylogenetic relationships. The

431

morphological–taxonomic procedure has led to a double systematics in which some taxa can be interpreted phyletically and others cannot (Bouché, 1972). Because similar characters indicate similar habits, it is possible on a morpho-functional basis independent of taxonomy to group species sharing more or less the same role or niche into ecological types (Bouché, 1971, 1977). Such ecological types show r (epigeics) and K (anecics, most endogeics) selection (Satchell, 1980) but also adaptations specific to soil life. Recognition of characters appropriate to each ecological type thus helps to distinguish convergent characters from both apomorphic (recent) and plesiomorphic (primitive) characters. Conversely, a phyletic interpretation, when possible, may help to explain the distribution of present ecological types. Interpretation of both phyletic and ecological data is a continuing process; nothing is definitive.

38.2.3 Biogeography

Observed distribution

The distribution of animals is independent of morphological or fossil data but the value of present distribution data is dependent on the quality of sampling. Only France has been sampled methodically. For other countries information is copious (Hungary, Belgium, Norway) or good (USSR, Italy, UK, Germany, Sweden, Australia). Elsewhere, some zones are fairly well documented while others are practically ignored. Even when information is copious, the density of sampling may be sufficient to establish the presence of species but not their absence. In France, a small territory, general though still incomplete sampling revealed three new families or subfamilies and numerous species new to science. This state of knowledge is general.

Present-day distribution is the result of three parameters: the origin of the founders, their migratory capacity and their ability to survive.

Ability to migrate

Endogeic earthworms have, in general, a poor capacity to migrate. Some very fragile species living in deep soil layers (oligohumic type of Lavelle) are very sensitive to desiccation and are unable to leave the soil. Other less typically endogeic species have a limited capacity for surface migration (Mazaud and Bouché, 1980) and may be caught in pitfall traps (Boyd, 1960; Bouché, 1976). Epigeic worms living in semi-permanent litter or organic matter accumulations are comparatively mobile and search on the surface for food or shelter. 'Anecic' worms are intermediate. Some leave their burrows during heavy rainfall and are able to migrate and colonize a few metres per year (Mazaud and Bouché, 1980).

For most earthworms, migration is only by dispersal into adjoining land capable of supporting them. There is no active dispersal, but passive migration has distributed some species over wide areas. Species which are active on the surface during rainfall are transported by running water as are cocoons and worms from eroded soil (Schwert and Dance, 1979). Water dispersal, leading to a pseudo-riparian distribution, is limited by watersheds which form migration barriers. Most megadrile worms cannot survive marine environments (Piearce and Piearce, 1979) but some genera, e.g. *Pontodrilus, Rhododrilus* (Jamieson, 1980), *Microscolex*, secondarily adapted to sea water, have been able to colonize oceanic islands more than 1000 km from continental land masses (Bouché, 1982). Transport by animals, e.g. moles, is generally limited to a few decametres, or longer distances by birds (Schwert, 1980). Transport by man in water affects freshwater species and species of cultivated land are commonly transported in soil attached to plants. Large worms or endogeic species which are fragile and unable to survive in disturbed soils are rarely successfully transported by these means.

Survival ability

The success of anthropochoric species in colonizing environments disturbed by man is well known (e.g. Ljungström, 1972) and empty niches may be created by deforestation, pesticides, waste disposal and destruction of existing habitats. They may also arise naturally from glaciation, marine emergence and sea recession. Empty niches in such virgin environments may also be occupied by migrants as in the case of the epigeic *Dendrodrilus rubidus tenuis* var. *norvegicus* in the sub-antarctic Kerguelen Islands (Bouché, 1982) where the only other earthworm present is the endogeic species *Microscolex kerguelarum*.

Overpopulation by one ecological type can have an adverse effect on the entire earthworm community (Mazaud and Bouché, 1980). The various populations of a local community adapt themselves to changes in the environment by sharing the ecological resources and filling the available niches. A precise adaptation of populations to their environment has resulted from limited gene flow; consequently immigrant allochthonal species are generally less fitted to the locality and fail to become established.

38.2.4 Geology and ecology

Interpretations must be capable of counter-proof (Claude Bernard) or to be falsifiable (Popper). Palaeogeographical information and ecological data provide independent sources for refuting or confirming the bio-

Table 38.1 Names and hierarchy of taxa cited. Underlined taxa are followed by subordinate taxa.

Class: Annelida
Subclass: Oligochaeta
Orders: Lumbriculida, Tubificida, Haplotaxida
Haplotaxida suborders: Haplotaxina, Alluroidina, Moniligastrina, Lumbricina
Haplotaxina family: Haplotaxidae
    species: *Haplotaxis gordioides*
Lumbricina superfamilies: Criodriliodea, Lumbricoidea, Biwardriloidea,
    Glossoscolecoidea, Megascolecoidea

Families:

Lumbricoidea	Glossoscolecoidea	Megascolecoidea
Sparganophilidae	Kynotidae	Ocnerodrilidae
Ailoscolecidae	Microchaetidae	Megascolecidae
Hormogastridae	Glossoscolecidae	Acanthodrilidae
Lumbricidae	Almidae	Octochaetidae
		Eudrilidae

Sparganophilidae genus: *Sparganophilus*
Hormogastridae genera: *Hemigastrodrilus, Vignysa popi, Hormogaster praetiosa, Hormogaster redii, Hormogaster samnitica.*
Lumbricidae genera: *Lumbricus, Prosellodrilus, Ethnodrilus, Scherotheca, Allolobophora, Eophila, Aporrectodea, Dendrobaena, Dendrodrilus*
Scherotheca subgenera: *Opothedrilus, Scherotheca*
Scherotheca subgenus group 1: *S. monspessulensis, S. gigas, S. dugesi,*
                                                  S. dugesi brevisella
                        group 2: *S. guipuzcoana, S. coineaui, S. corsicana*
Allolobophora pereli, A. muldali
Eophila januae-argenti
Aporrectodea (= Nicodrilus) longa, A. longa ripicola, A. velox, A. gogna, A. balisa,
    A. giardi (= terrestris), A. giardi voconca
Dendrobaena jeanneli
Dendrodrilus rubidus tenuis var. *norvegicus*
Acanthodrilidae genus: *Microscolex kerguelarum*

---

geographical interpretations and phyletic reconstructions presented in this chapter.

38.3 INTERPRETATION

38.3.1 The Precambrian primary annelids

Valentine (1980) traces the origins of the main animal groups to the vermiform coelomates of 700 million years ago on the basis of fossil tracks and morpho-functional arguments concerning the coelom and hydro-

Table 38.2 Summary of: (1) proposed relations between geological stages; (2) time in millions of years; (3) selected main events; (4) taxonomic hierarchy (not synchronous between families); (5) palaeopedological events (Kubiena, 1948) and terrestrial adaptation.

Stages (1)	Time (2)	Selected Main Events (3)	Taxa (4)	(5)
Quaternary		Post-ice-age migrations / Extinction in ice age	Varieties	Gyttja
Pliocene	2	*Scherotheca dugesi brevisella* / *Hormogaster sammitica lirapora*	Subspecies	Moder
Miocene	5	Continent ↔ Corsica–Sardinia / *S. dugesi / S. gigas* / Massif Central ↔ Alps / *H. praetiosa, S. gigas*	Species	Active mull
Oligocene	23			Aquatic terrestrial earthworms
Eocene	36	*Prosellodrilus, Ethnodrilus, Opothodrilus, Scherotheca*	Subgenus and genus	Marine worms
Palaeocene	53.5	*Hemgastrodrilus, Hormogaster Vignysa?*		
Cretaceous	65	Kynotidae, Microchaetidae, Almidae, Lumbricidae, Hormogastridae, Ailoscolecidae, Eudrilidae, Malabrinae	Families	
Jurassic / Triassic	120	Lumbricoidea, Megascolecoidea, Glossoscolecoidea	Superfamilies	
Palaeozoic	225	Lumbricina / Oligochaeta	Suborder / Order	
Ediacarian	570	Annelida	Class	
Precambrian	700	Burrow tracks: coelomates		
	1400	Metazoa / Eucaryotes?		

Only South of France (bracket spanning the Pliocene–Eocene main events).

static skeleton. These coelomates differentiated into various lines, among them the Annelida, which appear to be represented in the fossil Ediacara fauna by *Marywadea* (680 million years), a metamerized animal. The Annelida were probably well established at the junction of the Precambrian and Cambrian (570 million years).

The evolution of the Oligochaeta by adaptation of the Annelida to epicontinental habitats can be interpreted in morpho-functional terms. The conquest of land began with an aquatic fauna which became adapted to freshwater streams. Two linked fundamental traits mark this ecological adaptation. First, the loss of all very mobile stages, both zygotes and diaspores. Oocytes and spermatozoa are no longer released into water, instead exchange of sperm occurs between partners. This sperm is contained in spermatophores or spermathecae to await fertilization which remains external in the ootheca where oocytes are layed. The free-living trochophore larva disappears and instead the embryo grows in the ootheca. Correlated with this, the clitellum spreads out to produce oothecae. This loss of mobility is a response to the need to avoid being carried away in water currents.

The second trait is a genetical consequence of the first. Copulation between animals with very limited mobility leads to considerable potential inbreeding. This handicap is corrected, at least partly, by hermaphroditism, polyploidism and parthenogenesis.

This differentiation of Oligochaeta from the sea annelids probably occurred in the early Palaeozoic. Among freshwater Oligochaeta, the Haplotaxida today inhabit very humid oxygen-deficient soils or penetrate far into deep soils by natural drains where they live as endogeic rheophiles. Their distribution is world wide, corresponding to their origin in Pangaea.

38.3.2 The Mesozoic: adaptation to terrestrial life

At the end of the Palaeozoic or during the early Mesozoic, the Oligochaeta acquired the characters adapted to true terrestrial life. First the oesophageal crushing organs, the progastral gizzard, appeared, then, freeing the anterior part, the postgastral gizzard. Earthworms obtained nourishment from various sources distributed in the soil (endogeics) then, later, from rich accumulations on the soil surface (epigeics), finally they took to mixing organic and mineral soil layers (mainly anecics).

This evolution affected the characters of the entire animal. An earthworm population living in firm burrows was subjected to selective pressure which affected organs in a linear way (pholeoiptomy); expansion of an organ in the anterior part brought about the contraction of another organ, a feature of burrow-living animals. Development of the oesophageal gizzard led to posterior displacement of some genital organs and the postgastral organization was associated with a backward movement of clitellum and puberculum. This posterior displacement was, in turn, linked with indirect copulation.

Adaptation to the soil environment also entails greater development of the body wall and of the anterior septal muscles, particularly in the postgastra, better skin protection, dispersal of nephridiopores (holonephridia) with solfeggio-like distribution of nephridiopores, meronephridia, coelomic pores, and diverse arrangements of setae.

The end of the Palaeozoic and the beginning of the Mesozoic was the period of the single continent, Pangaea. Since Wegener's classic work (1920) various reconstitutions of this land mass and its evolution to the present day have been proposed. Plate tectonics and choice of absolute geographic co-ordinates have led to a modern reconstruction (Dietz and Holden, 1970) and to computer simulations (Smith and Briden, 1978).

The present classification–distribution of earthworms may be superimposed on to Pangaea and its various stages of dissociation. Some difficulties result from the need to include all earthworms whereas one can only be confident of the phylogeny of the best-known groups. Here I use the arrangement of Sims (1980a) but the position of some families or superfamilies having only one or two species seems uncertain (Criodriloidea, Biwadriloidea, Lutodrilidae) from lack of information.

Following Jamieson (1978), I consider Alluroidina and Moniligastrina as independent from Lumbricina. These three suborders probably have their origin in haplotaxine ancestors.

If the present distribution of endemic Lumbricina is projected on to a map of Pangaea (Fig. 38.1), the distribution of the Megascolecoidea will be found to show a discrepancy. This group, well established in 'south' Pangaea, the future Gondwana, also spread into Southeast Asia where the Megascolecidae, also present in Australia, were well established. Perhaps the Acanthodrilidae (Australia, Africa and Americas) are also endemic in Southeast Asia.

This situation at the time of the Trias (220 million years) may be interpreted by various hypotheses:

(a) The present taxonomy is incorrect: this is doubtful for this level in the hierarchy;
(b) The two families spread into Southeast Asia after the junction of the Australian continent with Asia: this event was probably too late;
(c) The animals were translocated on fragments of Pacifica which drifted through Panthalassa, today's Pacific Ocean, from the Australian region to Asia and the Americas (Nur and Ben-Avraham, 1977, interpretation in Sims, 1980a). This hypothesis seems chronologi-

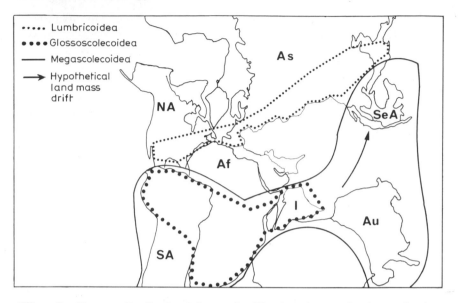

Fig. 38.1 Present distribution of superfamilies based on endemic species in relation to Pangaea in the early Triassic (220 million years). After Smith and Briden (1978) and present area of superfamilies based on endemic species. Present continents projected on Pangaea: Af, Africa; NA, North America; As, Asia; SA, South America; Au, Australia; SeA, Southeast Asia; I, India.

cally incompatible with the total evolution of Lumbricina if Pacifica broke away from Pangaea in the mid-Permian;

(d) An inadequate interpretation of continental drift. A Southeast Asian fragment could have migrated as India, from the Southern zone (Gondwana) to present-day Asia. It must be remembered that sometimes zoological arguments are better than the accepted geological theses. Continental drift was accepted by Michaelsen (1922) and Cernosvitov (1935), while geologists generally rejected it (Wilcke, 1955). The 'impossible' migration of *Scherotheca dugesi (brevisella)* from the continent to Corsica in mid-Miocene (Bouché, 1972) provides another example. The geotectonics of Southeast Asia remain poorly interpreted and I consider this hypothesis to be probable.

Gondwana was occupied only by the Glossoscolecoidea and Megascolecoidea. The Ocnerodrilidae spread into America, Africa, Madagascar and India. The Acanthodrilidae and Octochaetidae with similar territories extend eastward to Australia. The Megascolecidae, however, settled only in the eastern part, Southeast Asia and Australia.

The Eudrilidae, a highly evolved family, seem to have differentiated from this stock at a later date, after the isolation of Africa. Laurasia, which separated from Gondwana in the Jurassic, harbours the Lumbricoidea.

38.3.3 End of the Mesozoic: differentiation of families

Differentiation of the families occurred, generally speaking, around the end of the Jurassic or during the Cretaceous and the disruption of Gondwana (Fig. 38.2). The Glossoscolecoidea gave rise to the Glossoscolecidae in South and Central America, the Microchaetidae in South Africa and the Kynotidae in Madagascar. Finally, the Almidae, a group with various genera which are perhaps ecologically convergent, all living in muddy soils, evolved in South America, Central Africa and India to Sulawasia.

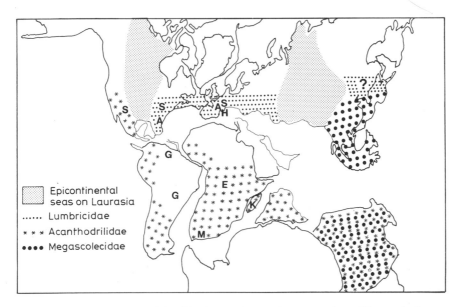

Fig. 38.2 Distribution of families in relation to the separation of Laurasia from Gondwana in the early Cretaceous (120 million years). After Smith and Briden (1978). A, Ailoscolecidae; E, Eudrilidae; G, Glossoscolecidae; H, Hormogastridae; K, Kynotidae; S, Sparganophilidae.

The Lumbricoidea probably spread from an epicontinental sea in 'North America' to Siberia. The maximum of endemism is observed in families with different genera in America and Europe (Ailoscolecidae, Lumbricidae). The Sparganophilidae, with probably a single genus on both Atlantic shores, became established in very humid soils, and two families are very localized, the Lutodrilidae in America and the Hormogastridae in Western Europe.

The Megascolecoidea gave rise to the Eudrilidae, peculiar to Africa, and the Acanthodrilidae spread into Western North America, bounded eastward by an epicontinental sea. Sims (1980a) provides another interpretation. The Ocnerodrilidae formed a subfamily peculiar to India. In general, the Gondwanian disruption gave rise to families or subfamilies on its fragments while the Laurasian breakup, which took place later, is today characterized by genera. This difference in hierarchical levels appears again at lower levels for earthworms living in mud (Almidae, Sparganophilidae).

38.3.4 The Tertiary: intrageneric differentiation

At the end of the Mesozoic and during the Tertiary, the present-day genera, subgenera and species became differentiated. The north of France

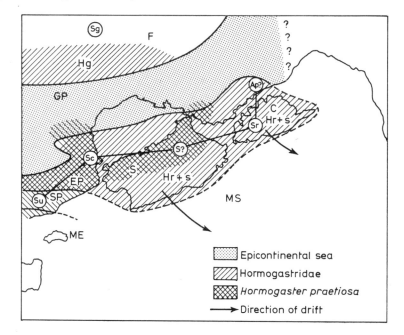

Fig. 38.3 Approximate position of Sardinia and Corsica in the Palaeocene during separation from the continent. GP, Aquitaine–Provence gulf; SP, South-Pyrenean furrow; C, Corsica; EP, East-Pyrenean ridge; F, Continental France; S, Sardinia; Su, *Scherotheca guipuzcoana*; Sc, *S. coineaui*; S?, *hypothetical Scherotheca* sp.; Sr, *S. corsicana*; Ap?, *Allolobophora pereli?*; Sg, *S. gigas* group; Hg, *Hemigastrodrilus*; Hr, *Hormogaster redii*; Hs, *H. samnitica*; MS, Mediterranean Sea; ME, Menorca.

was deprived of fauna during Quaternary glaciations and consequently gives no information on Tertiary history but the area now forming the south of France illustrates some of the factors involved. In this zone, widely submerged by epicontinental seas in mid and late Jurassic, a great emergence occurred in the Jurassic–Cretaceous when some of the Lumbricoidea were probably able to spread into the new terrestrial territory. This land, before and during the Palaeocene, split into three zones, a 'Massif Central' in the North separated by a Lower-Provence–Aquitaine gulf from a ridge running from the East-Pyrenees to Sardinia, Corsica and the Maures (EPSACOM). This ridge was in turn probably separated from an 'Ebro continent' by a sea occupying a South Pyrenean furrow (Fig. 38.3).

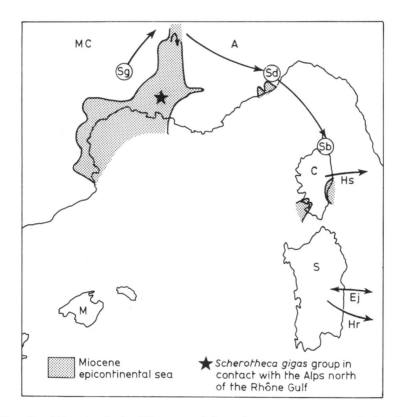

Fig. 38.4 Migration in the Miocene and through temporary passages during the drying up of the Mediterranean during the Pliocene. Sg, *S. gigas*; Sd, *S. dugesi*; Sb, *S. dugesi brevisella*; Hs, *Hormogaster samnitica*; Hr, *H. redii*; Ej, *Eophila januae-argenti*; MC, Massif Central; A, Alps; M, Majorca; C, Corsica; S, Sardinia.

Knowledge of the Iberian fauna is poor but in the Massif Central and EPSACOM three lumbricoid groups seem to have evolved: Hormogastridae, *Prosellodrilus/Ethnodrilus*, *Scherotheca*. In the Massif Central land mass live: 1 *Hemigastrodrilus* (and possibly *Vignysa*); 2 *Ethnodrilus;* 3 *Scherotheca*, branch *gigas-monspessulensis*. In EPSACOM: 1 *Hormogaster* (and possibly *Vignysa*); 2 *Prosellodrilus*; 3 *Opothedrilus* and the species assemblage *S. guipuzcoana, S. coineaui, S. corsicana* with a relatively anterior puberculum (30)31–40(42) (+ *Allolobophora pereli?*).

Before the Oligocene, an island which later became Sardinia and Corsica broke away from the continent (Auzende *et al.*, 1973). Today, the distribution of *Hormogaster praetiosa* is perhaps a testimony to the former link between Catalonia-Sardinia and West Provence.

During the Miocene, contact between the Alps and the Massif Central land mass occurred allowing the *Scherotheca gigas* group to invade this new territory where it became differentiated into the species *S. dugesi*. At the end of the Miocene, the Mesogea, the present-day Mediterranean, fell to a low level (Hsii, 1972). *S. dugesi* invaded North Corsica where it is represented today by a peculiar subspecies: *S. dugesi brevisella*. *Hormogaster redii* and *H. samnitica* which were differentiated in Sardinia and Corsica, formerly the eastern part of EPSACOM, spread around and through the present Ligurian-Thyrenean sea. *Eophila januae-argenti*, today occupying Macedonia, Campania and Sardinia (Omodeo, 1961), had probably a single area during this period (Fig. 38.4).

38.3.5 THE QUATERNARY: DESTRUCTION OF THE FAUNA AND RECOLONIZATION

Quaternary glaciations eliminated earthworms over a large part of middle and North Europe (Michaelsen, 1903; Cernosvitov, 1935). In western Europe, and especially in France, we can see endemics as discrete isolated populations which have only been able to maintain themselves in warm micro-biotopes, including caves (*Dendrobaena jeanneli*) (Bouché, 1972). This zone of isolated population refuges lies in the middle of France with a more favourable seashore zone around Charente and Périgord (*Ethnodrilus*). There are no endemics from the area of inland ice in the Alpine mountains. Southward, the biogeographical areas reflect tertiary events and were not greatly affected by glaciation.

The recolonization of land after the ice age can sometimes be traced. For example, the group of closely related species *Aporrectodea (= Nicodrilus) longa* (Ude), *A. gogna, A. velox* inhabits a funnel-shaped zone in the high Rhône watershed which opens on to the great North-European Plain extending from the British Isles to the USSR. Two species are confined to the zone of refuges and did not leave the Rhône watershed while *A. longa longa* and *A. longa ripicola* spread northward (Fig. 38.5).

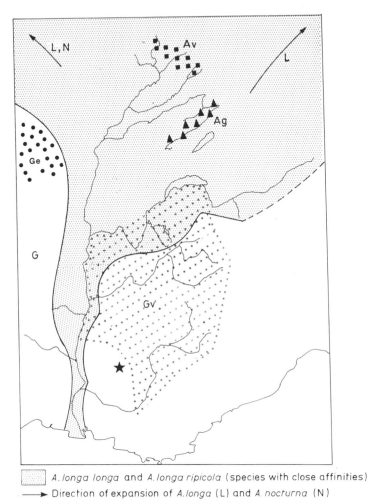

A. *longa longa* and *A. longa ripicola* (species with close affinities)

⟶ Direction of expansion of *A. longa* (L) and *A. nocturna* (N)

Fig. 38.5 Post-glacial migration of *Apporrectodea* from the Rhône watershed. Av, *A. velox* (■); Ag, *A. gogna* (▲); G, *A. giardi giardi*; Ge, var. *eudensis* (●); Gv, *A. giardi voconcus* (·); *A. balisa* ★ (the last six are all competing species).

The territory of this group is bounded by other *Aporrectodea* which were probably competitive (*A. giardi giardi*, *A. giardi voconca*, *A. balisa*), all also anecics, and the area is shared with another anecic, *A. nocturna*, in the west. This partial co-habitation of two anecic species may be explained by food sharing (Ferrière, 1980). While boundaries are not clearcut and the areas overlap to some extent, the species *A. longa*, *A. giardi* and *A. nocturna* are usually mutually exclusive (Bouché, 1972).

The north European fauna is poor and although there are some unexplained cases, for example, *Allolobophora limicola* from the Rhine watershed, including its former British tributaries, one can recognize the origin of the species in the south.

Recolonization after the ice age was followed in modern times by transport by man.

38.3.6 Ecological types and distribution

The evolution of the taxa was accompanied by adaptation to often precise ecological niches. It is difficult to retrace this but over such a long period it is probable that groups occupied various niches successively and alternately. This has led to morphological types which have descended from various ancestors and are more or less convergent. Direct taximetric classification of morphological data thus gives information both on present ecological similarities and on phyletic relations (Sims, 1980b; Bouché, 1980). Starting with the Haplotaxida of the endogeic type, earthworms were more or less specialized in relation to food sources. Small and narrow polyhumic worms tend to establish themselves near the surface of roots, or in the zone where organic matter is carried by drainage into pore spaces (shrinkage cracks: *Allolobophora muldali*; natural drainage channels: *Haplotaxis gordioides*). The mesohumic endogeics are in an intermediate position and ingest moderately organic soils. The oligohumic endogeics live in soils poor in organic matter of which they consume large quantities. The latter are important in tropical zones with *Vignysa popi* present in humid mediterranean soils.

These adaptations are in equilibrium with organic sources, metabolic needs and microbial life (Lavelle *et al.*, 1980) which are in turn dependent on climate. Surface organic horizons are invaded by epigeic earthworms or can be buried by anecics which contrast with other earthworms by feeding and casting in different layers. The anecic ecological type has been well established only in the Lumbricidae. Anecics create the conditions for microbial maturation of their food for, after burying, the casts are reingested (Bouché, 1981). Their burrows and cast systems favour the penetration of organic compounds by solution, by gravity in burrows, by ingestion, by traction and by excretion of metabolites (Dietz, 1979; Bouché, 1981).

The different ecological adaptations of earthworms lead to co-existence of various species which are in actual or partial competition and occupy distinct or partly overlapping niches. This scheme is probably general and explains why the total earthworm niche is generally filled and why the areas of the various endogeics and anecics are stable. The mobile epigeics are in equilibrium with their temporary media. This scheme also explains why palaeogeographic phenomena can be observed in present population distributions and, conversely, how earthworms continued to occupy their total niche as the epicontinental ecosystems evolved. The filling of the annelid niche seems to have coincided with ecosystem development, and fossil humus types (Kubiena, 1948) are a testimony to the emergence of earthworms (Wilcke, 1955). The humus type gyttja is associated with freshwater groups, Lumbriculida, Tubificida, Haplotaxida; moder and poorly active mull are associated with endogeics while active mull is linked with anecics (Table 38.2).

38.4 CONCLUSION

Since the time of the Ediacara fauna, earthworm ancestors have occupied a fundamental position in ecosystems using plant organic matter which is subjected to complex transformations through earthworm/microflora interactions. Analogous interactions occur in grassland (Loquet *et al.*, 1977) and estuaries (Loquet and Dupont, 1977). The Annelida are energetically efficient with a hydrostatic skeleton, poikilothermy, reduced search for food and sexual partners, and protection by the soil environment, and have therefore filled a wide niche. One group, the Oligochaeta, has become well adapted to epicontinental life, the most successful branch being the Lumbricina with its diverse terricolous adaptations. Various ecological types representing the adaptations of different earthworm groups fill the annelid niche in emerged soils. Historical events such as continental drift, tectonic movements and glaciation, remain imprinted in many rather immobile groups. The imprint is to some extent indelible as a result of the loss of mobile stages during adaptation to terrestrial life and of specialization and competitive interactions (Bouché, 1972). This does not exclude a diversification of communities nor adaptation of their structure to macroclimatic conditions and consequent availability of food.

There are nevertheless various difficulties in interpretation, in particular, dyssynchronism of families in the taxonomic hierarchy although these anomalies will doubtless be resolved as taxonomy improves. On the broad time and space scales used in this study a coherent pattern can be seen which is consistent with the independent studies of Jamieson (1974).

We can conclude that the vermiform coelomates became established very early in the development of the Metazoa and that the Annelida have played a dominant role in the porosphere since the Precambrian.

38.5 ACKNOWLEDGEMENT

I wish to thank Brenda Healy for critical contributions to the manuscript and for undertaking the considerable task of translating it.

38.6 REFERENCES

Auzende, J. M., Bonnin, J. and Olivet, J. L. (1973) The origin of the western Mediterranean basin. *J. Geol. Soc., London*, **129**, 607–620.

Bouché, M. B. (1971) Relations entre les structures spatiales et fonctionnelles des écosystèmes illustrées par le rôle pédobiologique des vers de terre. In *La Vie dans les Sols* (ed. P. Pesson), Gauthier-Villars, Paris, pp. 187–209.

Bouché, M. B. (1972) Lombriciens de France. Ecologie et systematique. *Ann. Zool. – Écol. Anim.*, **72**, INRA, 1–671.

Bouché, M. B. (1976) Étude de l'activité des invertébrés épigés prairiaux. I. Résultats généraux et géodrilologiques (*Lumbricidae: Oligochaeta*). *Rev. Écol. Biol. Sol*, **13**, 261–281.

Bouché, M. B. (1977) Stratégies lombriciennes. In *Soil Organisms as Components of Ecosystems* (eds. U. Lohm and T. Persson), *Proc. 6th Int. Soil Zool. Coll., Ecol. Bull. (Stockholm)*, **25**, 122–132.

Bouché, M. B. (1980) L'interprétation morphologique des lombriciens: un commentaire de l'évaluation numérique de R. W. Sims. *Pedobiologia*, **20**, 227–229.

Bouché, M. B. (1981) Contribution des lombriciens à la migration des éléments dans les sols en climats tempérés. *C. R. Coll. Int. C.N.R.S. Migrations Organo-minérales dans les Sols Tempérés*. Nancy, 1979. C.N.R.S., 145–153.

Bouché, M. B. (1982) Les lombriciens des terres australes et antarctiques françaises. In *Colloque sur les Ecosystèmes Subantarctiques*, Paimpont, 1981 (eds. T. Jouventin, L. Masse and P. Trehen), CNFRA, **51**, pp. 175–180.

Bouché, M. B. (in press) Observations sur les lombriciens: une nouvelle espéce (*Sparganophilus langi*) de la famille amphiatlantique Sparganophilidae (Oligochaeta).

Boyd, J. M. (1960) Studies of the differences between the fauna of grazed and ungrazed grassland in Tiree, Argyll. *Proc. Zool. Soc. London*, **135**, 33–54.

Cernosvitov, L. (1935) Monograph on Czechoslovakian earthworms. *Arch. Prirod. Vyzleum Cech.*, **19**, 1–86. [In Czech.]

Dietz, S. (1979) *Étude de l'incorporation de la litière en système herbacé à l'aide de materiel végétal marqué au C^{14}*. Thèse 3ème cycle écologie terrestre, univ. sci. techn. Languedoc, Montpellier, pp. 1–78.

Dietz, R. S. and Holden, J. C. (1970) The breakup of Pangaea. *Sci. Am.*, **223**, 30–40.

Ferrière, G. (1980) Fonctions des lombriciens. VII. Une méthode d'analyse de la matière organique végétale ingérée. *Pedobiologia*, **20**, 263–273.

Hsii, K. J. (1972) When the Mediterranean Sea dried up. *Sci. Am.*, **227**, 27–36.

Jamieson, B. G. M. (1974) VIII. The zoogeography and evolution of Tasmanian Oligochaeta. In *Biogeography and Ecology of Tasmania* (ed. W. D. Williams), Dr. W. Junk, The Hague, pp. 195–228.

Jamieson, B. G. M. (1978) Phylogenetic and phenetic systematics of the Opisthoporous Oligochaeta (Annelida: Clitellata). *Evol. Theor.*, **3**, 195–233.

Jamieson, B. G. M. (1980) Preliminary discussion of an Hennigian analysis of the phylogeny and systematics of Opisthoporous Oligochaetes. *Rev. Ecol. Biol. Sol*, **17**, 261–275.

Kubiena, W. (1948) *Entwicklungslehre des Bodens*. Springer-Verlag, Vienna, pp. 1–215.

Lavelle, P., Sow, B. and Schaefer, B. (1980) The geophagous earthworm community in the Lamto savanna (Ivory Coast): niche partitioning and utilization of soil nutritive resources. In *Soil Biology as Related to Land Use Practices* (ed. D. L. Dindal), E. P. A. Washington D.C., pp. 653–672.

Ljungström, P. O. (1972) Introduced earthworms of South Africa. On their taxonomy, distribution, history of introduction and on the extermination of endemic earthworms. *Zool. Jb. Syst.*, **99**, 1–81.

Loquet, M. and Dupont, J. P. (1977) Étude morphologique et microbiologique des terriers de Nereis dans un faciès sablo-vaseux (Baie de Somme, France). In *Soil Organisms as Components of Ecosystems* (eds. U. Lohm and T. Persson), *Proc. 6th Int. Soil Zool. Coll., Ecol. Bull. (Stockholm)*, **25**, 496–500.

Loquet, M., Bouché, M. B., Bhatnagar, T. and Rouelle, J. (1977) Essai d'estimation de l'influence écologique des lombriciens sur les micro-organismes. *Pedobiologia*, **17**, 400–417.

Mazaud, D. and Bouché, M. B. (1980) Introduction en surpopulation et migrations de lombriciens marqués. In *Soil Biology as Related to Land Use Practices* (ed. D. L. Dindal), E. P. A. Washington, D.C., pp. 687–701.

Michaelsen, W. (1903) *Die Geographische Verbreitung der Oligochaeten*. R. Friedlander und Sohn, Berlin. 186 pp.

Michaelsen, W. (1922) Die Verbreitung der Oligochaten im Lichte der Wegener'schen Theorie der Kontinentenverschiebung und andere Fragen zur Stammesgeschichte und Verbreitung diese Tiergruppe. *Verhandl. Naturwiss. Ver. Hamburg*, **29**, 45–79.

Nur, A. and Ben-Avraham, Z. (1977) Lost Pacifica continent. *Nature (London)*, **270**, 41–43.

Omodeo, P. (1961) Le peuplement des grandes iles de la Méditerranée par les Oligochètes terricoles. In *Le Peuplement des Iles Méditerranéennes et le Problème de l'Insularité*, C.N.R.S., pp. 127–133.

Piearce, T. G. and Piearce, B. (1979) Responses of Lumbricidae to saline inundation. *J. Appl. Ecol.*, **16**, 461–474.

Satchell, J. E. (1980) r and K worms: a basis for classifying lumbricid earthworm strategies. In *Soil Biology as Related to Land Use Practices* (ed. D. L. Dindal), *Proc. 7th Int. Soil Zool. Coll., EPA Washington, D.C.*, pp. 848–864.

Schwert, D. P. (1979) Description and significance of a fossil earthworm (Oligochaeta: Lumbricidae) cocoon from postglacial sediments in southern Ontario. *Can. J. Zool.*, **57**, 1402–1405.

Schwert, D. P. (1980) Active and passive dispersal of lumbricid earthworms. In *Soil Biology as Related to Land Use Practices* (ed. D. L. Dindal), *Proc. 7th Int. Soil Zool. Coll. EPA Washington D.C.*, pp. 182–189.

Schwert, D. P. and Dance, K. W. (1979) Earthworm cocoons as a drift component in a southern Ontario stream. *Can. Fld. Nat.*, **93**, 180–183.

Sims, R. W. (1980a) A classification and the distribution of earthworms, suborder Lumbricina (Haplotaxida: Oligochaeta). *Bull. Br. Mus. (Nat. Hist.)*, **39**, 103–124.

Sims, R. W. (1980b) A preliminary numerical evaluation of the taxonomic characters of *Allolobophora* auct. and some allies (Lumbricidae: Oligochaeta) occurring in France. *Pedobiologia*, **20**, 212–226.

Smith, A. G. and Briden, J. C. (1978) *Mesozoic and Cenozoic Paleocontinental Maps*. Cambridge University Press, Earth Science Series, pp. 1–63.

Valentine, J. W. (1980) L'origine des grands groupes d'animaux. *La Recherche*, **112**, 666–674.

Wegener, A. (1920) Die Enstehung der Kontinente und Ozeane. In *Die Wissenschaft*, Vieweg Verlag, Braunschweig.

Wilcke, D. E. (1955) Bemerkungen sum Problem des erdzeitlichen Alters der Regenwürmer (Oligochaeta Opisthopora). *Zool. Anz.*, **154**, 149–156.

Chapter 39

The structure of earthworm communities

P. LAVELLE

39.1 INTRODUCTION

In most terrestrial ecosystems, soil macrofauna (i.e > 4 mm, Bachelier, 1963) communities are dominated by earthworms. Their density may reach 10^6 ha^{-1} and their biomass 2t ha^{-1}. They are present everywhere except in arid and frozen regions.

Because of the rapidly changing conditions of their environment, soil invertebrates must have a flexible and quickly adaptable organization at the individual, population and community levels. The hypothesis of this chapter is that the success of earthworms rests on the organization and plasticity of their communities. It is analysed in terms of their ecological amplitude, i.e. the distribution of biomass between the different ecological categories defined by Bouché (1971) and Lavelle (1981), their organization by niche separation and their plasticity in relation to environmental changes.

39.2 ECOLOGICAL AMPLITUDE

39.2.1 Distribution of ecological categories in earthworm communities

Earthworms can be divided into five ecological categories, epigeic, anecic, oligo-, meso- or polyhumic endogeic. The ecological amplitude of a community reflects the distribution of its population among these categories.

A factorial analysis has been made of 42 earthworm communities in diverse temperate and tropical environments (Table 39.1). Thirty-five variables were used to describe them, including density, biomass, species

Table 39.1 Attributes of the 42 sites investigated.

	Country	Locality	Vegetation	Latitude	Authors
1	France	Citeaux	Pasture	47°5 N	Bouché (1978)
2	France	Sivrite	Conifer plantation	48°5 N	Bouché (1978)
3	France	Le Rouquet	Oak wood	43°5 N	Bouché (1978)
4	France	Brunoy	Ash–Birch wood	49° N	Bouché (1978)
5	France	La Madeleine	Oak wood	43°5 N	Bouché (1978)
6	France	Bellefontaine	Beech wood	48°5 N	Bouché (1978)
7	Sweden	Spiboke	Abandoned field	60° N	Persson and Lohm (1977)
8	Sweden	Lund	Spruce plantation	56° N	Nordström and Rundgren (1974)
9	Sweden	Lund	Spruce plantation	56° N	Nordström and Rundgren (1974)
10	Sweden	Lund	Spruce plantation	56° N	Nordström and Rundgren (1974)
11	Sweden	Lund	Beech wood	56° N	Nordström and Rundgren (1974)
12	Sweden	Lund	Alder–pine wood	56° N	Nordström and Rundgren (1974)
13	Sweden	Lund	Abandoned grassland	56° N	Nordström and Rundgren (1974)
14	Sweden	Lund	Permanent pasture	56° N	Nordström and Rundgren (1974)
15	Sweden	Lund	Pine plantation	56° N	Nordström and Rundgren (1974)
16	Sweden	Lund	Alder–birch wood	56° N	Nordström and Rundgren (1974)
17	Sweden	Lund	Tall herb meadow	56° N	Nordström and Rundgren (1974)
18	Sweden	Lund	Pine plantation	56° N	Nordström and Rundgren (1974)
19	Sweden	Lund	Elm–ash wood	56° N	Nordström and Rundgren (1974)
20	Sweden	Lund	Juniperus pasture	56° N	Nordström and Rundgren (1974)
21	Sweden	Lund	Elm–ash wood	56° N	Nordström and Rundgren (1974)
22	Sweden	Lund	Spruce plantation	56° N	Nordström and Rundgren (1974)
23	Sweden	Lund	Beech wood	56° N	Nordström and Rundgren (1974)
24	Sweden	Lund	Elm wood	56° N	Nordström and Rundgren (1974)
25	Sweden	Lund	Corynephorus heath	56° N	Nordström and Rundgren (1974)
26	Sweden	Lund	Alder wood	56° N	Nordström and Rundgren (1974)
27	Ivory Coast	Lamto	Fringing forest	6°1 N	Lavelle (1978)

28	Ivory Coast	Lamto	Forest edge	6°1 N	Lavelle (1978)
29	Ivory Coast	Lamto	Unburned savanna	6°1 N	Lavelle (1978)
30	Ivory Coast	Lamto	Dense shrub savanna	6°1 N	Lavelle (1978)
31	Ivory Coast	Lamto	Open shrub savanna	6°1 N	Lavelle (1978)
32	Ivory Coast	Lamto	Grass savanna	6°1 N	Lavelle (1978)
33	Ivory Coast	Lamto	Grass savanna	6°1 N	Lavelle (1978)
34	Ivory Coast	Foro Foro	Shrub savanna	7°0 N	Lavelle, unpublished
35	England	Wytham Woods	Beech wood	51°5 N	Phillipson et al. (1978)
36	Ivory Coast	Foro Foro	Shrub savanna	7°0 N	Lavelle, unpublished
37	India	Berhampur	Ungrazed pasture	21°3 N	Senapati (1980)
38	India	Berhampur	Grazed pasture	21°3 N	Senapati (1980)
39	France	Borculo	Permanent pasture	48°4 N	Bouché (1978)
40	Zaire	Lwiro	Savanna	2°1 S	Kanyonyo, unpublished
41	Ivory Coast	Ouango Fitini	Shrub savanna	10° N	Lavelle, unpublished
42	Mexico	Laguna Verde	Pastures	22° N	Lavelle et al. (1981)

richness and diversity, ecological categories, weight distribution of adults of the different species, the major climatic variables (temperature, annual rainfall, drought, frost), soil characteristics (pH, organic matter content, C/N, mechanical analysis) and vegetation type.

As described by these characteristics, the communities fell into eight groups in a latitudinal vegetational sequence: (1) coniferous forest, (2) heath, (3) cold grassland, (4) cold deciduous forest, (5) temperate deciduous forest, (6) temperate grassland and mediterranean woodland, (7) moist savanna, (8) dry savanna. Examples of (1)–(4) were drawn from the cold temperate zone of Southern Sweden, (6) from the milder temperate zone of England and France, (7) with its adjacent gallery-forests from the Southern Ivory Coast, and (8) from the Northern Ivory Coast, Mexico, Zaire and India (Fig. 39.1). In the factorial analysis, temperature difference related to latitude was the main factor explaining 31 % of the calculated variance, followed by litter characteristics such as quantity, quality and decomposability (16 %) and drought (8 %).

The earthworm communities in the eight groups of sites show marked differences in the representation of different ecological categories

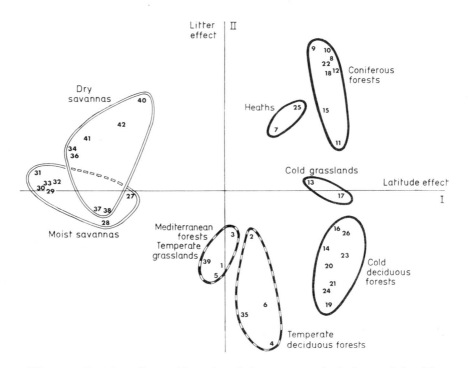

Fig. 39.1 Position of assemblages in relation to two principal axes defined by principal components analysis.

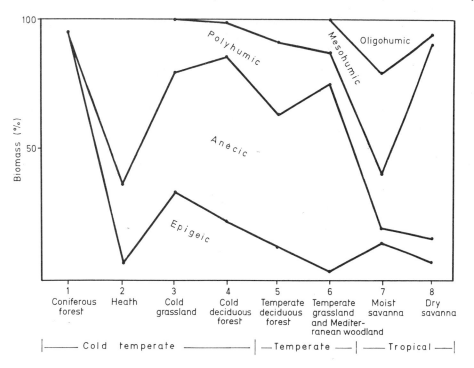

Fig. 39.2 Life form in relation to the latitudinal sequence of vegetation.

(Fig. 39.2). In the sites from Southern Sweden the epigeic group contributes nearly all the biomass in coniferous forests, less in cold deciduous forests and grasslands, and only 5.6% in heather. They contribute 2–13% of the biomass in temperate and tropical environments but their importance varies according to the habitat, whether forest or grassland, and with the degree of exposure to drought or heat.

In the two grassland sites in southern Sweden anecic species comprised 28 and 46% of the biomass. They reached their maximal development in the cold (64%) or temperate (52%) forests and in temperate grasslands and mediterranean forest (72%). In savannas and dry tropical forests, they are less abundant and contributed < 10% to the earthworm biomass.

Mesohumic endogeic species were dominant (41%) in moist tropical savannas but were not found in cold temperate ecosystems except in deciduous forests where they constituted < 1% of the biomass. They were only minimally represented in other temperate environments, and in dry savannas also their role is secondary.

Oligohumic species have not been reported from temperate environments but have been found in the mediterranean region (Bouché, 1972).

They are most abundant (21 %) in moist tropical areas, but drought quickly reduces their populations in savannas with an extended dry season.

In the polyhumic category such diverse types as top soil feeders, small filiform species of the rhizosphere and even carnivores constitute a relatively stable fraction of most communities (10–20 %). They can be even more important when drought or frost reduces the populations that otherwise would have been dominant. They formed 67 % of the biomass in the two Swedish heaths and 42 % in the dry savannas.

Thus, from the most northern latitudes to the equator, the successively dominant categories of earthworms are epigeic, anecic, meso- and oligo-humic endogeic. Polyhumic species proliferate opportunistically when any of the normally dominant groups are limited by environmental factors. Thus, the ecological amplitude of the earthworm community changes towards the tropics as the populations occupy the deeper soil, exploiting the humic reserves and depending less on surface litter.

39.2.2 Environmental determinants of earthworm community structure

The variation observed in the ecological amplitude of earthworm communities is a consequence of the interactions between soil organic matter, microflora and fauna. Most earthworm species apparently have poor enzyme systems and rarely have cellulase (Laverack, 1963; Mill, 1978). They therefore feed on simple soluble organic matter such as glucosides, amino acids or fatty acids with low molecular weights (Satchell and Lowe, 1967; Lavelle et al., 1980). They extract these simple compounds from the soil by feeding on rich substrates such as leaf litter, dead roots or micro-organisms with high contents of organic compounds. There can be little doubt that their intestinal microflora plays a symbiotic role and assists in their nutrition by transforming complex humic compounds into simpler more readily assimilated substances.

In cold regions, litter decomposes slowly and constitutes an abundant and reliable food source of high energy content. It contains hydrosoluble organic compounds and energy substrates that can be transformed into simple substances by a low level of microbial activity (Wright, 1972; Cooke and Luxton, 1980). By contrast, soil humic reserves though abundant are rarely accessible and relatively little soil is ingested ($1.2–1.6\,g\,g^{-1}$ fresh wt. day^{-1} for A. rosea; Bolton and Phillipson, 1976). Moreover microbial activity appears to be insufficient to produce enough primary substances from stable humus to sustain earthworm nutrition. The communities are consequently composed predominantly of de-tritivorous species (epigeic and above all anecic), whereas the endogeic group is only represented by polyhumic species that feed on surface organic concentrations (H horizon) or in the rhizosphere.

In tropical areas the situation is very different. Rapidly decomposed litter is no longer an abundant and reliable food source and, in any case, is soon deprived of its hydrosoluble energetic content by heavy warm rains. Consequently epigeic and anecic populations are reduced. In tropical grasslands the microclimatic conditions close to the surface of the soil can be very harsh and thus can reduce the size of detritivore populations. Soil humic reserves however constitute an abundant food source. In warm temperatures earthworms are able to ingest greater amounts of soil. The daily soil consumption of five geophagous species of the Lamto Savanna has been estimated as $6-30\,g\,g^{-1}$ (Lavelle, 1978) and $3-6\,g\,g^{-1}$ for the glossoscolecid *Pontoscolex corethrurus* in the tropical Mexican pastures of Laguna Verde (Lavelle and Cruz, in press.). Microbial activity is also greater, producing more primary organic matter on which earthworms can feed (Lavelle *et al.*, 1980).

Temperature thus appears to be the primary factor controlling the structure of earthworm communities, determining the metabolic rate of individuals, the level of microbial activity and so the availability of litter and organic matter. Frost, drought and the chemical composition of litter are also locally important.

39.3 COMMUNITY ORGANIZATION

39.3.1 Species richness and diversity

In Europe, the number of species increases between Iceland and the Mediterranean. Iceland has 8 (Bengtson *et al.*, 1975), Sweden 13 (Nordström and Rundgren, 1973), Denmark 19 (Bornebusch, 1930), England 28 (Gerard, 1967), Belgium, Luxemburg and Northern France 29 and the whole of France 97 (Bouché, 1972). These numbers reflect a variety of factors including quaternary glaciation, island phenomena, altitude and perhaps the predilections of particular taxonomists for proliferation or simplification of taxa. The number of species recorded in studies of individual sites is however generally low compared with that of some other soil groups, e.g. Arachnida, ranging from 2 to 11 in the 42 studies cited. It shows little variation with latitude, comprising 8–11 species from Sweden to the Ivory Coast unless reduced by soil acidity, frost or drought.

Specific diversity as measured by the Shannon Index does not show significant variations, so that earthworms seem to be an exception to the rule expressed by MacArthur and Wilson (1967) and since widely verified in various communities. This is however only an apparent exception.

While the number of species remains stable, their average size increases towards the Equator. In cold and acid Swedish soils, most of the species are small, and adult individuals weigh less than 1 g fresh weight. In temperate deciduous forests and grasslands some larger species weigh between 3 and 10 g.

The giant species live no further north than the Mediterranean region and reach their maximal development in the moist tropical savannas. At Lamto, three species have adults weighing over 10 g and two of them reach a maximum individual weight of 30 g. Savannas also have species with very small individuals whose maximum weight only slightly exceeds 100 mg. The Amazonian forest contains species with the most extreme weights, ranging between 10 mg and 20 g (Nemeth, 1981). The diversification of the earthworm community associated with this range in size is paralleled by an equivalent diversity in the functions of the community in the ecosystem.

39.3.2 Niche separation

Niche overlap can be measured with the symmetric index devised by Pianka (1974), a similarity coefficient with values between 0 and 1. For each type of resource, e.g. space and food, a matrix of similarity between species is calculated. The mean coefficients presented in Table 39.2 show how strictly the resource is partitioned.

Numerous studies aimed at characterizing the ecological niches of earthworms show that the main separation factors are spatio-temporal. The spatial distribution can be vertical and horizontal in the same ecosystem or horizontal between different ecosystems in the same region. The temporal distributions are measured as seasonal variations in population density. Further, species with similar spatio-temporal distributions can occupy different trophic niches defined by food substrates, the size of particles which can be ingested as determined by the size of the worms, and the metabolic rates and demographic profiles of the species (Lavelle et al., 1980; Satchell, 1980).

Vertical distribution

The vertical distributions of earthworm populations have been widely studied (e.g. Gerard, 1967; Lavelle, 1973; Nemeth, 1981) and have been used in different attempts to construct ecological classifications (e.g. Rosa, 1882; Lee, 1959; Perel, 1975).

Similarity indexes calculated for the vertical distribution of the nine species studied by Nordström and Rundgren range from 0.69 to 0.999 with a mean of 0.90. The five species studied by Gerard range between

Table 39.2 Mean values of Pianka niche overlap indexes for various earthworm communities.

	Vertical distribution	Horizontal distribution	Temporal distribution		Alimentary regime	Size spectrum
			Density	Production		
Three facies of moist savannas (Ivory Coast) (Lavelle, 1978)	0.47	0.69	0.90	0.73	—	0.24
Three facies of Amazonian forest (Venezuela) (Nemeth, 1981)	0.50	0.70	—	—	—	—
Five different environments (Sweden) (Nordstrom and Rundgren, 1974)	0.90	0.39	—	—	—	—
Twelve cultivated fields (England) (Evans and Guild, 1947)	—	0.47	—	—	—	—
Pastures (England–France) (Gerard, 1967; Ferriere, 1980)	0.92	—	—	—	0.21/0.71	—
Beechwood (England) (Phillipson et al., 1976)	—	—	0.81	—	—	—

0.84 and 0.94 with a mean of 0.92. The vertical distribution patterns of the species in these communities are thus quite similar. In the moist Lamto savannas, however, the seven commonest species have much more diverse vertical distribution patterns, the similarity index ranging from 0.008 to 0.96 with a mean of 0.47. The seven species in the Venezuelan Amazon forest studied by Nemeth also had a wide vertical separation, the index ranging from 0.01 to 0.99 with a mean of 0.50. In cold and temperate regions, then, the preferential use of litter and top soil as food resources apparently limits the possibility of vertical niche separation and so increases the competition for space.

Horizontal distribution

The horizontal distribution patterns of earthworms are difficult to evaluate as they have been described on different scales: 'homogeneous' sampling plots (Phillipson et al., 1976; Lavelle, 1978), different facies of one environment (e.g. Evans and Guild, 1948a; Nemeth, 1981), different environments within a region (Nordström and Rundgren, 1973), and the geographical scale of Bouché's (1972) sites.

Earthworms show marked preferences for different facies of the Amazonian forest at San Carlos de Rio Negro and the Lamto savannas. Their similarity indexes range respectively from 0.43 to 0.95 around a mean of 0.70 and from 0.40 to 0.96 with a mean of 0.69. In temperate areas, the calculated indexes are significantly lower: 0.47 on average (0.22–0.86) in 12 cultivated fields in England (Evans and Guild, 1947) and 0.39 (0–0.95) for five different environments in Southern Sweden (Nordström and Rundgren, 1973). Niches thus seem to be better separated horizontally in cold and temperate regions than in tropical ones.

Time distribution

Niches can be separated temporally in populations that are active and disappear as others that occupy the same niche increase. The density and biomass of earthworm populations show marked seasonal variation (e.g. Evans and Guild, 1948b; Lavelle, 1978; Senapati, 1980). Nevertheless, variations in density tend to synchronize, populations fluctuating together in relation to soil temperature, moisture and organic input and cycles. The similarity indexes between monthly values for density ranged from 0.80 to 0.99 ($\bar{x} = 0.90$) for the seven species of the Lamto savannas and from 0.44 to 0.98 ($\bar{x} = 0.78$) for the ten species of an English beechwood (Phillipson et al., 1978). Similar calculations, made with monthly values of tissue production, a better measure of real activity, give a lower similarity index of 0.73 from the Lamto species.

Food partitioning

Earthworms can be roughly divided into the detritivores, which eat litter and roots, and geophages, which feed on soil humic reserves. The depth distribution of geophagous populations gives some indication of their diet. The soil in which three meso- and oligohumic species of the Lamto savannas grow best was found to be precisely the stratum in which the worms are most frequent (Lavelle *et al.*, 1980). This is especially true for the oligohumic species living in poor soil at deeper levels.

Detritivorous species are more difficult to analyse. Ferrière (1980) showed that all the ten species in pastures at Citeaux, Dijon, even the polyhumic species, eat some litter. Similarity indexes calculated from recognizable fragments of different plant species in the gizzards ranged from 0 to 0.97 ($\bar{x} = 0.21$). Including the unidentified fragments which constituted 48 % of the total, the similarity coefficients, probably over-estimated, ranged from 0.27 to 0.99 ($\bar{x} = 0.71$).

Size distribution

Another important dimension of a trophic niche is the size distribution of the individuals comprising the population. Size affects metabolic rates (Fenchel, 1974; Blueweiss *et al.*, 1978), the size of the particles ingested and ability to move in the soil. The energy cost to small animals of crawling in soil slits and fissures is considerably less than that to larger animals of active burrowing.

The ability of two species of different size to exploit the same resource in different ways is illustrated by two detritivorous species in the Lamto savanna. The small *Dichogaster agilis* (max. fresh wt. 0.6 g) feeds at the soil litter interface on small plant debris already well decomposed, and *Millsonia lamtoiana*, a giant species that reaches a length of 50 cm and 32 g fresh wt., drags larger and less-decomposed pieces of litter into the soil like anecic lumbricids. There is little overlap in the size of the species in the Lamto savanna. In temperate areas the size range is undoubtedly less.

Conclusion

Earthworm communities are separated into ecological niches by various vertical and horizontal distributions, alimentary specialization and species sizes. These factors differ in importance in different ecosystems. Vertical distribution and size separation are better distinguished in tropical regions whereas temperate areas have greater horizontal and temporal separation and greater alimentary specialization. Competition is greater in temperate communities because they depend heavily on litter as a food source and the depth of soil they can exploit is restricted by their smaller size.

39.4 PLASTICITY OF COMMUNITIES

39.4.1 Individual plasticity

Young *Millsonia anomala* usually live closer to the soil surface and withstand both drought and heat better than the adults do (Fig. 39.3). They ingest three to four times more soil per unit weight than the adults and can feed on soil from the upper 25 cm. However, the adults continue to grow only if they feed on soil from the 2–5 cm stratum (Lavelle *et al.*, 1980). Thus, young and adult *M. anomala* seem to occupy quite different ecological niches.

Functional plasticity* can also be demonstrated on a shorter time scale, for example in the rate of soil ingestion which varies in many species with

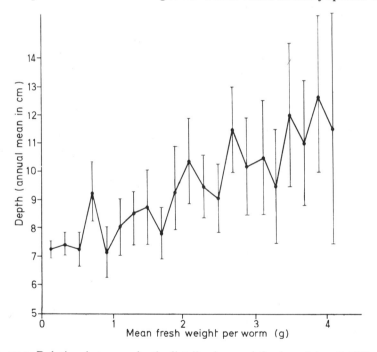

Fig. 39.3 Relation between depth distribution and fresh weight of *Millsonia anomala* ($N = 3365$; $R = 0.25$; $\alpha < 0.001$).

* The term 'function' used by the ecological group of the École Normale Supériéure is defined by the author as: 'La fonction des vers de terre est l'ensemble des activitiés qu'ils accomplissement dans le sol consideré en tant que

the soil organic matter content. Individual functional plasticity can also appear instantaneously. In *M. anomala*, soil ingestion has been shown to vary inversely with the soil hydrosoluble organic content (Lavelle *et al.*, in press). Earthworm behaviour also varies in relation to environmental factors, particularly soil moisture and temperature as reviewed by Satchell (1967). Activity is sometimes stopped by quiescence or diapause.

39.4.2 Functional plasticity of populations

Populations are composed of individuals which behave differently in relation to their age and size. The function of a population thus depends on its weight structure as well as on environmental factors. A factorial analysis showed that earthworm populations in the Lamto savanna pass through seven phenological stages characterized by weight structure, degree of activity and depth distribution (Lavelle and Meyer, 1976; Lavelle, 1978) (Fig. 39.4). The ecological effect of the population, for example on the soil ingested at different depths, the amount of organic matter directly mineralized through respiration, and the production of surface castings, thus varies directly with the seasonal cycle. In turn, the variations change the population structure.

At the end of the dry season, in February, when bush fires have destroyed the grass layer, the populations are inactive, biomass is reduced and the vertical distribution is relatively deep. They have a great proportion of young individuals that are better adapted than adults to withstand the high temperatures and irregular hydric regime as well as being able to feed on soil impoverished by the burning of the litter. Populations in this condition are represented in Fig. 39.4 by the phenological types 4, 2 and 1, differing according to the intensity of drought.

After a few months, the herb layer is restored and the soil microclimate becomes much more favourable. Litter accumulates and soluble organic

systeme écologique. Cette fonction globale peut se diviser en fonctions élémentaires d'ou la notion de structure fonctionelle = distribution de la fonction globale entre les differentes fonctions élémentaires, ici répresentées par les catégories écologiques. Example la fonction 'endogé oligohumique' consiste a ingérer de la terre en profondeur et a la rejeter dans le sol au même niveau aprés avoir remélangé les substrats organiques et la microflore et activé l'activité microbienne de façon sélective (activation des 'cellulolytiques' et inhibition des consommateurs d'hydrosolubles).

Ces fonctions, élémentaires ou globales, peuvent varier suivant les conditions du milieu entre certaines limites et avec un temps de reponse plus ou moins long, d'ou la notion de plasticité fonctionelle qui caractérise la vitesse et les capacités d'adaptation d'un individu, d'une population ou d'un peuplement au changement intervenu.' – ed.

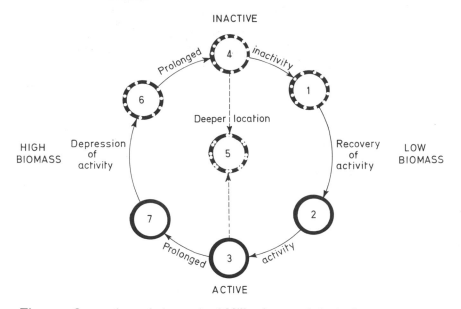

INACTIVE

ACTIVE

HIGH BIOMASS

LOW BIOMASS

Prolonged inactivity

Depression of activity

Recovery of activity

Deeper location

Prolonged activity

Fig. 39.4 Seasonal population cycle of *Millsonia anomala* in the Lamto savanna.

compounds enrich the topsoil. The earthworms are predominantly adults concentrated in the upper 10 cm, and the biomass is at a maximum, as is metabolic activity as measured by the quantity of soil ingested. The population is then in stage 3 or 5. Reproduction begins and intensifies with the coming of the dry season (stage 7). When drought comes, mortality increases, especially among adults, and the population passes into stage (6) with reduced activity, low biomass and a high proportion of juveniles located deeper in the soil. As the drought is established, the population passes again through stages 4, 2 or 1 according to the moisture conditions. This detail of cycle varies according to the type of savanna. Some of the environmental factors that synchronize these changes are known (e.g. Tsukamoto and Watanabe, 1977; Lavelle and Meyer, 1977; Hartenstein *et al.*, 1979; Neuhauser *et al.*, 1980). Earthworm populations thus appear to adapt to environmental conditions by changing both their activities and their demographic composition by variations in such developmental parameters as fecundity, growth rate and mortality. On the wider scale of the geographical area a species occupies, the mean demographic composition can also vary in relation to the environment. Niches filled by a species in one area may be left partially vacant in others and available for occupation by competing species. As each population adapts to the local environmental conditions it may acquire morpho-logical characteristics that are slightly different from those of other populations which differ in their average demographic constitution. The diversity of *Scherotheca* species, all with a quite restricted distribution (Bouché, 1972), may have arisen, at least partially, by this process.

39.5 DISCUSSION

The problem initially posed of the reasons for the success of megadrile earthworms can be explained by their great plasticity as individuals, populations and communities. Arthropods, which are divided into castes and stages, are prisoners of external skeletons which prevent any possibility of changing their size or alimentary regime. They must therefore resort to greater specialization. Microarthropod communities include tens to hundreds of species whose small size and niche specialization restrict them to relatively rich substrates. Thus, unlike earthworms, they do not function as litter movers or soil mixers.

Consideration of the factors that determine the function, composition and degree of activity of earthworm communities leads us to reassess competition and predation, the forces traditionally recognized as primordial in determining the organization of communities. To earthworms, predation is of quite minor importance for, although it may stimulate the reproduction of epigeic populations, its role is negligible when compared with the lethal effects of physical extremes in the environment.

The pressure of competition seems to be great in temperate regions where the amplitude of the corporate earthworm niche is quite narrow. As litter is the primary food, populations tend to concentrate close to the soil surface and a reduced range of size limits behavioural diversification. The result is greater food specialization and a wider niche separation temporally and in horizontal space. In tropical regions, the earthworm niche is enlarged by a greater food diversity and there is consequently a wider range in earthworm size and a reduction in competition pressure.

Besides these competitive pressures there is another organizing force – the system of interactions between fauna, microflora and organic matter which has evolved over millennia. This appears to be the principal source of evolution at the species and community levels. It is not so much the interactions between species that guide their evolution, as the relationships that each species maintains within the small ecosystem subunits, which include micro-organisms, nematodes, microarthropods and other soil animals. These units are organized around organic matter in various stages of mineralization and humification. Such small composite units seem to be the basic functional units of the soil system and should constitute the object of future research.

39.6 REFERENCES

Bachelier, G. (1963) *La Vie Animale dans les Sols*. O.R.S.T.O.M., Paris. 279 pp.

Bengtson, S. A., Nilsson, A., Nordström, S. and Rundgren, S. (1975) Habitat selection of lumbricids in Iceland. *Oikos*, **26**, 253–263.

Blueweiss, L., Fox, H., Kudzma, V., Nakashima, D., Peters, R. and Sams, S. (1978) Relationships between body size and some life history parameters. *Oecologia, (Berl.)*, **37**, 257–272.

Bolton, P. J. and Phillipson, J. (1976) Burrowing, feeding, egestion and energy budgets of *Allolobophora rosea* (Savigny) (*Lumbricidae*). *Oecologia (Berl.)*, **23**, 225–245.

Bornebusch, C. H. (1930) The fauna of forest soil. *Det Forstlige Forsogsvaesen*, **11**, 1–158.

Bouché, M. B. (1971) Relations entre les structures spatiales et fonctionelles des ecosystemes, illustrées par le rôle pédobiologique des vers de terre. In *La Vie dans les Sols, Aspects Nouveaux, Études Experimentales* (ed. P. Pesson), Gauthier-Villars, Paris, pp. 187–209.

Bouché, M. B. (1972) Lombriciens de France. Ecologie et systematique. *Ann. Zool. Ecol. Anim.*, No. hors serie, 671 pp.

Bouché, M. B. (1978) Fonction des Lombriciens. I. Recherches françaises et résultats d'un programme forestier cooperatif (RCP 40). *Bull. Sci. Bourgogne*, **20**, 143–227.

Cooke, A. and Luxton, M. (1980) Effects of microbes on food selection by *Lumbricus terrestris*. *Rev. Ecol. Biol. Sol*, **17**, 365–370.

Evans, A. C. and Guild, W. J. McL. (1947) Studies on the relationships between earthworms and soil fertility. I. Biological studies in the field. *Ann. Appl. Biol.*, **34**, 307–330.

Evans, A. C. and Guild, W. J. McL. (1948a) Studies on the relationships between earthworms and soil fertility. IV. On the life cycles of some British Lumbricidae. *Ann. Appl. Biol.*, **35**, 471–484.

Evans, A. C. and Guild, W. J. McL. (1948b) Studies on the relationships between earthworms and soil fertility. V. Field populations. *Ann. Appl. Biol.*, **35**, 485–493.

Fenchel, T. (1974) Intrinsic rate of natural increase: the relationship with body size. *Oecologia, (Berl.)*, **14**, 317–326.

Ferrière, G. (1980) Fonctions des Lombriciens. VII. Une methode d'analyse de la matiere organique vegetale ingerée. *Pedobiologia*, **20**, 263–273.

Gerard, B. M. (1967) Factors affecting earthworms in pastures. *J. Anim. Ecol.*, **36**, 235–252.

Hartenstein, R., Neuhauser, E. F. and Kaplan, D. L. (1979) Reproductive potential of the earthworm *Eisenia foetida Oecologia (Berl.)*, **43**, 329–340.

Lavelle, P. (1973) Peuplement et production des vers de terre dans les savanes de Lamto. *Ann. Univ. Abidjan*, Ser. E, **VI**, 79–98.

Lavelle, P. (1978) Les vers de terre de la savane de Lamto (Côte d'Ivoire): peuplements, populations et fonctions dans l'écosystème. Thèse Doctorat, Paris VI. Publ. Lab. Zool. ENS, **12**, 301 pp.

Lavelle, P. (1981) Stratégies de reproduction chez les vers de terre. *Acta Oecol./Oecol. Gen.*, **2**, 117–133.

Lavelle, P. and Cruz, I. (in press) Soil ingestion, growth and fecundity of the earthworm *Ponthoscolex corethrurus* (Glossoscolecidae) in relation to soil water content. In *New Trends in Soil Biology* (eds. Ph. Lebrun, H. M. Andŕe, A. de Medts, C. Grégoire-Wibo and G. Wauthy).

Lavelle, P., Maury, M. E. and Serrano, V. (1981) Estudio cuantitativo de la fauna del suelo en la región de Laguna Verde, Vera Cruz. Epoca delluvias, *Inst. Ecol. Pub.* **6**, 75–105.

Lavelle, P. and Meyer, J. A. (1976) Les populations de *Millsonia anomala* (Acanthodrilidae-Oligochetes): structure, variations spatio-temporelles et pro-duction. Application d'une analyse multivariée (programme Constel). *Rev. Ecol. Biol. Sol.*, **13**, 561–577.

Lavelle, P. and Meyer, J. A. (1977) Modelisation et simulation de la dynamique, de la production et de la consommation des populations du ver de terre geophage *Millsonia anomala* (Oligochetes-Acanthodrilidae) dans la savane de Lamto (Cote d'Ivoire). In *Soil Organisms as Components of Ecosystems* (eds. U. Lohm and T. Persson), *Proc. VIth Int. Soil Zool. Coll., Ecol. Bull. (Stockholm)*, **25**, 420–430.

Lavelle, P., Sow, B. and Schaefer, R. (1980) The geophagous earthworm community in the Lamto savanna (Ivory Coast): Niche partitioning and utilization of soil nutritive resources. In *Soil Biology as Related to Land Use Practices* (ed. D. L. Dindal), E. P. A., Washington D.C., pp. 653–672.

Lavelle, P., Zaidi, Z. and Schaefer, R. (in press) An interaction system among earthworms, soil organic matter and microflora in an African savanna (Lamto, Ivory Coast). In *New Trends in Soil Biology* (eds. Ph. Lebrun, H. M. André, A. de Medts, C. Grégoire-Wibo and G. Wauthy).

Laverack, M. S. (1963) *The Physiology of Earthworms*. Pergamon, Oxford. 206 pp.

Lee, K. E. (1959) The earthworm fauna of New Zealand. *Bull. N.Z. Dep. Sci. Ind. Res.*, **130**, 1–486.

MacArthur, R. H. and Wilson, E. O. (1967) Some generalized theorems of natural selection. *Proc. Natl. Acad. Sci. U.S.A.*, **48**, 1893–1897.

Mill, P. J. (1978) *Physiology of Annelids*. Academic Press, London. 638 pp.

Nemeth, A. (1981) Estudio ecológico de las lombrices de tierra (Oligochaeta) en ecosistemas de bosque humedo tropical en San Carlos de Rio Negro, Territorio Federal Amazonas. Thesis, Universidad Central de Venezuela. 92 pp.

Neuhauser, E. F., Hartenstein, R. and Kaplan, D. L. (1980) Growth of the earthworm *Eisenia foetida* in relation to population density and food rationing. *Oikos*, **35**, 93–98.

Nordström, S. and Rundgren, S. (1973) Associations of lumbricids in Southern Sweden. *Pedobiologia*, **13**, 301–326.

Nordström, S. and Rundgren, S. (1974) Environmental factors and lumbricid associations in Southern Sweden. *Pedobiologia*, **14**, 1–27.

Perel, T. S. (1975) Lifeforms of Lumbricidae. *Zh. Obshch. Biol. Moscou*, **36**, 189–202. [In Russian.]

Persson, T. and Lohm, U. (1977) Energetic significance of the annelids and arthropods in a Swedish grassland soil. *Ecol. Bull. (Stockholm)*, **23**, 211 pp.

Phillipson, J., Abel, R., Steel, J. and Woodell, S. R. J. (1976) Earthworms and the factors governing their distribution in an English beechwood. *Pedobiologia*, **16**, 258–285.

Phillipson, J., Abel, R., Steel, J. and Woodell, S. R. J. (1978) Earthworm numbers, biomass and respiratory metabolism in a beech woodland – Wytham Woods, Oxford. *Oecologia (Berl.)*, **33**, 291–309.

Pianka, E. R. (1974) Niche overlap and diffuse competition. *Proc. Natl. Acad. Sci. U.S.A.*, **71**, 2141–2145.

Rosa, D. (1882) Sulla distribuzione verticale dei Lumbricidi. *Boll. Mus. Torino*, **2**.

Satchell, J. E. (1967) Lumbricidae. In *Soil Biology* (eds. A. Burgess and F. Raw), Academic Press, London, pp. 259–322.

Satchell, J. E. (1980) r worms and K worms: a basis for classifying earthworm strategies. In *Soil Biology as Related to Land Use Practices* (ed. D. L. Dindal), E.P.A., Washington D.C., pp. 848–864.

Satchell, J. E. and Lowe, D. G. (1967) Selection of leaf litter by *Lumbricus terrestris*. In *Progress in Soil Biology* (eds. O. Graff and J. E. Satchell), North Holland Publishing Co., Amsterdam, pp. 102–119.

Senapati, B. K. (1980) Aspects of ecophysiological studies on tropical earthworms (Distribution, population dynamics, production, energetics and their role in the decomposition process). Ph.D., Sambalpur University, India. 154 pp.

Tsukamoto, J. and Watanabe, H. (1977) Influence of temperature on hatching and growth of *Eisenia foetida* (Oligochaeta, Lumbricidae). *Pedobiologia*, **17**, 338–343.

Wright, M. A. (1972) Factors governing ingestion by the earthworm *Lumbricus terrestris* with special reference to apple leaves. *Ann. Appl. Biol.*, **70**, 175–188.

Chapter 40

The scientific names
of earthworms

R. W. SIMS

'The object . . . is to promote stability and universality in the scientific names of animals, and to ensure that each name is unique and distinct.' – *International Code of Zoological Nomenclature. Preamble, paragraph 2.*

40.1 INTRODUCTION

The taxonomy of earthworms has undergone a profound reorganization in recent years due partly to reappraisals of established data and concepts but more importantly to the accumulation of new information. Some changes have been startlingly novel, immoderately subjective or seemingly inconsistent, creating an impression of esoteric confusion. The situation should not however be seen as chaotic but indicative of the labours of dedicated workers striving to advance our knowledge of the group. Unfortunately little of this progress has been reported in compilatory works, and, as the information has not been sifted, the rejected and the accepted, the old and the new co-exist to the confusion of the non-specialist.

40.2 GENERAL

Recent changes include the new hierarchy of names of earthworm groups with five superfamilies containing 20 or more families and subfamilies (Sims, 1980). Major advances in North American studies have led to many western species previously contained within a heterogeneous Indo–Australasian genus, *Plutellus*, being assigned to the smaller but more homogeneous genus *Argilophilus* (Gates, 1977), while the new genera *Komarekiona* and *Lutodrilus* have been erected to accommodate newly described lumbricoids (Gates, 1974; McMahan, 1976). In Australia, the research of Jamieson has wrought long overdue name changes (Jamieson, 1974), as have the revisionary studies on the Asio–Australasian megascolecid genus *Pheretima* by Sims and Easton (1972) and Easton (1979, 1984). This huge taxon contained nearly eight hundred nominal species of which less than half are now regarded as valid and assigned to eight different genera with many of the common allochthonous (peregrine) species contained within the genus *Amynthas* Kinberg, 1867. Even greater changes have occurred in the taxonomy of the Lumbricidae, possibly because it occurs throughout the eastern Nearctic and western and central Palaearctic, an area containing more workers than any other. Research has progressed at different rates from one country to another with the emphases varying between taxonomy, distribution and ecology. Some reports thus retain an outmoded taxonomy while others incorporate untested and perhaps transitory proposals. Added to this, non-specialists have tended to consult literature principally in their own language regardless of its age, or have taken miscellaneous advice, so producing the now familiar nomenclative and taxonomic discrepancies.

40.3 NAME CHANGES OF COMMON LUMBRICIDAE

At one time, the lob worm or night-crawler was known by two names, *Lumbricus terrestris* Linnaeus, 1758 and *Enterion herculeum* Savigny, 1826. This problem disappeared (Bouché, 1976) when it became apparent that Savigny misapplied the Linnaean name and gave it to a species now recognized as *giardi* Ribaucourt, 1901. Perhaps one of the more troublesome names has been the generic cognomen *Allolobophora*. Eisen (1873) did not indicate a type species in the original description but simply included seven heterogeneous species nowadays accommodated in several different genera. Over the years, other species were added to *Allolobophora* until it became a 'catch-all' and a source of confusion until Omodeo (1956) designated *chlorotica* Savigny, 1826 (one of the 'odd men out' of the included species) as the type species. This made it impossible for workers any longer to ignore the problem, as many species previously assigned to the taxon were excluded. Bouché (1972) grasped the nettle and redefined *Allolobophora sensu stricto* for a small group of species allied to *chlorotica*. He then placed the related species *caliginosus* (Savigny, 1826) and *longus* (Ude, 1886) in a new genus *Nicodrilus* with *terrestris* (Savigny) = *giardi* (Ribaucourt, 1901) as the type species. This action had the virtue of bringing together several allied species from the 'catch-all' genus although many more still remained in a generic limbo which he categorized as '*Allolobophora sensu lato*'. Unfortunately this particular solution to a difficult problem did not bring stability to the nomenclature. In the first place it stimulated further research into the inter-relationships of the species comprising the residual groups in *Allolobophora sensu lato* with the recognition of further taxa, for example *Perelia* Easton, 1983 (see below) [nom. nov. pro *Allolobophora* (*Svetlovia*) Perel, 1976; non *Svetlovia* Chekanovskay, 1975 (Tubificidae)]. Then the new genus *Nicodrilus* Bouché, 1972 was placed as a junior synonym of *Aporrectodea* Örley, 1885 because the two type species are congeneric, the type species of *Aporrectodea* being *trapezoides* Dugès, 1828, a junior synonym of *caliginosa* Savigny, 1826, a species originally included in the genus *Nicodrilus*.

The species *caliginosa* Savigny, 1826 also causes confusion as several well-known species are placed in its synonymy. Some were proposed in ignorance of Savigny's name but others were recognized as representing phases within the full range of morphological variation of the species caused by growth, sexual and seasonal changes, with each stage possibly further affected by phenotypic modifications. Nowadays (Zicsi, 1982), *caliginosa* Savigny, 1826 includes *trapezoides* Dugès 1828; *tuberculata* Eisen, 1874 (non *tuberculata* Cernosvitov, 1935); *nocturna* Evans, 1946 and *turgida* Eisen, 1873. The last name is sometimes resurrected by North American authors on the grounds that Savigny did not clearly define his species (Reynolds, 1977). If this practice were followed then many eighteenth and nineteenth century names would have to be suppressed and replacements provided. It is unnecessary since *caliginosa* has been established by usage and long ago was fully characterized by later authors – moreover there is no nomenclatural confusion. (A similar situation has arisen over the use of the name *pygmaea* Savigny, 1826 with a few authors uncertain whether to apply to this species the names of the junior synonyms *minima* Rosa, 1889 or *cognettii* Michaelsen, 1903 (nom. nov. pro *ribaucourti* Cognetti, 1901; non Bretscher, 1901).)

40.4 NAMES OF NEW LUMBRICIDAE

Other new lumbricid genera have been described either to recognize groups detected during taxonomic studies, for example *Fitzingeria* Zicsi, 1978, or to accommodate new species discovered in seldom-sampled localities or habitats, e.g. forests instead of cultivated land, or obtained by employing different collecting techniques, e.g. extracting by chemical means where digging is impracticable. Thus Bouché described *Orodrilus, Ethnodrilus, Prosellodrilus* and *Kritodrilus* from parts of France largely unaffected by the Pleistocene ice-sheets, while Zisci (1981) has widened our knowledge of the Lumbricidae by describing new species of

Octodrilus Omodeo, 1956 with inconspicuous male pores located in the clitellar region far behind the customary lumbricid position on segment 15.

40.5 THE *Octolasion* PROBLEMS

40.5.1 Orthography

Perhaps one of the most confusing tangles of names centres around the group usually known by the generic name *Octolasion*. One cause has been a misunderstanding on the part of a few nineteenth century zoologists about the latinization of scientific names. Instead of latinizing only new names, Michaelsen (1900) in *Das Tierreich* retrospectively latinized established Greek-derived names in error, in particular *Octolasion* to *Octolasium*, *Bimastos* to *Bimastus* and in the genus *Eisenia*, he emended the species name *fetida* to *foetida*. Under Articles 32 and 33 of the *Code of Zoological Nomenclature*, Michaelsen's emendations are clearly invalid and moreover were contrary to the nomenclatural procedures of his own day. Unfortunately Michaelsen's invalid emendations have been perpetuated.

40.5.2 Nomenclature

An oversight of the literature (including the *International Code of Zoological Nomenclature*) in studies on species associated with the genus *Octolasion* has left an unresolved problem akin to a ticking time-bomb that has the potential to destroy nomenclatural stability. It began in 1867 when Kinberg defined several genera including *Alyattes* and described the species *Lumbricus alyattes* which Michaelsen (1899) validly designated as the type species of *Alyattes* by virtual tautonymy (*Int. Code Zool. Nomencl.* Article 69 and Recommendation 69B) and recognized it as a junior synonym of *cyaneum* Savigny, 1826. Strangely, Michaelsen in *Das Tierreich* (1900) did not apply priority in this instance and *Octolasion* Örley, 1885 was given precedence while *Alyattes* Kinberg, 1867 was buried in synonymy. This situation is serious but worse was to follow. Omodeo (1952) reviewed the genus *Octolasion* (employing Michaelsen's improper emendation of *Octolasium*) and designated *complanatum* Dugès, 1828 as the type species. He then went on to describe the subgenera *Octolasion* (*Incolore*) with *lacteum* Örley, 1881 as the type species and *Octolasion* (*Purpureum*) with *lissaense* Michaelsen, 1891 as its type species. Next, Omodeo (1956) reviewed the family Lumbricidae and had (invalid) second thoughts about *Octolasion* (spelt *Octolasium*) because now he designated *lacteum* Örley, 1881 as the type species in place of *complanatum*

Dugès, 1828. This became instead the type species of a new subgenus *Octolasion* (*Octodrilus*) while his 1952 subgenera were ignored.

His actions in attempting to change the type-species were invalid, clearly contravening Article 61 of the 'Code' and contributed to a tangle subsequently exacerbated by authors following the 1956 proposals. However, authors should *not* now start applying correct nomenclatural procedures as application is being made to the International Commission on Zoological Nomenclature to use their plenary powers to deal with this problem. Clearly in the interests of stability it is best to continue to follow recent usage, e.g. Bouché (1972). The name *Alyattes* Kinberg, 1867 needs to be suppressed and the designations and new subgenera proposed by Omodeo in 1952 should be set aside while this author's actions in 1956 should be validated. Without action by the Commission all species now in *Octolasion* would be transferred to *Alyattes* and all species currently in *Octodrilus* would be accommodated in a new concept *Octolasion*, undesirable actions likely to cause instability, confusion and lack of universality of nomenclature.

40.5.3 Taxonomy

In the genus *Octolasion* a multiplicity of names has arisen from a lamentable reluctance on the part of a few oligochaete taxonomists to seek and examine type material or read the original descriptions of species. Thus in *Octolasion* s.s. two species are common, *cyaneum* with the lips of the male pores confined to segment 15, and its allied form with the lips extending across segments 14, 15 and 16. The latter species was named *tyrtaeum* from material collected around Paris (Ile de France) by Savigny (1826). This name was long overlooked and the species was often known by the name *lacteum* given to Hungarian material by Örley (1881). Recently, Bouché (1972) separated French populations of this species into two subspecies, principally on the occurrence of the tubercula pubertatis: an eastern element, chiefly east of the Rhône and a southern element, mainly Pyrenean, which he associated with the central European (Hungarian) populations, and a central, north and western element associated with British and other western European populations. Bouché (1976) now recognizes the central European subspecies as *Octolasion tyrtaeum tyrtaeum* (Savigny, 1826) and the western European subspecies as *O. t. gracile* (Örley, 1885). Thus the Hungarian *lacteum* Örley, 1881 has become a junior synonym of the Parisian *tyrtaeum* Savigny, 1826 although only *gracile* Örley, 1885 is now recorded around Paris. Clearly this problem has not yet been fully resolved so there is the possibility of further subspecific name changes. Fortunately the specific name should remain stable.

40.6 BIOSPECIES: PROBLEMS AND LIMITATIONS OF NOMENCLATURE

Some problems cannot be readily solved, for example how and whether it is desirable, to reflect in the formal scientific nomenclature, the genetic distinction reported by Satchell (1967) between the green phase of *Allolobophora chlorotica* Savigny, 1826 and the red phase, for which Savigny also provided the name *virescens*. This phenomenon is not isolated and a parallel is once again attracting attention in the atypical lumbricid *Eisenia fetida* Savigny, 1826. This commercially important species may be biparental, self-fertilizing or parthenogenetic but whatever the mechanism, the reproductive rate can be high under favourable conditions. Uniformly pigmented forms are commonly encountered under intense culturing and since these seem to be derived from normally striped forms, many observers believe them to be phenotypic variants. Other workers, however, regard the uniformly pigmented phase as a separate species (Avel, 1937; Jaenike *et al.*, 1982). Despite presenting evidence of the apparent biological isolation of these two phases, André (1963) made them conspecific by naming the uniformly pigmented phase 'var. *unicolor*' (the striped phase he named '*typica*'). As varietal names described after 1961 are regarded as infrasubspecific and have no status in nomenclature, Bouché (1972) validly provided the uniformly pigmented worm with the new subspecific name *Eisenia fetida andrei*. These actions were taken despite the fact that subspecies are usually regarded as morphologically distinct, potentially or actually interbreeding conspecific populations, whereas a species can be defined as a reproductively isolated group of interbreeding populations. Nevertheless it is difficult to apply the scientific nomenclature to these forms since taxonomy is ultimately based on preserved type-specimens in which the distinguishing characters can be observed, and, unfortunately, pigments in earthworms do not permanently survive preservation. Before attempting to apply a name, we need to be certain of the necessity of employing the formal system of zoological nomenclature. There is too a precedent for employing permanent characters: a physiological subspecies of the starling, *Sturnus vulgaris*, was once rejected because blood groups were used to differentiate between migratory mainland European and non-migratory British populations.

40.7 FUTURE TRENDS

Taxonomic research on the family Lumbricidae appears to be reaching the end of an era. The long period of taxa description, alpha taxonomy, is nearly over and is giving way to an age of revision, beta taxonomy. With this change in emphasis, the kind of turmoil outlined above will cease and permit one of the aims of the International Organization of Oligochaete Taxonomists to be realized. This organization, formed in Braunschweig in 1976, was founded by taxonomists anxious to obtain universality and remove confusion in earthworm taxonomy, especially in the family Lumbricidae for which the production of a modern checklist was seen as the best method of achieving stability. Seven years later much work still remains to be done before a widely accepted, definitive checklist can be produced; it is encouraging therefore that a start is made below with the production of a provisional list intended mainly as a discussion document for taxonomists. The list of valid species by E. G. Easton is clearly a synthesis and summarizes the studies of many others, especially Bouché (1972), Perel (1979) and Zicsi (1982).

The publication of this list is not intended to freeze or restrict taxonomic research, in fact the contrary, but more important it can provide ecologists, physiologists, geneticists and other non-taxonomists with a simple guide to the combinations of names currently applied to valid species of Lumbricidae. (The generic groups recognized are often units of convenience reflecting a broad concensus of agreement but doubtlessly some conflict, or are inconsistent in part with firmly held views of a few workers.) Therefore we should be grateful to Mr Easton for providing this guide, for its use will remove doubt and prevent confusion for non-taxonomists and provide the taxonomist with a document for discussion as well as a comprehensive list of the valid and nominal species currently included in the family Lumbricidae. Hopefully it will stimulate constructive discussions and lead to the early production of a widely accepted definitive checklist to attain a greater stability in lumbricid taxonomy than we have known hitherto.

40.8 REFERENCES

André, F. (1963) Contribution à l'analyse expérimentale de la reproduction des lombriciens. *Bull. Biol. Fr. Belg.*, **97**, 1–101.

Avel, M. (1937) *Titres et travaux scientifiques*. Delmas, Bordeaux.

Bouché, M. B. (1972) Lombriciens de France. Écologie et systématique. *Ann. Zool. Écol. Anim.*, (Num. spec.) **72-2**, 671 pp.

Bouché, M. B. (1976) Contribution à la stablisation de la nomenclature des Lumbricidae, Oligochaeta I. Synonymies et homonymies d'espèces du Bassin Parisien. *Bull. Mus. Natn. Hist. Nat. (Ser. 3, No. 354) Zool.*, **247**, 81–88.

Easton, E. G. (1979) A revision of the *Pheretima* group (Megascolecidae: Oligochaeta): *Archipheretima, Metapheretima, Planapheretima, Pleinogaster* and *Polypheretima. Bull. Br. Mus. Nat. Hist. (Zool.)*, **35**, 1–126.

Easton, E. G. (1984) The allochthonous, perichaetine earthworms (Megascolecidae: Oligochaeta) of the world. *Bull. Br. Mus. Nat. Hist. (Zool.)*, (in press).

Eisen, G. (1873) Om Skandinaviens Lumbricider. *Ofvers K. VetenskAkad. Förh. Stockholm*, **30**, 43–56.

Gates, G. E. (1974) On a new species of earthworm in a southern portion of the United States. *Bull. Tall Timbers Res. Stn.*, **No. 15**, 1–13.

Gates, G. E. (1977) On the correct generic name for some west coast native earthworms, with aids for a study of the genus. *Megadrilogica* **3(4)**, 54–60.

Jaenike, J., Ausubel, S. and Grimaldi, D. A. (1982) On the evolution of clonal diversity in parthenogenetic earthworms. *Pedobiologia*, **22**.

Jamieson, B. G. M. (1974) The indigenous earthworms (Megascolecidae: Oligochaeta) of Tasmania. *Bull. Br. Mus. Nat. Hist. (Zool.)*, **26**, 204–328.

Kinberg, J. G. H. (1867) Annulata nova. *Öfvers K. VetenskAkad. Förh. Sotckholm.*, **23**, 97–103 and 356–357.

McMahan, M. L. (1976) Preliminary notes on a new megadrile species, genus and family from the southeastern United States. *Megadrilogica*, **2(11)**, 6–8.

Michaelsen, W. (1899) Revision der Kinbergschen Oligochäten-Typen. *Öfvers K. VetenskAkad. Förh. Stockholm*, **56**, 413–448.

Michaelsen, W. (1900) Oligochaeta. *Tierreich*, **10**, 1–575.

Omodeo, P. (1952) Materiali zoologici raccolti dal Dr. Marcuzzi sulle Alpi Dolomitiche. *Arch. Zool. Ital.*, **37**, 29–59.

Omodeo, P. (1956) Contributo alla revisione dei Lumbricidae. *Arch. Zool. Ital.*, **41**, 131–212.

Örley, L. (1881) A Magyarorszagi Oligochaetak Paunaja. I. Terricolae. *Math. Term. Közlem. Magyar Akad.*, **16**, 563–611.

Perel, T. S. (1979) *Range and Regularities in the Distribution of Earthworms of the USSR Fauna.* Moscow Nauka, (Academy of Sciences of the USSR, Laboratory of Forest Science). 272 pp. [In Russian, English summary].

Reynolds, J. W. (1977) The earthworms (Lumbricidae and Sparganophilidae) of Ontario. *Life Sci. Misc. Publs. R. Ont. Mus.*, 141 pp.

Satchell, J. E. (1967) Colour dimorphism in *Allolobophora chlorotica* Sav. (Lumbricidae) *J. Anim. Ecol.*, **36**, 623–630.

Savigny, J. C. (1826) La multiplicité des espèces de ver de terre . . . *Mem. Acad. Sci. Inst. Fr. (Phys.)*, **5**, 176–184.

Sims, R. W. (1980) A classification and the distribution of earthworms, suborder Lumbricina (Haplotaxida: Oligochaeta). *Bull. Br. Mus. Nat. Hist. (Zool.)*, **39**, 103–124.

Sims, R. W. and Easton, E. G. (1972) A numerical revision of the earthworm genus *Pheretima* auct. (Megascolecidae: Oligochaeta) with the recognition of new genera and an appendix on the earthworms collected by the Royal Society North Borneo Expedition. *Biol. J. Linn. Soc.*, **4**, 169–268.

Zicsi, A. (1981) Weitere Angaben zur Lumbricidenfauna Italiens (Oligochaeta: Lumbricidae). *Opusc. Zool. Budapest*, **17/18**, 157–180.

Zicsi, A. (1982) Verzeichnis der bis 1971 beschriebenen und revidierten Taxa der Familie Lumbricidae. *Acta Zool. Hung.*, **28/34**, (in press).

Chapter 41

A guide to the valid names of Lumbricidae (Oligochaeta)

E. G. EASTON

41.1 PROVISIONAL LIST OF VALID NAMES UP TO DECEMBER 1981

Family LUMBRICIDAE Rafinesque-Schmaltz, 1815

Subfamily DIPORODRILINAE Bouché, 1970

Diporodrilus Bouché, 1970
Type species: *Diporodrilus pilosus*
Bouché, 1970

D. omodeoi Bouché, 1970
D. pilosus Bouché, 1970

Subfamily LUMBRICINAE Rafinesque-Schmaltz, 1815

Allolobophora Eisen, 1874 (sensu Perel, 1976)
Type species: *Enterion chloroticum*
Savigny, 1826

A. carpathica Cognetti, 1927
  (*omodeoi* Zajonc, 1963)
A. chlorotica chlorotica (Savigny, 1826)
  (*virescens* Savigny, 1826; *anatomicus* Dugès, 1828; *riparius* Hoffmeister, 1843; *viridis* Johnston, 1865; *riparius rufescens* Eisen, 1871; *riparius pallescens* Eisen, 1871; *neglecta* Rosa, 1882; *cambrica* Friend, 1892; *chlorotica curiosa*

Ribaucourt, 1896; *hortensis* Bretscher, 1901)
A. chlorotica postepheba Bouché, 1972
A. chlorotica waldensis Ribaucourt, 1896
  (*chlorotica morgensis* Ribaucourt, 1896; *nusbaumi* Ribaucourt, 1896)
A. chlorotica kosovensis Sapkarev, 1975
A. dacica (Pop, 1938)
A. eiseni (Levinsen, 1884)
  (*rubra* Bretscher, 1900; *eiseni gracilis* Friend, 1911; *merciensis* Friend, 1911; *oltenicus* Pop, 1938)
A. leoni Michaelsen, 1891
A. parva Eisen, 1874
  (*beddardi* Michaelsen, 1894; *parva udei* Ribaucourt, 1896; *constricta geminata* Friend, 1897)
A. pseudonematogena (Perel, 1967)

Allolobophora (s.l.)

A. aharonii (Stephenson, 1922)
A. andreinii (Baldasseroni, 1907)

475

A. apuliae (Baldasseroni, 1913)

A. atlantica (Bouché, 1969)

A. balcanica balcanica (Cernosvitov, 1942)
(*macedonica* Sapkarev, 1971)

A. balcanica plavensis (Karaman, 1972)

A. balcanica sineporis (Omodeo, 1952)

A. bartolii bartolii (Bouché, 1970)
[*icterica* forma *dicystis* Bartoli, 1963 non Cernosvitov, 1942 (= *Aporrectodea icterica dicystis*)]

A. bartolii alpemarea (Bouché, 1970)

A. bartolii meougensis (Bouché, 1970)

A. burgondiae Bouché, 1972

A. chaetophora Bouché, 1972

A. cuginii Rosa, 1905

A. cupulifera Tétry, 1937

A. demirkapiae Karamen, 1969

A. eurytanica (Tzelepe, 1943)

A. fernandae Graff, 1957

A. festae Rosa, 1892

A. gestroides Zicsi, 1970

A. getica Pop, 1947

A. graffi Bouché, 1972

A. haymozi Zicsi, 1977

A. hrabei (Cernosvitov, 1935)

A. kosowensis kosowensis Karamen, 1968

A. kosowensis montenegrina Sapkarev, 1975

A. kuzuvensis Sapkarev, 1971

A. lanzai (Omodeo, 1961)

A. ligra (Bouché, 1969)

A. lopezi Bouché, 1979

A. madeirensis Michaelsen, 1891

A. micella Bouché, 1972

A. mozsaryorum Zicsi, 1974

A. oliveirae Rosa, 1894

A. opisthosellata Graff, 1961

A. orionse Zicsi, 1977

A. osellai Zicsi, 1981

A. paratuleskovi Sapkarev, 1975

A. pereli Bouché, 1972

A. putricola putricola Bouché, 1972

A. putricola tebra Bouché, 1972

A. sapkarev **nom. nov.**
(*A. zicsi* Sapkarev, 1975 non Bouché, 1972)

A. satchelli Bouché, 1972

A. segalensis Bouché, 1972

A. sarnensis (Pierantoni, 1904)

A. sturanyi sturanyi Rosa, 1895

A. sturanyi dacidoides Bouché, 1973

A. smaragdina Rosa, 1892

A. tiginosa Bouché, 1972

A. tuleskovi (Cernosvitov, 1934)

A. vardarensis Sapkarev, 1971

A. vasconensis Bouché, 1979

A. virei (Cognetti, 1902)

A. zarandensis Pop, 1978

A. zicsi Bouché, 1972
[non: *A. zicsi* Sapkarev, 1975 (= *A. sapkarev*)]

Aporrectodea Örley, 1885 (sensu Perel, 1976)

Type species: *Lumbricus trapezoides* Dugès, 1828 (= *A. caliginosa trapezoides*)

Syn: *Eiseniona* Omodeo, 1956 (type species: *Allolobophora handlirschi* Rosa, 1897)
Nicodrilus (Nicrodrilus) Bouché, 1972 [type species: *Enterion terrestre*: Savigny, 1826 (= *A. giardi* Ribaucort, 1901)]
Nicodrilus (Rhodonicus) Bouché, 1972 (type species: *Allolobophora arverna* Bouché, 1969)

A. arverna (Bouché, 1969)

A. balisa (Bouché, 1972)

A. caliginosa caliginosa (Savigny, 1826)
(*carneum* Savigny, 1826; *lividus* Templeton, 1836; *gordianus* Templeton, 1836; *purus* Dugès, 1837; *hortensiae* Kinberg, 1867; *novaehollandiae* Kinberg, 1867; *communis olivaceus* Eisen, 1871: *communis pellucidus* Eisen, 1871; *turgida turgida* Eisen, 1874; *turgida tuberculata* Eisen, 1874; *levis* Hutton, 1877; *australiensis* Fletcher, 1886; *inflata* Michaelsen, 1900; *borellii* Cognetti, 1904; *similis* Friend, 1910; *remyi* Cernosvitov, 1929; *caliginosa hellenica* Tzelepe, 1943; *nocturna* Evans, 1946; *iowana* Evans, 1948; *arnoldi* Gates, 1952: *molita* Gates, 1952)

A. caliginosa alternisetosa (Bouché, 1972)

A. caliginosa meridionalis (Bouché, 1972)

A. caliginosa obscuricola (Cernosvitov, 1936)

A. caliginosa trapezoides (Dugès, 1828)
(*capensis* Kinberg, 1867; *caliginosa beddardi* Ribaucourt, 1896; *mariensis* Stephenson, 1917; *augilensis* Sciacchitano, 1931; *samarigera graeca* Cernosvitov, 1938)

A. dubiosa (Örley, 1881)
(*blinovi* Cernosvitov, 1938; *dubiosa pontica* Pop, 1938)

A. georgii (Michaelsen, 1890)
(*georgii transylvanica* Pop, 1938)

A. giardi giardi (Ribaucourt, 1901)
(*terrestre*: Savigny, 1826; *bretscheri* Cognetti, 1903; *longus occulatus* Cognetti, 1904)

A. giardi voconea (Bouché, 1972)

A. gogna (Bouché, 1972)

A. handlirschi handlirschi (Rosa, 1897)
(*rhenani* Bretscher, 1899; *vejdovskyi* Bretscher, 1899; *pallida* Bretscher, 1900; *aporata* Bretscher, 1901; *rosea alpina* Vedovini, 1967)

A. handlirschi mahnerti (Zicsi, 1973)

A. icterica icterica (Savigny, 1826)
(*claparedi* Ribaucourt, 1896; *capilla* Ribaucourt, 1901)

A. icterica dicystis (Cernosvitov, 1942) [non Bartoli, 1963 (= *Allolobophora* (sl) *bartolii bartolii*)]

A. icterica occidentalis (Graff, 1957)

A. icterica pannonica (Cognetti, 1906)

A. jassyensis jassyensis (Michaelsen, 1891)
(*jassyensis orientalis* Michaelsen, 1897; *cavaticus* Michaelsen, 1910)

A. jassyensis phoebea (Cognetti, 1913)

A. limicola (Michaelsen, 1890)

A. longa longa (Ude, 1885)
(*lactea* Friend, 1892; *intermedium* Friend, 1909)

A. longa ripicola (Bouché, 1972)

A. rosea rosea (Savigny, 1826)

(*mucosa* Eisen, 1874; *aquatilis* Vejdovsky, 1875; *mediterranea* Örley, 1881; *rosea macedonica* Rosa, 1893; *danielirosai* Ribaucourt, 1896; *alpestris* Bretscher, 1899; *nobilii* Cognetti, 1903; *diomedaeus* Cognetti, 1906; *rosea glandulosa* Friend, 1910; *prashadi* Stephenson, 1922; *rosea storkani* Cognetti, 1934; *dairenensis* Kobayashi, 1940; *harbinensis* Kobayashi, 1940; *hataii* Kobayashi, 1940; *jeholensis* Kobayashi, 1940; *rosea paucipartita* Tzelepe, 1943; *rosea dendrobaenoides* Omodeo, 1950; *jenensis* Fuller, 1953; *moderata* Csekanovszkaja, 1959; *rosea interposita* Plisko, 1965)

A. rosea balcanica (Cernosvitov, 1942)

A. rosea bimastoides (Cognetti, 1901)
(*kulagini* Malevics, 1949; *rosea troglodyta* Alvarez, 1971)

A. rosea budensis (Szuts, 1909)

A. rubra rubra (Vedovini, 1969)
(*rubicundus rubicundus* Bouché, 1972)

A. rubra acidicola (Bouché, 1972)

A. velox (Bouché, 1967)

Bimastos Moore, 1893 (sensu Zicsi, 1981)
Type species: *Bimastos palustris* Moore, 1895 (= *Bimastos* sp. Moore, 1893)
Syn: *Bimastus*: Michaelsen, 1900 (invalid emendation of *Bimastos*) *Spermophorodrilus* Bouché, 1975 [type species: *S. albanianus* Bouché, 1975 (= *B. antiquus michalisi*)]

B. antiquus antiquus (Cernosvitov, 1938)

B. antiquus bouchei Zicsi and Michalis, 1981

B. antiquus michalisi (Karaman, 1972)
(*albanianus* Bouché, 1975)

B. antiquus tuberculatus (Tzelepe, 1943)

B. baloghi Zicsi, 1981
B. gieseleri gieseleri (Ude, 1895)
B. gieseleri hempeli (Smith, 1915)
B. hauseri (Zicsi, 1973)
B. heimbergeri Smith, 1928
B. palustris Moore, 1895
    (*B. sp.* Moore, 1893)
B. samarigerus (Rosa, 1893)
B. schweigeri Zicsi, 1981
B. syriacus (Rosa, 1893)
    (*atheca* Cernosvitov, 1940; *atheca kosswigi* Omodeo, 1952)
B. tumidus (Eisen, 1874)
    (*ducis* Stephenson, 1932)
B. welchi (Smith, 1917)
B. zeteski (Smith and Gittins, 1915)

Cernosvitovia Omodeo, 1956 (sensu Zicsi, 1981)
Type species: *Allolobophora rebelii* Rosa, 1897

C. biserialis (Cernosvitov, 1937)
C. bulgarica (Cernosvitov, 1934)
C. dobrogeana (Pop, 1938)
C. dudichi Zicsi and Sapkarev, 1982
C. rebelii (Rosa, 1897)
C. schweigeri (Zicsi, 1973)

Dendrobaena Eisen, 1874
Type species: *Dendrobaena boeckii* Eisen, 1874 (= *D. octaedra*)

D. adaiensis adaiensis (Michaelsen, 1900)
    [*Helodrilus (Eophila) montanus* Michaelsen, 1910 (non *Helodrilus (Eisenia) venetusmontanus* Michaelsen, 1910 = *D. montana*); *schmidti* Michaelsen, 1907; *schmidti violacea* Michaelsen, 1910]
D. adaiensis surbiensis (Michaelsen, 1910)
D. adaiensis tellermanica Perel, 1966
D. aegea (Cognetti, 1913)
D. alpina alpina (Rosa, 1884)
    (*alpina armeniaca* Rosa, 1893; *alpina tatrensis* Nusbaum, 1895; *octaedra irregularis* Ribaucourt, 1896; *veneta cognettii* Cernosvitov, 1935; *alpina alteclitellata* Pop, 1938; *hypogea* Malevics, 1947)

D. alpina graffi **nom. nov.**
    (*Allolobophora rosea lusitana* Graff, 1957 non *Dendrobaena lusitana* Graff, 1957)
D. alpina mavrovensis Sapkarev, 1971
D. alpina popi Sapkarev, 1971
D. alpina zeugochaeta Bouché, 1972
D. attemsi attemsi (Michaelsen, 1902)
D. attemsi decipiens (Michaelsen, 1910)
D. bokakotorensis Sapkarev, 1975
D. boneinensis (Michaelsen, 1910)
D. byblica (Rosa, 1893)
    (*ganglbaueri* Rosa, 1894; *ganglbaueri annectens* Rosa, 1895; *fedtschenkoi* Michaelsen, 1900; *ganglbaueri olympiacus* Michaelsen, 1902; *thyrrena* Baldasseroni, 1907; *ganglbaueri daghestanensis* Michaelsen, 1907; *schelkovnikovi* Michaelsen, 1907; *ganglbaueri meledaensis* Michaelsen, 1908; *schelkovnikovi veliensis* Michaelsen, 1910; *schelkovnikovi bakuensis* Michaelsen, 1910; *fedtschenkoi lenkoranensis* Michaelsen, 1910; *insularis* Chinaglia, 1913; *lacustris* Stephenson, 1913; *ganglbaueri bulgarica* Cernosvitov, 1937; *ganglbaueri cylindrica* Tzelepe, 1943; *ganglbaueri differentis* Tzelepe, 1943; *schelkovnikovi graeca* Omodeo, 1955; *galloprovincialis* Bartoli, 1962)
D. clujensis Pop, 1938
D. faucium (Michaelsen, 1910)
D. franzi Zicsi, 1965
D. hortensis (Michaelsen, 1890)
    (*hibernica* Friend, 1892; *veneta tepidaria* Friend, 1904; *veneta dendroidea* Friend, 1909; *birsteini* Malevics, 1947)
D. illyrica (Cognetti, 1906)
    (*illyricus hintzei* Michaelsen, 1907; *bohemicus* Cernosvitov, 1931)
D. indica (Michaelsen, 1907)
D. jeanneli Pop, 1948
D. juliana juliana Omodeo, 1954
D. juliana auriculifera Zicsi, 1969
D. juliana colloquia Bouché, 1973
D. kurashvilii Kvavedze, 1971

D. lusitana Graff, 1957
(non *Allolobophora rosea lusitana* Graff, 1957 = *D. alpina graffi*)

D. mahnerti Zicsi, 1974

D. metallorum (Tétry, 1926)

D. michalisi Karaman, 1972

D. montana (Michaelsen, 1910)
[*Helodrilus (Eisenia) venetus montanus* Michaelsen, 1910 (non *H. (Eophila) montanus* Michaelsen, 1910 = *D. adaiensis adaiensis*)]

D. montenigrina Karaman, 1972

D. mrazeki (Cernosvitov, 1935)

D. nassonovi Kulagin, 1889
(*mariupoliensis* Wyssotzky, 1898; *crassa* Michaelsen, 1900; *mariupoliensis monticolus* Michaelsen, 1910; *mariupoliensis relicta* Perel, 1967; *mariupoliensis adjarica* Kvavadze, 1973)

D. nicaensis Vedovini, 1971

D. octaedra (Savigny, 1826)
(*flaviventris* Leuckart, 1849; *boeckii* Eisen, 1874; *camerani* Rosa, 1882; *octaedra alpinula* Ribaucourt, 1896; *octaedra liliputiana* Ribaucourt, 1896; *octaedrus casterinensis* Chinaglia, 1911; *octaedra quadivesiculata* Pop, 1938)

D. olympica (Cernosvitov, 1938)

D. orientalis Cernosvitov, 1940

D. osellai Zicsi, 1970

D. pantaleonis pantaleonis (Chinaglia, 1913)

D. pantaleonis balagnensis Bouché, 1972

D. parabyblica Perel, 1972

D. pentheri (Rosa, 1905)
(*schemachaensis* Michaelsen, 1910)

D. pseudohortensis Sapkarev, 1977

D. pygmaea (Savigny, 1826)
(*Allolobophora minima* Rosa, 1884 (non Muldal, 1953 = *Murchieona minuscula*); *cognettii* Michaelsen, 1903; *ribaucourti* Cognetti, 1901)

D. ressli Zicsi, 1973

D. rhodopensis (Cernosvitov, 1937)

D. ruffoi Zicsi, 1970

D. semitica semitica (Rosa, 1893)

D. semitica michaelseni **nom. nov.**
[*D. semitica kervillei* Michaelsen,

1926 non *Eisenia veneta kervillei* Michaelsen, 1926 (= *D. veneta kervillei*)]

D. sketi Karaman, 1972

D. vejdovskyi (Cernosvitov, 1935)
(*octaedra filiformis* Pop, 1947)

D. veneta veneta (Rosa, 1886)
(*bogdanowi* Kulagin, 1889; *caucasica* Kulagin, 1889; *veneta zebra* Michaelsen, 1902; *veneta succinta* Rosa, 1905; *veneta robusta* Friend, 1909; *venetus concolor* Michaelsen, 1910; *veneta picta* Michaelsen, 1910; *veneta tumida* Friend, 1927; *austriaca* Michaelsen, 1936; *veneta balcanica* Cernosvitov, 1937; *veneta minuta* Malevics, 1947; *veneta crassa* Malevics, 1947; *svetlovia* Grieb, 1948)

D. veneta kervillei (Michaelsen, 1926)
[*Eisenia veneta kervillei* Michaelsen, 1926 non *Dendrobaena semitica kervillei* Michaelsen, 1926 (= *D. semitica michaelseni*)]

D. veneta meleica (Tzelepe, 1943)

D. veneta ochridana Sapkarev, 1977

Dendrodrilus Omodeo, 1956 (sensu Perel, 1976)
Type species: *Enterion rubidum* Savigny, 1826

D. rubidus rubidus (Savigny, 1826)
(*xanthurus* Templeton, 1836; *puter* Hoffmeister, 1845; *valdiviensis* Blanchard, 1849; *pieter* Udekem, 1865; *havaicus* Kinberg, 1867; *victoris* Perrier, 1872; *arborea* Eisen, 1874; *fraissei* Örley, 1881; *constricta* Rosa, 1884; *darwini* Ribaucourt, 1896; *putris subrubicunda* var. *helvetica* Ribaucourt, 1896; *subrubicunda papillosa* Pop, 1938; *magnesa* Tzelepe, 1943)

D. rubidus norvegicus (Eisen, 1874)

D. rubidus subrubicundus (Eisen, 1874)
[*putris dieppi* Ribaucourt, 1901; ?*arborea pygmaea* Friend, 1923; *rivulicola* Chandebois, 1958]

D. rubidus tenuis (Eisen, 1874)

Eisenia Malm, 1877 (sensu Perel, 1974)
Type species: *Enterion fetidum* Savigny, 1826
Syn: *Allolobophora (Notogama)* Rosa, 1893 (type species: *Enterion fetidum* Savigny, 1826: new designation)

[Note: Three species, *carpetana* (Alvarez, 1970), *hrabei* Cernosvitov, 1934 and *longicinctus* (Smith and Green, 1915), are excluded from *Eisenia* by Perel (1974). They are here retained in *Eisenia* since they have not yet been assigned to other genera.]
E. altaica Perel, 1968
E. balatonica Pop, 1943
  (*ukrainae* Malevics, 1950)
E. carpetana (Alvarez, 1970)
E. djungarica (Perel, 1969)
E. fetida (Savigny, 1826)
  (*foetida* auctorum (illegal emendation); *semifasciatus* Burmeister, 1835; *annularis* Templeton, 1836; *olidus* Hoffmeister, 1842; *luteus* Blanchard, 1849; *rubofasciatus* Baird, 1873; *annulatus* Hutton, 1877; *fetida fimetoria* Örley, 1881; *nordenskioeldi caucasica* Michaelsen, 1902; *fetida attica* Tzelepe, 1943; *fasciata* Backlund, 1948; *fetida* var *unicolor* André, 1963; *fetida andrei* Bouché, 1972)
E. grandis grandis (Michaelsen, 1907)
  (*perelae* Kvavadze, 1973; *perelae polysegmentica* Kvavadze, 1973)
E. grandis ebneri (Michaelsen, 1914)
E. gordejeffi (Michaelsen, 1899)
E. hrabei Cernosvitov, 1934
E. intermedia (Michaelsen, 1901)
  (*kazanensis* Michaelsen, 1910; *tanaitica* Malevics, 1953)
E. iverica (Kvavadze, 1973)
E. japonica (Michaelsen, 1891)
  (*japonica gigantica* Oishi, 1934; *japonica minuta* Oishi, 1934)
E. kattoulasi Zicsi and Michalis, 1981
E. koreana (Zicsi, 1972)

E. kucenkoi Michaelsen, 1902
E. longicinctus (Smith and Gittins, 1915)
E. lucens (Waga, 1857)
  (*submontanus* Vejdovsky, 1875; *fetida hungarica* Örley, 1881; *tigrina* Rosa, 1896; *latens* Cognetti, 1902; *croatica* Szuts, 1909; *gavrilovi* Cernosvitov, 1942)
E. magnifica (Svetlov, 1957)
E. malevici Perel, 1962
E. nordenskioeldi nordenskioeldi (Eisen, 1879)
  (*acystis* Michaelsen, 1902)
E. nordenskioeldi lagodechiensis (Michaelsen, 1910)
E. nordenskioeldi manshurica Kobayashi, 1940
E. nordenskioeldi polypapillata Perel, 1969
E. oltenica (Pop, 1938)
  (*colchidica* Perel, 1967)
E. salairica (Perel, 1968)
E. spelaea spelaea (Rosa, 1901)
  (*triglavensis* Pop, 1943)
E. spelaea athenica Cernosvitov, 1938
E. transcaucasica (Perel, 1967)
E. uralensis (Malevics, 1950)

Eiseniella Michaelsen, 1900
Type species: *Enterion tetraedrum* Savigny, 1826
Nom. nov. pro: *Allurus* Eisen, 1874 (type species: *Enterion tetraedrum* Savigny, 1826) [*non* Foerster, 1862 (Hymenoptera)]
  Tetragonurus Eisen, 1874 [type species: *T. pupus* Eisen, 1874 (= *E. tetraedra pupa*)] [*non* Risso, 1810 (Pisces)]

E. eutypica (Michaelsen, 1910)
E. ochridana ochridana Cernosvitov, 1931
  (*ochridana stankovici* Cernosvitov, 1931; *lacustris* Cernosvitov, 1931; *lacustris ochridana* Cernosvitov, 1931)
E. ochridana profunda Cernosvitov, 1931
  (*lacustris naumi* Cernosvitov, 1931)
E. paradoxoides (Alvarez, 1971)

E. peleensis Tzelepe, 1943

E. tetraedra tetraedra (Savigny, 1826)
(*quadrangularis* Risso, 1826; *amphisbaenus* Dugès, 1828; *agilis* Hoffmeister, 1843; *tetraedrus luteus* Eisen, 1871; *tetraedrus obscurus* Eisen, 1871; *dubius* Michaelsen, 1890; *tetragonurus* Friend, 1892; *macrurus* Friend, 1893; *flavus* Friend, 1893; *tetraedrus bernensis* Ribaucourt, 1896; *tetraedrus novis* Ribaucourt, 1896; *tetraedrus infinitesimalis* Ribaucourt, 1896; *tetraedra hammoniensis* Michaelsen, 1909; *mollis* Friend, 1911; *intermedia* Jackson, 1931; *tetraedra popi* Zicsi, 1960)

E. tetraedra intermedia Cernosvitov, 1934

E. tetraedra neapolitana (Örley, 1885)
(*ninnii* Rosa, 1886; *tetraedra sewelli* Stephenson, 1924)

E. tetraedra pupa (Eisen, 1874)
(*hercynius* Michaelsen, 1890; *tetraedra quadripora* Cernosvitov, 1942)

Eisenoides Gates, 1969
Type species: *Allolobophora loennbergi* Michaelsen, 1894

E. carolinensis (Michaelsen, 1910)
(*pearsei* Stephenson, 1932)

E. loennbergi (Michaelsen, 1894)

Eophila Rosa, 1893 (sensu Omodeo, 1956)
Type species: *Allolobophora tellinii* Rosa, 1886

E. asconensis asconensis (Bretscher, 1900)

E. asconensis silvatica (Zicsi, 1976)

E. dofleini (Ude, 1922)

E. gestroi (Cognetti, 1905)
(*laurentii* Chinaglia, 1910; *chinagliae* Baldscher, 1919)

E. haasi Michaelsen, 1925

E. hispanica (Ude, 1885)

E. januaeargenti januaeargenti (Cognetti, 1903)

E. januaeargenti sarda (Michaelsen, 1910)

E. januaeargenti stankovici (Sapkarev, 1971)

E. mehadiensis mehadiensis (Rosa, 1895)

E. mehadiensis boscaiui (Pop, 1948)

E. mehadiensis oreophila (Pop, 1978)

E. moebii moebii (Michaelsen, 1895)
(*moebii tenerifana* Cognetti, 1931)

E. moebii baeticae Sims, 1962

E. molleri (Rosa, 1889)

E. opisthocystis (Rosa, 1895)

E. pyrenaicoides Sapkarev, 1977

E. robusta (Rosa, 1895)

E. sardonica sardonica (Cognetti, 1904)

E. sardonica catalaunensis (Bouché, 1972)

E. tardionii (Baldasseroni, 1906)

E. tellinii (Rosa, 1886)

Ethnodrilus Bouché, 1972
Type species: *Ethnodrilus zajonci* Bouché, 1972

E. aveli Bouché, 1972

E. gatesi Bouché, 1972

E. lydiae Bouché, 1972

E. zajonci Bouché, 1972

Fitzingeria Zicsi, 1978
Type species: *Enterion platyurum* Fitzinger, 1833

F. platyura platyura (Fitzinger, 1833)
(*terrestris platyurus* Örley, 1881; *oerleyi* Horst, 1887; *fitzingeri* Beddard, 1895)

E. platyura depressa (Rosa, 1893)
(*platyura moravica* Proksova, 1955; *platyura panonica* Proksova, 1955)

F. platyura montana (Cernosvitov, 1932)

Helodrilus Hoffmeister, 1845 (sensu Perel, 1979)
Type species: *Helodrilus oculatus* Hoffmeister, 1845

Syn: *Helodrylus* Udekem, 1855
(lapsus pro *Helodrilus*)
Anagaster Friend, 1921 [type species: *Anagaster fontinalis* Friend,
1921 (= *H. oculatus*)]

H. antipae antipae (Michaelsen,
1891)
(*tyrtaea* Ribaucourt, 1896; *riparia*
Bretscher, 1901; *cuginii helod-
riloides* Chandebois, 1958)
H. antipae tuberculatus (Cernosvitov,
1935)
H. antipae vogesianus (Tétry, 1938)
H. cernosvitovianus (Zicsi, 1967)
H. oculatus oculatus Hoffmeister,
1845
(*hermanni* Michaelsen, 1890; *fontinalis* Friend, 1921)
H. oculatus samniticus Cognetti, 1914
(*oculatus dudichi* Pop, 1943;
massiliensis Bartoli, 1962)
H. patriarchalis (Rosa, 1893)
(*ariadne* Michaelsen, 1928)

Kritodrilus Bouché, 1972 (sensu
Perel, 1976)
Type species: *Octolasion calcarensis*
Tétry, 1944

K. auriculatus (Rosa, 1897)
(*skorikowi* Michaelsen, 1902)
K. calcarensis (Tétry, 1944)

Lumbricus Linnaeus, 1758
Type species: *Lumbricus terrestris*
Linnaeus, 1758
Syn: *Enterion* Savigny, 1824
(type species: *Lumbricus
terrestris* Linnaeus, 1758)
Omilurus Templeton, 1836
(type species: *Omilurus
omilurus* Templeton, 1836 = *L.
festivus*)

L. annulatus Perel, 1975
L. badensis Michaelsen, 1907
L. baicalensis Michaelsen, 1900
(*dueggelii* Bretscher, 1903: *pusillus*
Wessely, 1905)
L. castaneus (Savigny, 1826)

(*pumilum* Savigny, 1826; *triannu-
laris* Grube, 1851; *minor* Johnston,
1865; *josephinae* Kinberg, 1867;
purpureus Eisen, 1871; *castaneus
morelli* Ribaucourt, 1896; *castaneus
perrieri* Ribaucourt, 1896; *brunes-
cens* Bretscher, 1900; *castaneus dis-
junctus* Tétry, 1936; *castaneus
pictus* Chandebois, 1957)
L. centralis Bouché, 1972
L. festivus (Savigny, 1826)
(*omilurus* Templeton, 1836; *rubes-
cens* Friend, 1891)
L. friendi Cognetti, 1904
[nom nov pro *L. papillosus* Friend,
1893 non Müller, 1776
= *Arenicola marina* (Polychaeta)];
lumbricoides (Bretscher, 1901)
L. improvisus Zicsi, 1963
L. meliboeus Rosa, 1884
(*michaelseni* Ribaucourt, 1896)
L. polyphemus (Fitzinger, 1833)
L. rubellus rubellus Hoffmeister, 1843
(*campestris* Hutton, 1887; *rubellus
curticaudatus* Friend, 1892; *rubel-
lus tatrensis* Nusbaun, 1895; *her-
culeana* Bretscher, 1899;
ribaucourti Bretscher, 1901; *relictus*
Southern, 1909; *rubellus tristani*
Pickford, 1932)
L. rubellus castaneoides Bouché, 1972
L. rubellus friendoides Bouché, 1972
L. terrestris Linnaeus, 1758
(*herculeum* Savigny, 1826; *agricola*
Hoffmeister, 1842; *infelix*
Kinberg, 1867; *americanus* Perrier,
1872; *studeri* Ribaucourt, 1896)

Microeophila Omodeo, 1956
Type species: *Eophila marcuzzii*
Omodeo, 1952

M. alzonae (Cognetti, 1904)
M. cryptocystis (Cernosvitov, 1935)
M. kratochvili (Cernosvitov, 1937)
M. marcuzzii (Omodeo, 1952)
M. nematogena (Rosa, 1903)
(*meledaensis* Michaelsen, 1908; *be-
llicosus* Ude, 1922; *dudichiana*
Zicsi, 1966)
M. sotschiensis (Michaelsen, 1902)

Murchieona Gates, 1978
Type species: *Bimastos muldali*
Omodeo, 1956 (= *M. minuscula*)

M. minuscula (Rosa, 1905)
(*icenorum* Pickford, 1926;
Allolobophora minima Muldal,
1952 (non Rosa, 1884, = *Dendrobaena pygmaea*); *muldali* Omodeo,
1956)

Octodrilus Omodeo, 1956 (sensu
Zicsi and Sapkarev, 1982)
Type species: *Lumbricus complanatus*
Dugès, 1828
Syn: *Octolasion (Purpureum)*
Omodeo, 1952 (type species:
Allolobophora lissaensis
Michaelsen, 1891)

O. aelleni Zicsi, 1979
O. argoviensis (Bretscher, 1899)
(*croaticum eutypica* Pop, 1947)
O. benhami (Bretscher, 1900)
O. besncheti Zicsi, 1979
O. binderi Zicsi, 1979
O. boninoi (Omodeo, 1962)
O. bretscheri (Zicsi, 1969)
O. croaticus (Rosa, 1895)
O. complanatus (Dugès, 1828)
O. damianii (Cognetti, 1905)
O. eubenhami (Zicsi, 1971)
O. exacystis (Rosa, 1896)
O. frivaldszkyi (Örley, 1885)
O. gradinescui (Pop, 1928)
O. hemiandrus (Cognetti, 1901)
O. janetscheki (Zicsi, 1970)
O. kamnensis (Baldasseroni, 1919)
O. karawankensis (Zicsi, 1969)
O. kovacevici (Zicsi, 1970)
O. lissaensioides (Zicsi, 1971)
O. lissaensis (Michaelsen, 1891)
O. marenzelleri (Michaelsen, 1910)
O. mimus (Rosa, 1889)
O. minoris (Omodeo, 1952)
O. omodeoi Zicsi, 1982
O. ortizi (Alvarez, 1970)
O. phaenohemiandrus (Zicsi, 1971)
O. pseudocomplanatus (Omodeo,
1962)
O. pseudokovacevici (Zicsi, 1971)

O. pseudotranspadanus (Zicsi, 1971)
O. racovitzai (Pop, 1938)
O. robustus (Pop, 1973)
O. rucneri (Plisko and Zicsi, 1970)
O. ruffoi Zicsi, 1982
O. tergestinus (Michaelsen, 1910)
O. transpadanoides Zicsi, 1982
O. transpadanus (Rosa, 1884)
(*opimum* Savigny, 1826; *trans-
padana cinerea* Rosa, 1886; *cyanea
recta* Ribaucourt, 1896; *sulfurica*
Ribaucourt, 1896; *nivalis*
Bretscher, 1899; *transpadanum
alpinum* Bretscher, 1905)
O. vallorus (Baldasseroni, 1920)

Octolasion Örley, 1885
Type species: *Lumbricus terrestris
lacteus* Örley, 1881 (subsequent
designation: Omodeo, 1956; non
Omodeo, 1952)
Syn: *Alyattes* Kinberg, 1867 (type
species: *Lumbricus alyattes*
Kinberg, 1867. = *O. cyaneum*)
Octolasium: Michaelsen 1900
(invalid emendation of *Octolasion*)
Octolasion (Incolore) Omodeo,
1952 (type species: *Lumbricus
terrestris lacteus* Örley, 1881)

O. cyaneum (Savigny, 1826)
(*stagnalis* Hoffmeister, 1845; *alyattes* Kinberg, 1867; *studiosa*
Michaelsen, 1890; *kempi*
Stephenson, 1922)
O. lacteovicinum Zicsi, 1968
O. lacteum (Örley, 1881)
(*terrestris rubidus* Örley, 1881; *profuga* Rosa, 1884; *cyanea profuga
sylvestris* Ribaucourt, 1896; *hi-
malayana* Cernosvitov, 1937)
O. montanum (Wessely, 1905)
(non *O. montanum* Cernosvitov,
1932, = *Fitzingeria platyura
montana*)
O. tyrtaeum (Savigny, 1826)
(*gracile* Örley, 1885)

Orodrilus Bouché, 1972
Type species: *Helodrilus*
(*?Dendrobaena, ?Allolobophora*) *do-
deroi* Cognetti, 1904

O. doderoi (Cognetti, 1904)
O. gavarnicus (Cognetti, 1904)
O. paradoxus paradoxus (Cognetti, 1904)
O. paradoxus magnei Bouché, 1972

Perelia **nom. nov.**
Nom. nov. pro *Allolobophora*
(*Svetlovia*) Perel, 1976 (type
species: *Eophila arnoldiana* Perel,
1971);
non *Svetlovia* Chekanovskaya,
1975 (Tubificidae)

P. albicauda (Perel, 1977)
P. agatschiensis (Michaelsen, 1910)
P. arnoldiana (Perel, 1971)
P. bouchei (Perel, 1977)
P. brunnea (Perel, 1971)
P. chlorocephala (Perel, 1977)
P. diplotetratheca (Perel, 1967)
P. ferganae (Malevics, 1949)
P. ghilarovi (Malevics, 1949)
P. graciosa (Perel, 1977)
P. kaznakovi (Michaelsen, 1910)
  (*asiatica* Malevics, 1949)
P. kirgisica (Perel, 1971)
P. longoclitellata (Perel, 1977)
P. media (Perel, 1977)
P. microtheca (Perel, 1977)
P. muganiensis (Michaelsen, 1910)
P. ophiomorpha (Perel, 1977)
P. persiana (Michaelsen, 1900)
P. polytheca (Malevics, 1949)
P. sokolovi (Perel, 1969)
P. stenosoma (Perel, 1977)
P. schneideri (Michaelsen, 1900)
P. taschkentensis (Michaelsen, 1900)
P. tuberosa (Svetlov, 1924)
  (*baschirica* Malevics, 1950)
P. turkmenica (Malevics, 1941)
P. umbrophila (Perel, 1977)

Prosellodrilus Bouché, 1972
Type species: *Prosellodrilus idealis*
Bouché, 1972

P. alatus Bouché, 1972

P. amplisetosus amplisetosus Bouché, 1972
P. amplisetosus hexathecosus Bouché, 1972
P. biauriculatus Bouché, 1972
P. fragilis fragilis Bouché, 1972
P. fragilis biserialis Bouché, 1972
P. fragilis elisatus Bouché, 1972
P. fragilis polythecosus Bouché, 1972
P. idealis Bouché, 1972
P. praticola Bouché, 1972
P. pyrenaicus pyrenaicus (Cognetti, 1904)
P. pyrenaicus argonicus (Alvarez, 1971)

Satchellius Gates, 1975
Type species: *Enterion mammale*
Savigny, 1826

S. mammalis (Savigny, 1826)
  (*celtica* Rosa, 1886; *celtica rosea*
Friend, 1893)

Scherotheca Bouché, 1972
Scherotheca (Scherotheca)
Type species: *Lumbricus gigas* Dugès, 1828

S. (S.) coineaui Bouché, 1972
S. (S.) corsicana corsicana (Pop, 1947)
S. (S.) corsicana albomaculata
Bouché, 1972
S. (S.) corsicana simplex (Zicsi, 1981)
S. (S.) cyrnea (Michaelsen, 1926)
S. (S.) dollfusi (Tétry, 1939)
S. (S.) dugesi dugesi (Rosa, 1886)
S. (S.) dugesi brevisella Bouché, 1972
S. (S.) dugesi provincialis (Vedovini, 1971)
  (*dugesi porotheca* Bouché, 1972)
S. (S.) dugesi sanaryensis (Tétry, 1942)
S. (S.) gigas gigas (Dugès, 1828)
S. (S.) gigas aquitania Bouché, 1972
S. (S.) gigas dinoscolex Bouché, 1972
S. (S.) gigas mifuga Bouché, 1972
S. (S.) gigas orbiensis Bouché, 1972
S. (S.) gigas rhodana Bouché, 1972
S. (S.) guipuzcoana Bouché, 1972
S. (S.) hexatheca (Michaelsen, 1926)

S. (S.) monospessulensis monospessulensis Bouché, 1972

S. (S.) monospessulensis idica Bouché, 1972

Scherotheca (Opothedrilus) Bouché, 1972
Type species: *Allolobophora savignyi* Guerne and Horst, 1893

S. (O.) occidentalis occidentalis (Michaelsen, 1922)

S. (O.) occidentalis chicharia (Bouché, 1967)

S. (O.) occidentalis thibauti Bouché, 1972

S.(O.) savignyi savignyi (Guerne and Horst, 1893)

S. (O.) savignyi minor (Stephenson, 1931)

S. (O.) savignyi nivicola Bouché, 1972

41.2 REFERENCES

Only the more recent references are listed below. For earlier references see the *Zoological Record*, 1970–1977 and the bibliographies of Zicsi (1982), Perel (1979) and Bouché (1972).

Bouché, M. B. (1972) Lombriciens de France. Écologie et systématique. *Ann. Zool. Écol. Anim. Paris*, **72-2**, 671 pp.

Bouché, M. B. (1979) Observations sur les Lombriciens (5eme serie) XII. Lumbricidae (Oligochaeta) du Guipúzcoa. *Doc. Pedozool.*, **1**, 90–100.

Perel, T. S. (1979) *Range and Regularities in the Distribution of Earthworms of the USSR Fauna (with keys to Lumbricidae and other Megadrili)* Moscow: Nauka (Academy of Sciences of the USSR, Laboratory of Forest Science), 272 pp. [In Russian: English summary.]

Pop, V. V. (1978a) *Allolobophora zarandensis* sp. n. (Oligochaeta, Lumbricidae) from the Romanian western Carpathians. *Trav. Mus. Hist. Nat. Grigore Antipa*, **19**, 251–253.

Pop, V. V. (1978b) *Allolobophora mehadiensis* Rosa, 1895 and its subspecies (Oligochaeta, Lumbricidae). *Trav. Mus. Hist. Nat. Grigore Antipa*, **19**, 255–258.

Zicsi, A. (1978) Revision der art *Dendrobaena platyura* (Fitzinger, 1833) (Oligochaeta: Lumbricidae) *Acta Zool. Hung.*, **24**, 439–449.

Zicsi, A. (1979) Neue Angaben zur Regenwurm-Fauna der Schweiz (Oligochaeta: Lumbricidae). *Rev. Suisse Zool.*, **86**, 473–484.

Zicsi, A. (1981) Weitere Angaben zur Lumbriciden-fauna Italiens (Oligochaeta: Lumbricidae). *Opusc. Zool. Budapest*, **17/18**, 157–180.

Zicsi, A. (1981) Probleme der Lumbriciden-systematic sowie die revision zweir gattungen (Oligochaeta). *Acta. Zool. Hung.*, **27**, 431–442.

Zicsi, A. (1982) Verzeichnis der bis 1971 beschriebenen und revidierten taxa der family Lumbricidae. *Opusc. Zool. Budapest*, **28**, 421–454.

Zicsi, A. and Michalis, K. (1981) Übersicht der regenwurm-fauna Griechenlands (Oligochaeta: Lumbricidae). *Acta Zool. Hung.*, **27**, 239–264.

Zicsi, A. and Sapkarev, J. A. (1982) Eine neue *Cernosvitovia*-art aus Jugoslawien (Oligochaeta: Lumbricidae). *Acta Zool. Hung.*, **28**, 181–182.

41.3 CHANGES IN TAXONOMY AND NOMENCLATURE

Previous name	Emended name	Reason
Allolobophora (Notogama) Rosa, 1893	*Eisenia* Malm, 1877	Syn. nov.
Allolobophora (Svetlovia) Perel, 1976	*Perelia* **nom. nov.**	Preoccupied: *Svetlovia* Chekanovskaya, 1975 (Tubificidae)
Allolobophora mediterranea Örley, 1881	*Aporrectodea rosea* (Savigny, 1826)	Syn. nov.
Allolobophora parva udei Ribaucourt, 1896	*Allolobophora parva* Eisen, 1874	Syn. nov.
Allolobophora sulfurica Ribaucourt, 1896	*Octodrilus transpadanus* (Rosa, 1884)	Syn. nov.
Allolobophora zicsii Sapkarev, 1975	*Allolobophora sapkarevi* **nom. nov.**	Preoccupied: *Allolobophora zicsii* Bouché, 1972
Aporrectodea rubicunda rubicunda (Bouché, 1972)	*Aporrectodea rubra rubra* (Vedovini, 1969)	Priority
Aporrectodea rubicunda acidicola (Bouché, 1972)	*Aporrectodea rubra acidicola* (Bouché, 1972)	Priority
Dendrobaena alpina lusitana (Graff, 1957) (*Allolobophora rosea lusitana* Graff, 1957)	*Dendrobaena alpina graffi* **nom. nov.**	Preoccupied: *Dendrobaena lusitana* Graff, 1957
Dendrobaena auriculifera auriculifera Zicsi, 1969	*Dendrobaena juliana auriculifera* Zicsi, 1969	Priority
Dendrobaena auriculifera juliana Omodeo, 1954 (*Dendrobaena schmidti juliana* Omodeo, 1954)	*Dendrobaena juliana juliana* Omodeo, 1954	Priority
Dendrobaena mariupoliensis (Wyssotzky, 1898)	*Dendrobaena nassonovi* (Kulagin,1889)	Priority
Dendrobaena minima (Rosa, 1884)	*Dendrobaena pygmaea* (Savigny, 1826)	Priority (see Bouché, 1972:393)
Dendrobaena schmidti schmidti (Michaelsen, 1907) [*Helodrilus (Dendrobaena) schmidti* Michaelsen, 1907]	*Dendrobaena adaiensis adaiensis* (Michaelsen, 1900)	Priority
Dendrobaena schmidti surbiensis (Michaelsen, 1910) [*Helodrilus (Dendrobaena) schmidti surbiensis* Michaelsen, 1910]	*Dendrobaena adaiensis surbiensis* (Michaelsen, 1910)	Priority
Dendrobaena schmidti tellermanica Perel, 1966	*Dendrobaena adaiensis tellermanica* Perel, 1966	Priority
Dendrobaena semitica kervillei Michaelsen, 1926	*Dendrobaena semitica michaelseni* **nom. nov.**	Preoccupied: *Dendrobaena veneta kervillei* (Michaelsen, 1926)

Dendrobaena veneta tumida (Friend, 1927) [Allolobophora veneta tumida Friend, 1927]	Dendrobaena veneta veneta (Rosa, 1886)	Syn. nov.
Eiseniella tetraedra tetragonura (Friend, 1892) [Allurus tetragonurus Friend, 1892]	Eiseniella tetraedra tetraedra (Savigny, 1826)	Syn. nov.
Eiseniella tetraedra macrura (Friend, 1893) [Allurus macrurus Friend, 1893]	Eiseniella tetraedra tetraedra (Savigny, 1826)	Syn. nov.
Eiseniella tetraedra mollis (Friend, 1911) [Allurus mollis Friend, 1911]	Eiseniella tetraedra tetraedra (Savigny. 1826)	Syn. nov.
Lumbricus teres Dugès, 1828	Vignysa teres (Dugès, 1828) [family Hormogastridae)	Comb. nov. (Bouché, 1972:210)
Lumbricus valdiviensis Blanchard, 1849	Dendrodrilus rubidus rubidus (Savigny, 1826)	Syn. nov.
Lumbricus victoris Perrier, 1872	Dendrodrilus rubidus rubidus (Savigny, 1826)	Syn. nov.
Scherotheca (Scherotheca) dugesi porotheca Bouché, 1972	Scherotheca (Scherotheca) dugesi provincialis (Vedovini, 1971)	Priority
Vignysa popi Bouché, 1972 (Hormogastridae)	Vignysa teres (Dugès, 1828)	Priority

Systematic index

See also Chapter 41

Agastrodrilus dominicae, 425–428
Agastrodrilus multivesiculatus,
　425–428
Agastrodrilus opisthogynus, 425–428
Allolobophora/Apporrectodea/
　Nicodrilus, 468, 469
　antipae, 31
　balisa, 434, 443
　caliginosa, 29, 41–43, 52, 68, 69,
　　75, 76, 79, 80, 85, 86, 88–90,
　　93, 124–126, 128, 139–143,
　　145, 148, 154–160, 162, 187,
　　215–224, 244, 245, 249, 257,
　　258, 278–281, 254–356, 397,
　　417, 468, 469
　chlorotica, 85, 124–126, 132, 161,
　　162, 216, 217, 220–224, 250,
　　270, 278–281, 394, 397, 417,
　　468, 472
　cognetti, 469
　cupilifera, 85, 417
　giardi, 434, 443, 444, 468
　gogna, 434, 442, 443
　icterica, 62–65, 417
　limicola, 444
　longa/longus, 52, 56, 62–65, 68,
　　69, 80, 85, 86, 89, 107–121,
　　124, 125, 128, 132, 145, 148,
　　158, 159, 162, 171, 217,
　　220–223, 353, 417, 434, 444,
　　468
　minima, 469
　muldali, 434, 444
　nocturna/nocturnus, 124, 128, 417,
　　443, 444, 469
　pereli, 434, 440, 442
　prashadi, 216

　pygmaea, 469
　ribaucourti, 469
　(Eisenia) rosea, 52, 85, 89, 124,
　　141, 154, 158, 159, 162, 187,
　　198, 203, 204, 216, 217,
　　219–223, 278–281, 283, 417,
　　454
　trapezoides, 162, 198, 199, 203,
　　216, 469
　tuberculata, 52, 124, 217,
　　221–223, 248, 252, 253, 255,
　　394, 396, 469
　turgida, 52, 162, 469
　velox, 434, 442, 443
　virescens, 472
Alma emini, 189
Alma stuhlmanni, 189
Alyattes, 471
Amynthas, 204, 468
Argilophilus, 467

Bimastos (Lumbricus) eiseni, 123,
　124, 162, 220–222

Chuniodrilus zielae 426

Dendrobaena/Dendrodrilus, 478–480
　arborea, 162
　auriculata, 177
　hortensis, 278, 279
　jeanelli, 434, 442
　mammalis, 358
　octaedra, 123, 124, 162, 177, 216,
　　219–222, 244, 245, 248, 396
　rubida/rubidus, 124, 161, 167,
　　177, 220–223, 250, 252–254,
　　256, 258, 259, 396, 433, 434

subrubicunda, 124, 154, 158, 159, 358
veneta, 251, 254, 255, 354–356
Dichogaster agilis, 459

Eisenia fetida (foetida), 161, 247, 249, 252, 254, 256, 259, 278, 279, 286–292, 297–307, 323–329, 331, 339–348, 354–360, 375–380, 470
Eisenia lucens, 357
Eisenia nordenskioldii, 168, 352
Eisenia rosea, see A. rosea
Eiseniella tetraedra, 161, 198, 216, 221, 222, 251, 254, 255, 278, 279
Enterion herculeum, 468
Eophila januae-argenti, 434, 441, 442
Ethnodrilus, 434, 442, 469
Eudrilus eugeniae, 125, 185, 286, 287, 315–320, 331, 334, 335
Eukerria saltensis, 196, 198, 203

Fitzingeria, 469
Fitzingeria platyura, 172, 173, 175–177

Glyphidrilus, 189

Haplotaxis gordioides, 434
Hemigastrodrilus, 434, 435, 440, 442
Hippopera nigeriae, 185
Hoplochaetella affinis, 384
Hoplochaetella kempi, 384
Hoplochaetella suctoria, 383–391
Hormogaster, 442
Hormogaster praetiosa, 434, 440, 442
Hormogaster redii, 434, 440–442
Hormogaster samnitica, 434, 440–442
Hyperodrilus africanus, 125, 185

Komarekiona, 468
Kritodrilus, 469

Lumbricus castaneus, 52, 53, 56, 85, 124, 125, 219, 221, 222, 224, 244, 245, 248, 278, 279, 417, 421
Lumbricus festivus, 52, 53, 56, 124, 217

Lumbricus friendi, 219
Lumbricus polyphemus, 172–174, 176, 177
Lumbricus rubellus, 42, 53, 56, 68, 69, 75, 80, 85, 86, 89, 124, 157, 158, 160, 162, 166, 167, 177, 215–224, 242–245, 249–251, 253, 254, 257, 258, 268, 270–272, 353–355, 358, 396, 397, 417, 421
Lumbricus spencer, 293
Lumbricus terrestris, 52, 53, 56, 73, 85, 86, 89, 93, 107–121, 124, 125, 129–132, 143, 145, 148, 154, 156, 157, 159, 162, 166, 167, 168, 171–174, 176, 216, 217, 219–224, 230, 232, 234–240, 250, 252, 253, 255, 258, 260, 278, 279, 291, 331–338, 351, 357–360, 365–373, 375–380, 393–397, 402–406, 409, 417, 421, 468
Lutodrilus, 468

Megascolides australis, 290
Metapheretima jocchana, 185
Metaphire, 204
Microchaetus, 195, 197, 203
Microchaetus microchaetus, 200, 201, 204
Microchaetus modestus, 198–205
Microscolex, 196, 433
Microscolex kerguelarum, 433, 434
Millsonia anomala, 187, 460–462
Millsonia caecifera, 428
Millsonia hemina, 428
Millsonia lamtoiana, 426, 459

Nicodrilus, see Allolobophora

Octodrilus/Octolasion/Octolasium, 470
complanatus, 173
gradinescui, 172, 173, 176
mima, 173
rucneri, 173
transpadanus, 172, 173, 176
cyaneum, 68, 69, 85, 124, 125, 162, 187, 221–223, 417, 471
lacteum, 158–159, 168, 216, 352, 358, 417, 471
tyrtaeum, 222, 223, 471

Opothodrilus, 434, 435
Orodrilus, 469

Perichaeta communissima, 293
Perionyx excavatus, 309–313
Pheretima, 468
Pheretima alexandri, 209–212
Plutellus, 467
Pontodrilus, 433
Pontoscolex corethrurus, 197, 202, 455
Prosellodrilus, 434, 435, 442, 469

Rhododrilus, 433

Satchellius mammalis, 222
Scherotheca, 434, 435, 440, 442, 463
Scherotheca coineaui, 434, 440, 442

Scherotheca corsicana, 434, 440, 442
Scherotheca dugesi, 434, 435, 438, 441, 442
Scherotheca gigas, 434, 435, 440–442
Scherotheca guipuzcoana, 434, 440, 442
Scherotheca monspessulensis, 434, 442
Sparganophilus, 434
Standeria, 195
Stuhlmannia porifera, 425, 426

Tritogenia, 195, 197, 198, 203, 204

Udeina, 196, 198, 203–205

Vignysa, 435, 442
Vignysa popi, 434

General index

Aeration
  effects of earthworms on, 45, 73,
    74, 93, 94, 167
  effects on earthworms, 221
Age distribution, *see* Population
  dynamics
Aggregation
  of earthworm populations, 49,
    52, 56
  soil (*see also* Wormcasts),
    19–33, 217
Amino acids, 23, 74, 260
Anthrax, 15
Arable (*see also* Cultivation;
  Crops/crop rotation), 31,
  123–135, 139–149, 151–160,
  241–245
Assimilation efficiency, 78, 79, 184,
  297–307, 345, 353

Behaviour, 4, 10, 11, 191, 404, 405
Biomass, 52–54, 85, 142–146, 163,
  172, 182–184, 186, 285,
  299–307, 316, 325–327, 393,
  409, 417, 418, 449, 453, 454,
  462
Birds, *see* Predators
Burning 151–160, 209–212, 461
Burrows and burrowing, 20, 65, 68,
  71–74, 91, 93, 128, 130, 172,
  173, 190, 283, 353, 428, 436,
  444

Calciferous glands, 29, 31, 257, 425
Calcium, 134, 238, 251, 256, 257,
  368–372

Carbon, soil (*see also*
  Wormcasts/faeces), 35–48, 92,
  93, 133, 168, 181, 182,
  209–212, 216, 318, 319, 357
Carrying capacity, 215, 300, 301
Casts, *see* Wormcasts
Cellulose, as an earthworm food
  source, 301–307, 323–329, 454
Chloragogen cells, 257, 259, 260
Colloids, soil, 24–26, 28, 31, 32,
  319
Colonization (*see also* Dispersal),
  38, 88–90, 217–221, 237, 238,
  244, 245, 433, 444
Community structure, 431–445,
  449–463
Competition (*see also* Ecological
  niche), 125, 348, 443, 444
Coprophagy, 307, 444
Crops/crop rotation (*see also*
  Arable; Yield), 123–135,
  151–160, 200, 201
Cultivation (*see also* Arable; Crops),
  123–135, 151–160, 190, 197,
  205, 416–424

Darwin, Charles, xi, xii, 1–21,
  29–31, 107, 166, 171, 179, 185,
  187, 189, 211, 365, 379
Decomposition, 36, 40, 46, 67,
  76–78, 173–177, 181–184, 318,
  319, 328, 329, 356–358
Dehydrogenase, 320
Derelict land, *see* Reclamation
Detoxification, 247, 258–261
Diapause, 42, 202, 461

Digestion (*see also* Assimilation;
    Enzymes; Nutrition), 352, 353,
    358, 359, 426
Direct drilling, 124, 125, 128–130
Dispersal (*see also* Colonization),
    88–90, 215, 217, 432, 433,
    442–444
Diversity, 123, 163–165, 212, 224,
    455, 456
Dokuchaev, 2, 3, 59
Drainage
    effects of earthworms on, 128,
    216
    effects on earthworms (*see also*
    Aeration), 221
Drought, 42
Dung
    burial by earthworms, 49–57
    effect on growth rate, 298, 305,
    309, 323–329, 343
    effects on population density,
    124, 132, 139–149, 151–160,
    245, 248, 254, 406
    microflora of, 357, 390, 391
Dung beetles, 49–57

Ecological niche, 125, 191, 417,
    421, 422, 432–445, 449–463
Endemic species, 196, 205, 173,
    442
Energy flow, *see* Respiration;
    Tissue
Enzymes
    earthworm, 119, 167, 259, 454
    soil, 67, 74–77
Erosion, 21, 189
Evolution, 431–445, 463
Excretion, 142, 143, 257, 259

Faeces, earthworm, *see* Wormcasts
Feeding
    behaviour and ecology, 121, 166,
    173–177, 212, 220, 235, 306,
    307, 426, 454, 455
    effects of, 68, 75, 77
Feedstuffs, use of earthworms in,
    285–293, 309–313
Fertilizers, 124, 131–134, 197
Fire, *see* Burning
Fish bait, 331–338
Flooding, 217

Food chains, 247, 256, 267–270,
    277, 283, 393–410
Food selection, 119, 220, 261, 325,
    326, 358, 365–373
Food sources (*see also* Cellulose),
    126, 128, 130, 132, 133,
    156–160, 161, 162, 173–177,
    216, 220, 298–307, 358, 359,
    435, 444, 485, 459, 460
Fossils, 431, 434–436, 445
Fraud, in earthworm marketing,
    334–337
Fungi, 31, 72, 209–212, 357,
    365–373, 375
Fungicides, *see* Pesticides

Gleying, 95, 97
Growth rate, 139–142, 299–301,
    316, 317, 324–327, 341–343
Growth stimulators, plant, 79
Gut contents, 60–66, 115, 116, 187,
    247

Heavy metals, 247–261, 267–273,
    290
Hensen, Victor, 8, 9, 10, 14
Herbicides, *see* Pesticides
Humic acids, 21, 24, 167
Humus (*see also* Soil organic
    matter) 2, 9, 19–33, 148, 167,
    221, 234, 237, 445, 459

Infiltration, 72–74, 93, 95, 190
Ingestion (*see also* Feeding; Food
    selection; Food sources), 182,
    184, 185, 345, 456, 459–461
Inoculation, *see* Introduced
    earthworms
Introduced earthworms, 38, 45, 79,
    80, 85–105, 172, 179, 197, 198,
    200, 203, 205, 209, 215–225,
    230–239
Irrigation, 190, 197–201, 216, 221
Islands, 182, 433, 442

Kommetjies, 200, 201

Leaching, 181
Leys, 159, 160
Life form (*see also* Ecological
    niche), 453
Lignin, digestibility, 206, 358

Litter
  effects of earthworms on, 39–41,
    67, 68, 74–78, 172–177,
    209–212, 216–218, 235–237,
    248, 269–272, 409
  as food, 216, 452, 454, 455, 459,
    461
Litter fall, 179–181

Manure, *see* Dung
Marketing, earthworm, 331–338
Mat (thatch), 38, 43, 77, 86, 90, 91,
    93, 95–97, 102, 126, 128, 216
Microbial activity, 21, 28, 29, 77,
    139, 148, 149, 166, 319, 320,
    351–361, 444, 454, 455
Microflora, of casts, gut and soil,
    29, 148, 184, 209–212,
    351–361, 375–380, 383–391,
    454
Microflora, responses to, 166, 305,
    307, 365–373
Microrelief, 189
Mining, 218, 219, 229–239, 250
Moles, 102, 282, 283, 400, 401
Mortality, 243, 272, 304, 305, 307,
    311, 312, 341, 359, 418–424
Mucopolysaccharides, 28, 30
Mucoproteins, 30
Mucus, 72, 76, 139, 257
Mulching, 219
Müller, Paul, 9, 10, 59, 162
Mycorrhiza, 229

Nitrogen, 76–79, 92, 93, 133,
    139–149, 216, 238, 239, 318,
    319, 328, 353, 357, 365,
    368–372, 375
Nitrogen assimilation and
    excretion, 76, 142, 143, 305,
    353
Nomenclature, 467–485
Nutrient cycling (*see also* Nitrogen;
    Phosphorus; Potassium),
    67–81, 168, 181
Nutrition, 74, 75, 78, 79, 305–307,
    358, 359, 375–380

Orchards, 124–128, 131, 217, 218

Palatability, *see* Food selection
Parasites, 290

Pathogens, 15, 290, 357, 359–361,
    375–379, 383–391, 403
Peat, 215, 219–224
Pesticides, 68, 124, 128, 131, 217,
    248, 281, 291, 359
Pests, earthworms as, 6, 9, 13, 15
pH, of casts, food and soil, 134,
    162, 182, 209–212, 216, 218,
    220, 221, 237, 239, 319, 373,
    455
Pharmacology, 293
Phosphatases, 75, 76, 319, 320,
    354–356
Phosphorus, 67–81, 209–212, 238,
    239, 318, 319, 328, 354, 355
Phylogeny, 431, 432–442
Ploughing, 43, 126, 128, 129, 159,
    160, 190, 417–423
Podzols, 29, 37, 38, 46, 162, 217,
    352
Polders, 85–105, 217, 218
Pollution, 247–261, 267–273,
    275–283, 285, 315
Polymorphism, 472
Polysaccharides, 28
Population density, 42, 51–53, 55,
    85, 88–90, 104, 105, 125–134,
    143, 154–160, 163, 172,
    182–184, 186, 196, 198, 200,
    216, 217, 219–223, 242, 243,
    404, 417, 449
Population dynamics, 42, 52, 89,
    90, 151–160, 237, 242–245,
    339–348
Population metabolism, *see*
    Respiration
Pore space, 71–74, 94, 190, 216
Potassium, 209–212, 238, 239, 318,
    319, 328
Predators, 102, 128, 131, 203, 205,
    256, 267–270, 277–282, 360,
    393–410, 415–424, 425–429
Production, biomass, 139–145, 182,
    291, 316, 317, 319, 339–348
Profile development, 188
Protein, earthworm, composition
    of, 286, 287, 297, 311, 393
Protozoa, 352, 358–360, 375–380,
    383
Pulverized fuel ash, 219

Rainfall, 200, 201, 205

Rain forest, 38, 161, 182, 183
Reclamation, 85–105, 197, 215–225,
    229–240, 241–245, 250
Refuges, 189, 197, 442–444
Refuse, 219
Relict species, *see* Refuges;
    Endemics
Rendzina, 38
Reproduction, 205, 224, 225, 245,
    316, 317, 327, 328, 343–345,
    436, 462
Respiration, 45, 46, 141–145, 167,
    168, 182–184, 218, 258
Root penetration, 101

Savannas, 181, 188, 452, 454, 456–462
Sea water, 217
Seeds, 107–121
Selection, r and K, 162, 432
Sewage/sewage sludge, 132, 225,
    247–249, 253–255, 268–273,
    297–307, 343, 347
Size, 456, 459
Slurry, 132, 139–144, 419, 421, 422
Soil
    carbon in, *see* Carbon
    metabolism of, *see* Respiration
    organic matter in (*see also*
        Humus; Carbon), 33–46,
        60–66, 93, 123, 130, 160, 168,
        180–182, 269
    stability of, *see* Aggregation;
        Wormcasts
    structure of, 91, 94, 97, 197, 215
Spodosol, *see* Podzol
Stone collecting, 10
Succession, 244, 245
Sulphur, 74
Swamps, 189, 439, 440

Tanins, 163
Taxonomy, 467–485
Tectonics, 437–442
Temperature
    responses to, 200, 202–204, 342,
        346, 454, 455
    soil, effects of earthworms on,
        95–98

Termites, 187, 188
Tissue, composition of, 286, 287,
    393
Translocation
    of nutrients, 67
    of organic matter, 45, 49, 60–66,
        91, 93, 131, 166, 167, 234–237
    of pollutants, 131, 279–281
    of seeds, 107–121
Trituration 20, 21
Tropical soils, 123, 125, 179–191

Urine scorch 101

Vermiculture, 285–294, 309–313,
    314–320, 323–329, 331,
    334–338, 354–357, 360, 361
Vertical distribution, 142, 204–205

Waste disposal (*see also* Sewage),
    285–293, 315–321, 323–329,
    339–348, 354–361
Water relations, 73, 191, 202,
    368–373, 404, 405, 453–455,
    458, 461, 462
Wedgwood, Josiah, 5, 6, 10
Weeds/sward composition, 97, 99
White, Gilbert, 6, 7, 46
Wormcasts/faeces
    microflora of, 184, 211, 319, 320,
        351–361, 376–380, 444
    nutrient and organic content of,
        60–67, 71, 74–77, 143, 184,
        185, 209–212, 319, 328, 329
    other aspects of, 2, 5, 20, 21, 30,
        31, 115–121, 142, 143, 145,
        146, 149, 176, 184, 186,
        209–212, 224, 278, 297, 328,
        329
    pollutants in, 254, 255, 269–273,
        280, 281
    stability of, 19–33, 185, 187, 189

Yield, crop, 67, 79–81, 97, 100, 218

Zero tillage, 190